Geographic Visualization

Geographic Visualization

Concepts, Tools and Applications

Edited by

Martin Dodge, Mary McDerby and Martin Turner

The University of Manchester

John Wiley & Sons, Ltd

Telephone (+44) 1243 779777

Email (for orders and customer service enquiries): cs-books@wiley.co.uk
Visit our Home Page on www.wileyeurope.com or www.wiley.com

Other Wiley Editorial Offices

John Wiley & Sons Inc., 111 River Street, Hoboken, NJ 07030, USA

Jossey-Bass, 989 Market Street, San Francisco, CA 94103-1741, USA

Wiley-VCH Verlag GmbH, Boschstr. 12, D-69469 Weinheim, Germany

John Wiley & Sons Australia Ltd, 33 Park Road, Milton, Queensland 4064, Australia

John Wiley & Sons (Asia) Pte Ltd, 2 Clementi Loop #02-01, Jin Xing Distripark, Singapore 129809

John Wiley & Sons Canada Ltd, 6045 Freemont Blvd, Mississauga, Ontario, L5R 4J3

Wiley also publishes its books in a variety of electronic formats. Some content that appears in print may not be available in electronic books.

Library of Congress Cataloging-in-Publication Data

Geographic visualization : concepts, tools and applications / edited by Martin Dodge,
Mary McDerby and Martin Turner.
p. cm.
Includes bibliographical references and index.
ISBN 978-0-470-51511-2 (cloth)
1. Geography–Computer network resources. 2. Visualization.
3. Geographic information systems. I. Dodge, Martin, 1971– II. McDerby, Mary.
III. Turner, Martin (Martin John Turner), 1968–
G70.212.G463 2008
910.285–dc22

2008004951

British Library Cataloguing in Publication Data

A catalogue record for this book is available from the British Library

ISBN 978-0-470-51511-2

Typeset in 10/12pt Minion by Aptara Inc., New Delhi, India
Printed and bound by Printer Trento Srl., Trento Italy
This book is printed on acid-free paper.

Martin Dodge dedicates this book to Maggie
Mary McDerby dedicates this book to her mother
Margaret Murray (1944–2006)

Contents

Foreword

Encounters with (Geo) Visualization

David J. Unwin

Emeritus Chair in Geography, Birkbeck College, University of London and Visiting Chair in Geomatic Engineering, University College, University of London

This volume presents essays that collectively give an overview of the current state of the art in what has become to be known as 'geovisualization', the exploratory analysis by graphics of data in which their spatial location is used as an important and necessary part of the analysis. In this forword I examine the scientific, geographical and administrative contexts in which geovisualization has developed in UK, concluding with a series of concerns related to where it is now heading.

Contexts: scientific

The scientific background to visualization is both well known and well documented. Traditionally in science, graphical modelling is subservient to mathematical and even statistical analysis: although colloquially we might 'see' a result, invariably the preferred form of analysis is by mathematics or statistics. This started to change in the 1960s with the increasing use of computers to draw pictures for which a variety of stand-alone computer programs with names like GHOST, GINO and PICASSO were developed. However, it took a series of other necessary changes before arguments based on graphics became accepted, if not on equal terms with mathematical and statistical modelling, at least as part of the basic toolkit of scientific investigation.

First, throughout the sciences developments in sensor technology and automated data capture now provide data at rates faster than can easily be converted into knowledge. Second, some of the most exciting discoveries have been associated with non-linear dynamics where apparently simple equations like the finite difference form of the logistic conceal enormously complex, but real-world-like, behaviour that can only be appreciated when displayed graphically. Third, as science has progressed to produce ever more complex simulation models, so it became necessary to use graphics as the only practicable way to assimilate all the model outputs. Particularly relevant examples include the complex, large-scale environmental simulations provided by atmospheric general circulation models used in the verification of the carbon dioxide-induced greenhouse warming hypothesis and complex social simulations of the spatial behaviour of whole populations based on models of individual behaviour. Finally, there have been enormous changes in our computing environment, all of which promote graphics as the major communication medium. It is easy to forget how far we have come. Even as late as the 1980s, colour displays were expensive luxuries needing substantial computer power to drive them, most hard copy was by way of monochrome pen plotters, and software was still in the form of subroutine libraries such as GINO, GHOST and GKS or in visualization systems such as IBM Explorer, AVS and PV-WAVE. If you wanted to use a computer to draw maps, chances were that you were in for a difficult and expensive time. For example, in 1980 the Census Research Unit at Durham University published an atlas of maps of the 1971 UK Census of Population (Census Research Unit, 1980), which used the then-new laser printing technology to produce, at great difficulty and expense, maps with individual colour symbolism for each and every kilometre grid square over Britain. At the time, these were the most detailed population maps at this scale and resolution ever produced.

As our computing norm we now have 'point and click' graphic interfaces to very large and fast machines equipped with high screen and spectral resolution displays and, thanks to the World Wide Web, graphical communication has become easier and easier. Nowadays, if you want to draw a map, a few mouse clicks using some cheap and easy to use software is all that are required. If you want to explore the map content using visual methods, then the same software will provide all the necessary resolution, colour, linkage back to the data, and so on, that are required. Provided you have the data, high spatial resolution population maps of UK are relatively easy to create on the desktop with standard hardware. In this new environment, the software can take the role of a toolkit to enable the scientist to create data displays that enable the exploratory development and testing of ideas that may later form the basis of more formal hypotheses and mathematical models.

Contexts: geographical

The geographical context is perhaps a little less well-known, but given the general changes outlined above, it was inevitable that some cartographers would 'morph' into 'geovisualizers' and that the two traditions of cartography and scientific visualization using computer graphics would intersect to give what has become known as geovisualization. This union has taken place alongside the increasing use of 'location' in almost all walks of life and the increasingly widespread availability and use of geographical information systems software. A brief visit to almost any GIS trade exhibition, or, more to the point, a look at the 'map gallery' at the annual ESRI San Diego User Conference (www.esri.com/events/uc/results/map_gallery_results.html), will show that maps continue to be not only the main selling point for GIS and associated data products, but also are one of the principal outputs from such systems. One result of the democratization of mapping, in which every map user can also now become the cartographers, has been a huge increase in the use of maps. Sometimes, as at the ESRI conferences, these are fine examples of the art and science of cartography, but all too often the products show ignorance of quite basic cartographic design, and miss the potentials available had these same data been analysed using modern geovisualization techniques.

It is certainly true that spatial coordinates can be treated simply as just two additional variables added to an existing set to be visualized. From this perspective there is nothing particularly special about adding geographical space. Yet experience suggests that, although the techniques used might look much the same as those used in more general scientific visualization, there is actually something that is special about 'geo'. In part this is to do with the ubiquitous presence in the real world of spatial autocorrelation, but I suspect it is also to do with what for want of a better word I call 'context'. Consider the very simple set of 24 numbers located geographically by the eastings and northings of some geographical grid shown below. These data can be analysed as a simple problem in general scientific visualization by interpolation of a continuous surface passing through them. However, if I provide you with the context that these numbers are mean January temperatures across the Rocky Mountain foothills in Alberta, I strongly suspect that you will realize that your initial analysis is faulty. Adding the real world context provided by spatial location adds much more than just two or more additional columns of data, and I am not sure the same would be said, for example, if these numbers had been the rate of a chemical reaction visualized in a space provided by temperature and pressure coordinates.

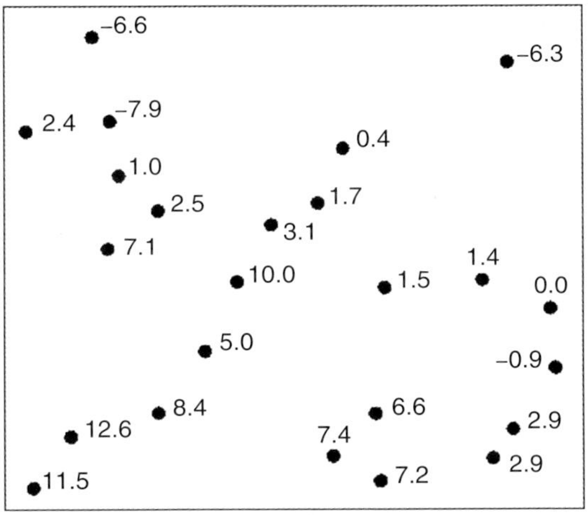

Figure F.1 Some numbers to be visualized?

Generations of cartographers have accumulated a great deal of knowledge of how these real world contexts can be addressed in their mappings, of what 'works' and what does not 'work'. Just because this knowledge has resisted formalization is no reason to ignore it. There is a continuing educational agenda here, in coupling those who have only recently discovered how useful maps can be to the community of cartographer/geovisualizers whose work is reported here and to the accumulated cartographic knowledge they bring into their work.

Contexts: organizational

This volume has a direct ancestral link back to a similar workshop sponsored by the (UK) Association for Geographic Information's Education and Research Committee and organized by the ESRC Midlands Regional Research Laboratory in Loughborough in 1992. The result then was an edited book which took the workshop title *Visualization in Geographical Information Systems* (Hearnshaw and Unwin, 1994; see also Unwin, 1994). In turn this spawned two further workshops on *Visualization in the Social Sciences* (1997, reported at www.agocg.ac.uk) and *Virtual Reality in Geography* (Fisher and Unwin, 2002). The second and third workshops were sponsored by a group set up by the UK academic research councils as their Advisory Group on Computer Graphics (AGOCG), a body I had been invited to join some time in the early 1980s as a representative of some hypothetical 'typical user' of graphic output from computers. As a geographer with a strong interest in cartography, sitting at meetings of this group was very instructive. First, as perhaps might have been expected, working with people very much at the cutting-edge of scientific research in computer graphics in UK, I became aware of the possibilities for cartography inherent in the newly developed and developing technology (see Brodlie *et al.*, 1992). Second, what perhaps was not so expected was the realization of the potential that cartography had to offer scientific visualization in areas such as what, for want of a better word, I will call the theory of graphics and graphical understanding, the use and perception of colour in graphics, symbology and so on. In the computer graphics community, the various works of Tufte were some sort of gospel and the work was conducted in almost complete ignorance of over a century of accumulated experience in the mapping and display of thematic spatial data. Much as I admire the spirit in which they were produced, the various books by Tufte are neither the only nor the last, word in graphical excellence. Dogmatic and essentially pre-computer he may be, but at the time even Bertin (1967, 1981) had much to offer this community (see Muller, 1981).

But ... what questions remain?

Reading the contributions to this volume makes it clear that much of what in the 1990s seemed to be at the cutting edge, such as the use of visually realistic displays, density estimation to visualize point patterns, area cartograms, linking and brushing, and map animation, have become commonplace. First and foremost, this volume reports immense progress in the further harnessing of the available technology to facilitate the visualization of geographic data. However, what I think stands out plainly from a comparison with the outputs from the history I have outlined is just how enduring some of the underlying themes have become. Examples in no special order include the balance between photo-realism and cartographic generalization in virtual reality, animation, projection (cartograms), conveying error/uncertainty graphically and temporal change. I suspect that there is more to this than at first meets the eye and that it is symptomatic of maybe three underlying problems in geovisualization.

The first concerns the interplay between the data that are being visualized, their geographical context and the technology used. Given that we have a research need to use visualization to generate, test and present ideas about some geographic data, three basic strategies might be recognized. The first is the geovisualization route, to provide affordances that enable interactive exploration of these data using object linking, brushing, and so on, but, by and large, leaving the data intact. The second is the spatial analytical route, to modify the numbers to be mapped by some form of arithmetic manipulation, for example by conversion into density estimates, probabilities against some hypothesized process or the derivation of 'local' statistics to isolate areas of specific research interest. The third, and least commonly adopted, is what I chose to call Tobler's way, which is to re-project these same data into a space, such as an area cartogram, in which some notion of geographic reality is better evident. For a ludicrously early and perceptive example of this see Tobler (1963).

Currently, work seems to be channelled down one or other of these three routes, yet it should be clear that most progress is likely to be made by combining them. The recent paper by Mennis (2006) provides an example of careful visualization of the results of the local statistical operation known as geographically weighted regression. Similarly, a classical spatial analytical tool, the Moran 'scatterplot' (Anselin, 1996), seldom makes much sense unless it is object-linked back to a choropleth map of the individual values. Doubtless the reader can provide other examples.

My second issue is that, despite the best efforts of some cartographers and members of the geographic information science community, as yet we seem to have little by way of 'well found' theory to enable us to answer basic visualization questions such as 'what works?', 'why does it work?' and even 'what's likely to be the best way of displaying these data?' What we have are some moderately well-articulated design rules, some interesting speculation based in, for example, communication theory or semiotics, some results from usability and perception experiments, and appeals to our instincts. The result is that in geovisualization we can be accused of finding tricky ways of turning one complex, hard-to-understand graphic and its associated data into another that is equally complex and hard to understand. It may well be that the basis for such theory exists and that what is lacking is the required synthesis, but it may also be that it cannot be formalized.

My third and final issue relates to the use of geovisualization and its relationship to the derivation, testing and presentation of social scientific theory. It was not the intention of any of the authors of the chapters in this volume to address this issue, but I doubt that in social science any hypothesis generation *ab initio* using graphics is either possible or even desirable. If this proposition seems unduly heretical in a forword to a book such as this, by way of evidence in its favour, I would point out that we do not to my knowledge have any published examples of pure hypothesis generation from graphics. Perhaps nobody is willing to come clean on the issue? The interplay between graphics, theory and prior domain knowledge seems to me to be always more complex than we usually recognize.

What I think we have in social science are examples of its use as a means of *testing* existing hypotheses. Nowhere is the relation of graphic to underlying theory better documented than in recent deconstructions of John Snow's 'iconic' 1854 map of cholera deaths in Soho, London, and its visual demonstration that a single polluted water supply pump was its cause and not the then popular notion of a 'miasma' in the air. Armed with the digitized data, numerous people have used statistical analyses to verify Snow's visual association (see Koch and Denke, 2004), but it is the role of the map that has attracted most attention. Many people – myself included – have cited Snow's mapping as a classic example of a geovisualization that in some sense led to the hypothesis that cholera is water-borne. What emerges from the more recent debates (see Brody *et al.*, 2000; Koch, 2004, 2005; Johnson, 2006) is that Snow already had his hypothesis and that the map was a very specific *test* against the 'air-borne' alternative.

Of course the entire episode remains a superb example of what this volume's editors refer to as the 'power in visualizing geography'. This power may have developed and have been best articulated in the physical and natural sciences, but it is of particular relevance to the social sciences where, like John Snow over 150 years ago, we have complex, multi-dimensional data with a variety of measurement scales from which it is necessary to test often contested and mutable theories. This volume not only shows how much progress has been made, it also points to many ways by which geovisualization will develop in the future.

References

Anselin, L. (1996) The Moran scatterplot as an ESDA tool to assess local instability in spatial association. In *Spatial Analytical Perspectives on GIS*, Fischer, M., Scholten, H. J. and Unwin, D. (eds). London, Taylor & Francis, pp. 111–125.

Bertin, J. (1967) *Semiologie Graphique. Les diagrammes – Les Reseaux – Les Cartes.* Paris, Mouton-Gauthier-Villars. (Translated and reprinted in 1984 as *Semiology of Graphics*, Madison, WI, University of Wisconsin Press.)

Bertin, J. (1981) *Graphics and Graphic Information Processing.* Berlin, Walter de Gruyter.

Brodlie, K. W., Carpenter, L. A., Earnshaw, R. A., Gallop, J. R., Hubbold, R. J., Mumford, A. M., Osland, C. D. and Quarendon, P. (1992) *Scientific Visualization Techniques and Applications.* Berlin, Springer.

Brody, H., Rip, M. R., Vinten-Johansen, P., Paneth, N. and Rachman, S. (2000) Map-making and myth-making in Broad Street: the London cholera epidemic, 1854. *The Lancet* 356(9223): 64–66.

Census Research Unit (1980) *People in Britain – a Census Atlas.* London, HMSO.

Fisher, P. F. and Unwin, D. J. (2002) *Virtual Reality in Geography.* London, Taylor and Francis.

Hearnshaw, H. M. and Unwin, D. J. (1994) *Visualization in Geographical Information Systems.* Chichester, Wiley.

Johnson, S. (2006) *The Ghost Map.* New York, Riverhead Hardcover.

Koch, T. (2004) The map as intent: variations on the theme of John Snow. *Cartographica* 39(4): 1–13.

Koch, T. (2005) *Cartographies of Disease: Maps, Mapping and Medicine.* Redlands, CA, ESRI Press.

Koch, T. and Denke, K. (2004) Medical mapping: the revolution in teaching – and using –maps for the analysis of medical issues. *Journal of Geography* 103: 76–85.

Mennis, J. (2006) Mapping the results of geographically weighted regression. *Cartographic Journal* 43(2): 171–179.

Muller, J. C. (1981) Bertin's theory of graphics: a challenge to North American thematic cartography. *Cartographica* 18: 1–8.

Tobler, W. R. (1963) Geographic area and map projections. *Geographical Review* 53: 59–78.

Unwin, D. J. (1994) ViSc, GIS and cartography. *Progress in Human Geography* 18: 516–522.

Acknowledgements

This book grew out of a selection of papers read and discussed at a workshop on *Geographic Visualization Across the Social Sciences* held at the University of Manchester in June 2006. The workshop was financially supported through an ESRC grant as an Agenda Setting Workshop administered by the National Centre for e-Social Sciences (NCeSS) and by the JISC-funded UK Visualization Support Network (vizNET). We also wish to acknowledge the support and encouragement from various groups within the University of Manchester, the University of Nottingham and University College London as well as the Cathy Marsh Centre for Census and Survey Research.

We are grateful for the encouragement of Mike Batty and Paul Longley in turning this into a book. Thanks also to the enthusiastic support of our editor Rachel Ballard, along with the help of Robert Hambrook, Liz Renwick and Fiona Woods, at John Wiley and Sons.

Martin, Mary and Martin

The University of Manchester, September 2007

Authors' Biographies

Anna Barford

Department of Geography, University of Sheffield, UK

Anna Barford is a researcher working in the Social and Spatial Inequalities Group. She has a BA from the University of Cambridge and an MA from the University of Nottingham. Anna has worked at the University of Leeds; interned in the Department of HIV/AIDS at the World Health Organization; and has undertaken independent research into participatory 'development' projects in Nepal.

Maurico Capra

Research Associate
Mixed Reality Laboratory, University of Nottingham

Mauricio Capra is an research associate in the Mixed Reality Laboratory at the University of Nottingham. His research interests include pervasive games, authoring, orchestration, evaluation, ethnographical studies, mobile technologies and augmented reality. He received his BSc in computers science from Universidade Luterana do Brasil, his MSc in cartographic engineering from Instituto Militar de Engenharia, Brazil, and is now waiting for his Viva at University of Nottingham. However, what he really likes to do is to ride his bicycle and fly over the moors in his paraglider.

William Cartwright

Professor of Cartography and Geographical Visualization
School of Mathematical and Geospatial Sciences, RMIT University, Australia

William Cartwright is Professor of Cartography and Geographical Visualization in the School of Mathematical and Geospatial Sciences at RMIT University, Australia. His major research interests are the application of New Media to cartography, the exploration of different metaphorical approaches to the depiction of geographical information and how the Art elements of cartography can be used to compose complementary views of the world to the views constructed by science and technology.

Stephanie Deitrick
School of Geographical Sciences, Arizona State University, USA

Stephanie Deitrick is a doctoral student in the School of Geographical Sciences at Arizona State University in Tempe, Arizona. She holds degrees in Geography and Mathematics. Her primary interests are the representation of uncertainty, the use of GIS in public policy decision-making, and the usability of geovisualization applications.

Martin Dodge
Lecturer in Human Geography
School of Environment and Development, University of Manchester, UK

Martin Dodge works at the University of Manchester as a Lecturer in Human Geography. His research focuses primarily on the geography of cyberspace, particularly ways to map and visualize the Internet and the Web. He is the curator of a web-based Atlas of Cyberspace (www.cybergeography.org/atlas) and has co-authored two books, *Mapping Cyberspace* (Routledge, 2000) and *Atlas of Cyberspace* (Addison-Wesley 2001), both with Rob Kitchin.

Danny Dorling
Professor of Human Geography
Department of Geography, University of Sheffield, UK

Danny Dorling worked on children's play schemes in the late 1980s, but has been trapped in universities since then. He has worked with several colleagues on a number of books, papers and reports. He is currently Professor of Human Geography at Sheffield University, Visiting Professor in Social Medicine at Bristol University and Adjunct Professor in the Department of Geography, University of Canterbury, New Zealand.

Robert Edsall
Assistant Professor
School of Geographical Sciences, Arizona State University, USA

Robert Edsall is an assistant professor in the School of Geographical Sciences at Arizona State University in Tempe. He holds degrees in Geography, Meteorology and Music and received his PhD in 2001 from Penn State University. His primary interests are in geovisualization tool and interface design, and the cognition and usability of cartographic applications.

Aude Esperbé
Geography Graduate Student
Department of Geography, San Diego State University, USA

Aude Esperbé is a graduate student pursuing an MS degree in Geography at San Diego State University. She holds an MS degree in Geology from the Institut Lasalle Beauvais, France. She worked for four years as a geological engineer in the oil and gas industry and previously in environmental consulting companies. Her interests are in the areas of visual representation of geologic and geographic phenomena, GIS and cartographic design.

Sara Fabrikant

Associate Professor of Geography
Department of Geography, University of Zurich, Switzerland

Sara Irina Fabrikant, a Swiss mapematician, is currently an associate professor of geography and head of the Geographic Information Visualization and Analysis (GIVA) group at the GIScience Center in the Geography Department of the University of Zurich, Switzerland. Her research and teaching interests lie in geographic information visualization, GIScience and cognition, designing cognitively adequate graphical user interfaces, and dynamic cartography. She holds an MS from the University of Zurich, Switzerland, and a PhD from the University of Colorado at Boulder, USA.

Michael F. Goodchild

Professor of Geography
Department of Geography, University of California, Santa Barbara, USA

Michael F. Goodchild is Professor of Geography at the University of California, Santa Barbara. His interests centre on geographic information systems and science, and he is the author of over 15 books and 400 papers. He is a member of the US National Academy of Sciences, and has played leading roles in the National Center for Geographic Information and Analysis, the Center for Spatially Integrated Social Science, and the Alexandria Digital Library.

Derek Hampson

Artist
School of Fine Art, University College for the Creative Arts, Canterbury, UK

Derek Hampson is a painter with an interest in problems of representation. His research interest centres on the processes of making the unseen seen through painting or drawing– that is, how we can visually represent that which we cannot see. These include abstract things such as thoughts, ideas, concepts but also existent things beyond the realm of standard vision, such as the very small or the very distant. He is the creator of painting cycles such as *Ulster in Albion, The Loves of the Plants* and *True*. His theoretical interests centre on philosophy, particularly twentieth century phenomenology as developed by Husserl and Heidegger and also to a lesser extent the transcendental philosophy of Kant. He is Course Leader of the University College for the Creative Arts undergraduate Fine Art course based at Canterbury, UK. He holds a BA and an MA in Fine Art from Nottingham Trent University.

Mark Harrower

Assistant Professor of Geography
Department of Geography, University of Wisconsin – Madison, USA

Mark Harrower is an assistant professor of geography and associate director of the Cartography Laboratory at the University of Wisconsin – Madison. His research interests include interactive and animated mapping systems, perceptual and cognitive issues in map reading, interface design and developing new tools for map production. Harrower has an MS and a PhD from the Pennsylvania State University, both in geography.

Andrew Hudson-Smith

Senior Research Fellow
Centre for Advanced Spatial Analysis, University College London, UK

Andrew Hudson-Smith is a Senior Research Fellow at the Centre for Advanced Spatial Analysis, University College London and author of the Digital Urban blog. His research is concentrated around the visualization of cities using digital means for both capture and representation. His latest work can be found at www.digitalurban.blogspot.com.

Menno-Jan Kraak

Professor and Chairman of the Department of Geo-Information Processing
International Institute for Geo-Information Science and Earth Observation

Menno-Jan Kraak is, together with F. J. Ormeling, the author of *Cartography: Visualization of Geospatial Data.* He works at the ITC – International Institute of Geo-Information Science and Earth Observation as Professor in Geovisualization. He is vice-president of the International Cartographic Association.

Mary McDerby

Visualization Support Officer
Research Computing Services, University of Manchester, UK

Mary McDerby is visualization support officer in Research Computing Services providing visualization, computer graphics, multimedia and image processing services to the University of Manchester. Her research is in the visualization of complex datasets within a virtual reality environment, as well as medical visualization. She is active in both national and international computer graphics/visualization communities such as Eurographics, and has been a co-editor of the proceedings of the UK chapter for the past three years.

Leif Oppermann

Researcher
Mixed Reality Laboratory, University of Nottingham, UK

Leif Oppermann is a researcher at the Mixed Reality Laboratory at the University of Nottingham. His main research interest is in pervasive games and the infrastructure, software tools and processes needed to get them running. He has been involved in the making of mixed reality games and is working towards a PhD about extending authoring tools for location-aware applications with completion anticipated for 2008. Prior to joining the MRL he studied Media Computer Science at the Hochschule Harz in Wernigerode, Germany. His degree in Interaction Surfaces in Augmented Reality was awarded the Best of Faculty prize in 2003.

Scott Orford

Lecturer in GIS and Spatial Analysis
School of City and Regional Planning, Cardiff University, UK

Scott Orford is a Lecturer in GIS and Spatial Analysis in the School of City and Regional Planning, Cardiff University. His research interests includes visualization, GIS and the statistical modelling of socio-economic spatial processes with a particular emphasis upon the built environment. He has a BSc in Geography from Lancaster University and a PhD from the School of Geographical Sciences, Bristol University.

Gary Priestnall

Associate Professor
School of Geography, University of Nottingham, UK

Gary Priestnall is an associate professor within the geographical information science research theme in the School of Geography, The University of Nottingham. His research draws upon interests in geography, art and computer science and focuses on the way digital geographic representations are constructed and how they are used in various contexts. His interests fuse research with teaching and learning, looking increasingly at the use of three-dimensional visualization and mobile mapping in the context of the field-based activities. He is director of the MSc in GIS at Nottingham and is the site manager for the Nottingham arm of the SPLINT (Spatial Literacy in Teaching) Centre for Excellence in Teaching and Learning.

Jonathan C. Roberts

Senior Lecturer
School of Computer Science, Bangor University, UK

Jonathan C. Roberts is a senior lecturer in the School of Computer Science, Bangor University, UK. He received his BSc and PhD from the University of Kent, both in Computer Science, and is a Fellow of the Higher Education Academy. His research interests focus around visualization, especially exploratory visualization, multiple views, visualization reference models, visualization in virtual environments, haptic interfaces and Web-based visualization. He is chair of the UK chapter of the Eurographics Association and sits on the editorial board of *Information Visualization.*

Ifan D. H. Shepherd

Professor of GeoBusiness
Middlesex University Business School, Middlesex University, UK

Ifan D. H. Shepherd is Professor of GeoBusiness at Middlesex University, where he teaches geodemographics, geographical information systems and e-business. His research interests include data visualization, the history of geodemographics, personal and religious marketing, and the transfer of learning. In his spare time, he is building a three-dimension virtual reality system for Victorian London with his son, who is a computer games programmer.

André Skupin

Associate Professor of Geography
Department of Geography, San Diego State University, USA

André Skupin is an associate professor of Geography at San Diego State University. He received a Dipl.-Ing. degree in Cartography at the Technical University Dresden, Germany, and a PhD in Geography at the State University of New York at Buffalo. Areas of interest and expertise include text document visualization, geographic visualization, cartographic generalization and visual data mining. Much of his research revolves around new perspectives on geographic metaphors, methods and principles, outside of traditional geographic domains.

Humphrey Southall

Reader in Geography
Department of Geography, University of Portsmouth, UK

Humphrey Southall is Director of the Great Britain Historical GIS Project and principal author of the web site *A Vision of Britain through Time.* His earliest research was on the historical development of labour markets in Britain and the origins of the north-south divide. He holds an MA and a PhD from Cambridge University.

Martin Turner

Visualization Team Leader
Research Computing Services, University of Manchester, UK

Martin Turner is the Visualization Team Leader within Research Computing Services at the University of Manchester. He gained his PhD in the Computer Laboratory, at Cambridge University. His research in visualization and image processing has resulted in a Fellowship with British Telecom, a published book, *Fractal Geometry in Digital Imaging* (Academic Press, 1998) as well as over 50 other publications, and he has supervised to completion seven successful MPhil/PhD students. Key activities and grants cover both local and nationally funded high-end visualization services as well as commercial contracts.

Michael Wright

Researcher
Mixed Reality Laboratory, University of Nottingham, UK

Michael Wright is a researcher at the Mixed Reality Laboratory at the University of Nottingham. His main research interest is in exploratory visualizations and pervasive games. He has previously worked on a visualization system for Hitchers and an authoring interface for Day of the Figurines. His current research interests are in the visualization of human activity in pervasive and ubiquitous environments.

1

The Power of Geographical Visualizations

Martin Dodge, Mary McDerby and Martin Turner

School of Environment and Development and Research Computing Services, University of Manchester

> Now when I was a little chap I had a passion for maps. I would look for hours at South America, or Africa, or Australia, and lose myself in all the glories of exploration. At that time there were many blank spaces on the earth and when I found one that looked particularly inviting on a map (but they all look that) I would put my finger on it and say, 'When I grow up I will go there'.
>
> (Joseph Conrad, *Heart of Darkness, 1902*)

> I believe we need a 'Digital Earth'. A multi-resolution, three-dimensional representation of the planet, into which we can embed vast quantities of geo-referenced data.... Imagine, for example, a young child going to a Digital Earth exhibit at a local museum. After donning a head-mounted display, she sees Earth as it appears from space. Using a data glove, she zooms in, using higher and higher levels of resolution, to see continents, then regions, countries, cities, and finally individual houses, trees, and other natural and man-made objects. Having found an area of the planet she is interested in exploring, she takes the equivalent of a 'magic carpet ride' through a 3-D visualization of the terrain. Of course, terrain is only one of the many kinds of data with which she can interact.... she is able to request information on land cover, distribution of plant and animal species, real-time weather, roads, political boundaries, and population.
>
> (Former Vice President Al Gore, *The Digital Earth: Understanding our Planet in the 21st Century, 1998*)

Geographic Visualization Edited by Martin Dodge, Mary McDerby and Martin Turner

1.1 Aims

'Geography' has the potential to provide the key to a whole raft of innovative means of information representation through the use of interactive spatial visualizations. This use is clearly seen in the rapid growth and uptake of geographic information systems (GIS), multimedia cartography, virtual globes and all manner of Web-based mapping tools that are currently available. *Geographic visualization* is a significant and growing area and for this book we take a necessarily broad view of what it constitutes. Drawing upon Harley and Woodward's (1987, xvi) definition of mapping, we see geographic visualization as the application of any graphic designed to facilitate a *spatial* understanding of things, concepts, conditions, processes or events in the human world.

The goal of this book is to explore the 'state of the art' of geographic visualization relevant to the social scientists, in particular, reviewing current and popular methods and techniques, examining software tools, and reporting on the development of new applications to support both research and pedagogy. In some senses this book represents a 10-year updating of the Advisory Group on Computer Graphics-sponsored 'Graphics, Visualization and the Social Sciences'[1] workshop held in May 1997. It is relevant now to see what has changed (for example, affordable mobile tracking and mass market in-car satellite navigation with increasingly sophisticated dynamic graphics) and what unexpected developments have occurred (powerful and accessible 'geoweb' tools for example, Google Maps, Google Earth, NASA World Wind, TerrainView-Globe and Microsoft Virtual Earth). It is also important to see where weaknesses and blockages still lie; so in the future these areas, through further research, can be resolved and social sciences then can exploit geographic visualizations to another level.

1.2 The nature of geographic visualization

To understand the power of visualization, one must grasp how it both stirs the imagination for exploration and works instrumentally in the exploitation of new spaces. As Joseph Conrad's narrator Marlow makes clear in the famous passage from the *Heart of Darkness*, maps (an archetype of geographic visualization) open up space to the imagination, even from a very early age. Furthermore, geographic visualization, primarily in the form of paper maps has, over millennia, provided uniquely powerful instruments by which to classify, represent and communicate information about spaces that are too large and too complex to be seen directly.

The ability to create and use geographic visualizations in the form of cartographic maps has been one of the most basic means of human communication, at least as old as the invention of language and arguably as significant as the discovery of mathematics. The recorded history of cartography, for example, clearly demonstrates the long pedigree and practical utility of geographic visualization in all aspects of Western society, being most important for organizing spatial knowledges, facilitating navigation and controlling territory. Some have gone further, to argue that spatial mapping processes are culturally universal, evident across

[1] Organized by Anne Mumford, Michael Batty and David Unwin; workshop report available at: http://www.agocg.ac.uk/wshop/33/33.pdf.

all societies (Blaut *et al.*, 2003), although the material and semiotic forms of the artefacts produced are diverse. At the same time as working pragmatically, geographic visualizations are also rhetorically powerful as graphic images that frame our understanding of the human and physical world, shaping our mental image of places, and constructing our sense of spatiality. So, in a very real sense, geographic visualization makes our world; and to an increasing extent may become our world.

Historically geographic visualizations visually represent a physical landscape often using the cartographic norms of a planar view – looking straight down from above – and a consistently applied reduction in scale. However, with the range of approaches available today it is impossible to neatly define geographic visualization according to a single type of mapping phenomena, the particular mode of presentation or their medium of dissemination. Geographic visualization has traditionally been in the form of printed maps and globes used as static storage devices for spatial data, but now they are much more likely to be interactive tools displayed on a computer screen. So today, we live in a map-saturated world (Wood, 1992), continually exposed to conventional geographic maps, along with many other map-like spatial images and media (for example three-dimensional city models, MRI scans of the brain, virtual globes on the news). Contemporary developments in computer-based geographic visualization are opening up unique ways to visually understand the complex, multivalent and intangible nature of space and society; some of this was considered within the now prophetic speech by former Vice President Al Gore when describing a possible future *Digital Earth* (Gore, 1988), which would be a geographically data-rich and selectively data-interactive globe representation.

Geographic visualization now covers a range of different scales from individual properties up to global scale visualization of vast landscapes of data and contains the graphical data-mining of the daily interactions of millions of people. Some of the geographic visualizations adhere to established conventions of cartographic design, but many more are employing quite different visual vocabularies. Some geographic visualization applications are beautiful; many more are really rather ugly in terms of aesthetic values. Some are actually quite useful as practical tools for social science analysis, but many visualization tools are not workable for real-world analysis.

This diversity in geographic visualization approaches can be usefully conceptualized into three broad epistemological classes. These classes we might call by the shorthand: 'looking', 'querying' and 'questioning'. Each class has increasing power to augment the human capacity in order to analyse and understand the world.

- '*Looking*': these are presentation graphics, thematic maps and charts that display data according to spatial coordinates. They have proved to be relatively easy to produce and are widely used in reports and papers. The next generation of presentation graphics (on websites with flash for example) go beyond just static images to utilize animation and interactive 2.5D data landscapes as well as fully 3D layered information spaces. The user is able to navigate through the landscape and animate it to display time-oriented information, make simple interactive queries on data items, and perhaps even calibrate display parameters (for example symbology, scales and classifications).

- '*Querying*': these are visual interfaces designed for information access. Based on database and data-mining techniques, they focus on enabling users to navigate through complex information spaces in order to locate and retrieve task-relevant subsets of information.

Supported user-tasks involve searching, backtracking and history-logging. User interface techniques attempt to preserve user-context and support smooth transitions between locations.

- '*Questioning*': these are full visual discovery and modelling structures. These systems combine the visual insights communicated by presentation graphics with an ability to probe, drill-down, filter and manipulate the display to answer the 'why' questions as well as the 'what' question. The difference between answering a 'what' and a 'why' question involves highly interactive operations and the capacity to simulate change. Interaction techniques let the user directly control the display of data, for example through projections, filtering, zooming and linked displays and brushing. Real-time response is necessary and often the linkage of multiple visualizations metaphors. Distortion techniques can also help in the interactive exploration process by providing a means for focusing while preserving an overview of the data (so-called 'focus + context').

1.3 The visualization process

In essence, geographic visualization exploits the mind's ability to more readily see complex relationships in images, and thus provide a clear understanding of a phenomenon, reducing search time and revealing relationships that may otherwise not have been noticed. It is a process which works essentially by helping people to see the unseen, premised on the simple notion that humans can reason and learn more effectively in a visual environment than when using textual or numerical description.

The ability of geographic visualization to elucidate meaningful patterns in complex data is clearly illustrated by some of the 'classics' from the pre-digital era, such as John Snow's 'cholera map' of 1854 (Johnson, 2006), Charles Joseph Minard's 'Napoleon map' of 1869 (see Chapter 15) or Harry Beck's 'Tube diagram' of 1933 (Garland, 1994). Even though these were all hand-drawn on paper, they are nonetheless still effective today and show the potential of geographic visualization to provide new understanding and compelling means of communicating to a wide audience. Their novel visual forms mean they also demonstrate the extent to which geographic visualization can be a creative design practice in and of itself. The best geographic visualizations go beyond merely representing to become a kind of cognitive shorthand for the actual places and processes themselves, as is illustrated in Beck's celebrated diagrammatic design of the London Underground (the Tube map), which has become such a powerful spatial template for the 'real' layout of London in the minds of many visitors and residents.[2]

Geographic visualization works by providing graphical ideation to render a place, a phenomenon or a process visible, enabling the most powerful human information-processing abilities – those of spatial cognition associated with the eye–brain vision system – to be directly brought to bear. Visualization is thus a cognitive process of learning through the active engagement with graphical signs that make up the display, and it differs from passive observation of a static scene, in that its purpose is also to discover unknowns, rather than

[2] The 'problem' is that, although Beck's visualization works well for underground movement, it can be confusing for surface navigation because it famously sacrifices geographic accuracy for topological clarity.

to see what is already known. Effective geographic visualization should reveal novel insights that are not apparent with other methods of presentation. In an instrumental sense, then, geographic visualization is a powerful prosthetic enhancement for the human body: '[l]ike the telescope or microscope, it allows us to see at scales impossible for the naked eye and without moving the physical body over space' (Cosgrove, 2003, p. 137).

1.4 Digital transition and geographic visualization

The development and rapid diffusion of information and communication technologies in the last three decades has affected all modes of geographic visualization, changing methods of data collection, cartographic production and the dissemination and use of maps. This has been termed the 'digital transition' in mapping (Pickles, 2004) and it is continuing apace (for example, developments in mobile communications and location-based services; see Chapter 16; Raper *et al.*, 2007). As such it is a vital component in understanding the milieu in which new modes of geographic visualization are emerging. Nowadays, most geographic visualizations are wholly digital and created only 'on demand' from geospatial databases for temporary display. The Web mapping portal MapQuest.com, for example, has already generated more digital maps than any other publisher in the history of cartography (Peterson, 2001); the popularity of Google map's API,[3] launched in the summer of 2005, has inspired an explosion of new online mapping tools and hacks (Gibson and Erle, 2006), and there is even the prospect that GIS itself will begin to adapt and evolve around such a Web services mapping model (Miller, 2006).

Cheap, powerful computer graphics on desktop PCs, and increasingly mobile devices, enable much more expressive and interactive geographic visualization that are potentially available to a growing number of people. The pervasive paradigm of hypertext as a way to structure and navigate information has also influenced digital geographic visualization and increasingly it is being used as a core component in larger multimedia information resources where locations and features on the map are hot-linked to pictures, text and sounds, to create distinctively new modes of geographic interaction (see Chapters 6 and 9). In design terms, the conventional planar map form itself is, of course, only one possible representation of geographic data and new digital technologies have enabled much greater diversity of forms, including pseudo 3-D landscape views, interactive panoramic photo image-maps, fully 3-D fly-through models and immersive VR space (see Chapters 9–11; Dykes, MacEachren and Kraak, 2005). Developments in computer graphics, computation and user interfaces have enabled visualization tools to be used interactively for exploratory data analysis (typically with the interlinking of multiple representations such as statistical charts, 3-D plots, tables, and so on).

Developments in networking and computer-mediated communications, and the rise of the World-Wide Web in the mid 1990s, means that now geographic visualizations are very easy to distribute at marginal cost and can be accessed 'on demand' by almost anyone, anywhere. The provision of Web mapping and online GIS tools is significantly shifting the accessibility to geographic visualization and spatial data, as well as altering the user

[3] An API (application programming interface) allows technically savvy users direct access to the database enabling sophisticated and novel third-party applications to be developed.

perception of what they can do. There are clear signs that geographic visualization will be seen as simply one of many available 'on demand' Web or portal services to the general public and integrated or 'mashed-up' within a multitude of other applications. As geographic visualization becomes more flexible and much more accessible, it is also, in some respects, granted a less reified status than the printed artefacts of the past. Visualizations will increasingly be treated as transitory information resources, created in the moment and discarded immediately after use. In some senses, this devalues the geographic visualization as it becomes just another form of ephemeral media, one of the multitude of screen images that barrage people everyday. Geospatial data itself is just another informational commodity to be bought and sold, repackaged and endlessly circulated.[4]

The production of geospatial information has always been dependent, to a large degree, on the available methods of data collection. These are being greatly augmented in the digital transition. The widespread importance of new digital measurement was noted by US National Science Foundation Director Rita Colwell (2004, p. 704): 'new tools of vision are opening our eyes to frontiers at scales large and small, from quarks to the cosmos'. Geographic visualization's ability to 'capture' the world has been transformed by digital photogrammetry, remote sensing, GPS-based tracking and distributed sensor network. Cartography can not only 'see' the world in greater depth (Pickles, 2004), but it can also 'see' new things (including virtual spaces), and with new temporalities.

Vast digital geospatial databases underlie many powerful geographic visualizations, such as the Ordnance Survey's Digital National Framework, comprising over 400 million features.[5] These are growing as part of the 'exponential world', being fed in particular by high-resolution imagery from commercial satellites. In the future, much of this growth will come from people gathering geospatial data as they go about their daily activity, automatically captured by location-aware devices that they will carry and use (see Chapter 16). From this kind of emergent mobile spatial data capture, it will be possible to 'hack' together new types of geographic visualization rather than be dependent on the products formally published by governments or commercial firms. Such individually made, 'amateur' mapping may be imperfect in many respects (not meeting the positional accuracy standards or adhering to the TOPO-96 surveying specifications, for example), but could well be better fit-for-purpose than professionally produced, generic visualization applications. There is also exciting scope for using locative media to annotate individual geographic visualization with ephemeral things, personal memories and messages for friends, which are beyond the remit of government agencies or commercial geospatial industry (Kwan, 2007).

1.5 The politics of visualization

The process of geographic visualization is also engendered because the objects are often visually appealing in their own right. The aesthetics of well-designed cartographic maps or

[4] However, the emergence of open-source cartography, as exemplified by the OpenStreetMap project, has the potential to challenge the commercial dominance over geospatial data by developing a 'bottom-up' capture infrastructure that is premised on a volunteerist philosophy.

[5] Source: www.ordnancesurvey.co.uk/oswebsite/media/news/2001/sept/masterchallenge.html.

globes, for example, is central to their success in rhetorical communication and means they are widely deployed as persuasive devices to present ideas, themes and concepts that are difficult to express verbally, as well as serving as decorative objects. The result, according to Francaviglia (1999, p. 155), is that '[c]artographers draw maps that have the power to both inform and beguile their users'. Most of the geographic visualization encountered on a daily basis (often with little conscious thought given to them) are maps used in the service of persuasion, ranging from the corporate marketing map to the more subtle displays such as states' claims to sovereign power over territory, implicitly displayed in the daily weather map seen on the news (Monmonier, 1996).

The production of geographic visualization involves a whole series of decisions, from the initial selection of what is to be measured, to the choice of the most appropriate scale of representation and projection, and the best visual symbology to use. The notion of 'visualization as decision process' is useful methodologically because it encourages particular ways of organized thinking about how to generalize reality, how to distil inherent, meaningful spatial structure from the data, and how to show significant relationships between things in a legible fashion. Geographic visualization provides a means to organize large amounts of, often multi-dimensional, information about a place in such a fashion as to facilitate human exploration and understanding. Yet, visualization practices are not just a set of techniques for information 'management'; they also encompass important social processes of knowledge construction. As scholars have come to realize, vision and culture are intimately entwined and inseparable (Pickles, 2004). Geographic visualization then is both a practical form of information processing and also a compelling form of rhetorical communication.

It must be recognized that geographic visualization is a process of creating, rather than revealing, spatial knowledge. Throughout the process of visualization creation a large number of subjective, often unconscious, decisions are made about what to include and, possibly more importantly, what to exclude, how the visualization will look, and what message the designer is seeking to communicate. In this fashion, geographic visualization necessarily becomes imbued with the social norms and cultural values of the people who construct them.

While contemporary geographic visualization developments, such as Google Earth and other online virtual globes (see Chapter 2), can be seen as a logical and even 'natural' evolution of 'flat' map representations, whose aim is to enhance users' knowledge of new spaces, making navigation and commerce more efficient and increasing the 'return-on-investment' in existing geospatial data by facilitating wider distribution on the Web, we would argue that the situation with geographic visualization is also more contestable. Only certain geographic visualizations get made and they show only certain aspects, in certain ways. They are not inherently 'good' and will certainly not be beneficial to all users. The visualization of geographic space is not a benign act, instead particular visualizations are made to serve certain interests. These interests may reflect dominant power relations in the society, especially when individuals and institutions with power commission a great deal of geographic visualization and control access to underlying data resources. Thus geographic visualizations are not objective, neutral artefacts but a political view point onto the world (Monmonier, 1996; Pickles, 2004). The prospect of propaganda and deceptive manipulation with geographic visualization always at times may rear its ugly head.

1.6 The utility of geographic visualization

Geographic visualizations have long been used in scholarly research of social and physical phenomena. They are, of course, a primary technique in geography but they are also widely used in other disciplines such as anthropology, archaeology, history and epidemiology, to store spatial data, to analyse information and generate ideas, to test hypotheses and to present results in a compelling, visual form. Geographic visualization as a method of enquiry and knowledge creation also plays a growing role in the natural sciences (for example, biological brain functional sensor mappings). This work is not limited to cartographic mapping; many other spatial visualization techniques, often using multi-dimensional displays, have been developed for handling very large, complex spatial datasets without gross simplification or unfathomable statistical output (for example, volumetric visualization in atmospheric modelling).

Within many social science disciplines there are growing signs of a 'spatial turn' as research questions and modes of analysis centre around geographic location and understanding of spatial relations and interactions come to greater prominence. Many social science disciplines are exploiting the spatial components of large data that they have collected or generated to facilitate their analysis. Interactive geographic visualization is then a crucial research tool.

Outside of academia there is also a great deal of excitement in so-called neogeography (Turner, 2006) and geographic visualization. This is most evident in terms of popular online mapping of data and new types of application – such as map 'mash-ups' using Google maps, the map hacking and open-source mapping activists using cheap GPS to visualize the world afresh. The value of high-resolution aerial photography and satellite imagery for 'backyard' visualization is being unlocked into easy (and fun) browsable interfaces such as Google Earth and Microsoft Local Live.

1.7 Conclusions

Although geographic visualization is a powerful research method for exploration, analysis, synthesis and presentation, it is not without its problems. Three particular issues are worth discussing: practical limitation, ethical concerns and political interests. Firstly, there are many practical issues to be faced and it is important to acknowledge the investment of time and effort necessary to make effective and appropriate geographic visualizations. Certain processes are now much easier to do today than in the past, but it is not necessarily a quick fix. Like any chosen research technique, the potential of geographic visualization has external practical constraints, including data quality and the level of user knowledge.

There are also issues to consider relating to the ethics and responsibility of researchers producing geographic visualization. As noted above, the processes of data selection, generalization and classification and the numerous design decisions mean that one can never remove the subjective elements in the process. Accordingly, Monmonier (1993, p. 185) argues in relation to thematic maps (and equally applicable to other geographic visualization):

> ... any single map is but one of many cartographic views of a variable or a set of data. Because the statistical map is a rhetorical device as well as an analytical tool, ethics require that a single map not impose a deceptively erroneous or carelessly

incomplete cartographic view of the data. Scholars must look carefully at their data, experiment with different representations, weigh both the requirements of the analysis and the likely perceptions of the reader, and consider presenting complementary views with multiple maps.

Further, some of these new forms of geographic visualizations open up society to a new kind of surveillance, revealing interactions that were previously hidden in unused transactions and databases. The act of visualization itself may constitute an invasion of privacy (Monmonier, 2002). If the appeal of some spaces is their anonymity, then people may object to them being placed under wider scrutiny, even if individuals are unidentifiable. Here, public geographic visualization and analysis may well represent an infringement of personal rights, especially if the individuals were not consulted beforehand and have no means to opt out. Thus, it is important to consider the ways in which, and the extent to which, geographic visualizations of social spaces are 'responsible artefacts' that do not destroy what they seek to represent or understand.

Lastly, it should be recognized that geographic visualization is also a cultural process of creating, rather than merely revealing, knowledge. All the sophisticated, interactive geographic visualizations have politics, just the same as any other form of representation, and we must be alert to their ideological messages. Geographic visualization for social scientists can prove to be very valuable, but at the same time they can never be value-free.

The future still requires research in many key unknown and under-explored areas, including uncertainty mapping, true temporal understanding and limits of human visual perception – but the future also is becoming more socially connected and geographic visualization may indicate a way forward.

References

Blaut, J. M., Stea, D., Spencer, C. and Blades, M. (2003) Mapping as a cultural and cognitive universal. *Annals of the Association of American Geographer* 93(1): 165–185.

Colwell, R. (2004) The new landscape of science: a geographic portal. *Annals of the Association of American Geographers* 94(4): 703–708.

Conrad, J. [1902] (1994) *Heart of Darkness.* London, Penguin.

Cosgrove, D. (2003) Historical perspectives on representing and transferring spatial knowledge. In *Mapping in the Age of Digital Media*, Silver, M. and Balmori, D. (eds). New York, Wiley-Academy, pp. 128–137.

Dykes, J., MacEachren, A. M. and Kraak, M.-J. (2005) *Exploring Geovisualization.* London, Elsevier.

Francaviglia, R. 1999. Walt Disney's Frontierland as an allegorical map of the American West. *The Western Historical Quarterly* 30(2): 155–182.

Garland, K. (1994) *Mr Beck's Underground Map.* Middlesex, Capital Transport Publishing.

Gibson, R. and Erle, S. (2006) *Google Mapping Hacks.* Sebastopol, CA, O'Reilly & Associates.

Gore, A. (1998) *The Digital Earth: Understanding our Planet in the 21st Century.* California Science Center, Los Angeles, CA, 31 January. Available at: www.isde5.org/al_gore_speech.htm.

Harley, J. B. and Woodward, D. (1987) *The History of Cartography, Volume 1. Cartography in Prehistoric, Ancient and Medieval Europe and the Mediterranean.* Chicago, IL, University of Chicago Press.

Johnson, S. (2006) *The Ghost Map*. London, Allen Lane.
Kwan, M.-P. (2007) Affecting geospatial technologies: toward a feminist politics of emotion. *The Professional Geographer* 59(1): 22–34.
Miller, C. C. (2006) A beast in the field: the Google maps mashup as GIS/2. *Cartographica* 41(3): 187–199.
Monmonier, M. (1993) *Mapping It Out*. Chicago, IL, University of Chicago.
Monmonier, M. (1996) *How To Lie With Maps*, 2nd edn. Chicago, IL, University of Chicago Press.
Monmonier, M. (2002) *Spying with Maps: Surveillance Technologies and the Future of Privacy*. Chicago, IL, University of Chicago Press.
Peterson, M. P. (2001) The development of map distribution through the Internet. *Proceedings of the 20th International Cartographic Conference*, Vol. 4, pp. 2306–2312.
Pickles, J. (2004) *A History of Spaces: Cartographic Reason, Mapping and the Geo-Coded World*. London, Routledge.
Raper, J., Gartner, G., Karimi, H. and Rizos, C. (2007) A critical evaluation of location based services and their potential. *Journal of Location Based Services* 1(1): 5–45.
Turner, A. 2006. *Introduction to Neogeography*. Sebastopol, CA, O'Riley Media.
Wood, D. 1992. *The Power of Maps*. New York, Guilford Press.

2

What does Google Earth Mean for the Social Sciences?

Michael F. Goodchild

National Center for Geographic Information and Analysis, and Department of Geography, University of California, Santa Barbara

Google Earth is an easy-to-use service for visualizing the surface of the Earth, and is readily extended to act as an output medium for a wide range of products of social-science research. It and other currently available geobrowsers closely approximate the vision of Digital Earth. Its designers overcame several apparently daunting technical problems, including the need for a hierarchical data structure and clever level-of-detail management. While the service is available to all, its use relies on fundamental spatial concepts, some of which are highly technical. Besides acting as an output medium, Google Earth presents a subject for social research in its own right, and there is a pressing need to address some of the issues identified in the earlier social critiques of cartography and geographic information systems. Several issues are identified that might, if addressed, lead to future geobrowsers that better meet the needs of social scientists.

2.1 Introduction

Google Earth was launched in early 2005, and quickly captured the popular imagination. Anyone with minimal computing skills, a basic personal computer and an Internet connection was able to download the software, see an image of the Earth, zoom from global to local scales, and simulate flight over prominent features such as the Eiffel Tower. The service came

Geographic Visualization Edited by Martin Dodge, Mary McDerby and Martin Turner

close to meeting the vision sketched by US Vice-President Al Gore in a speech prepared for delivery in early 1998:

> Imagine, for example, a young child going to a Digital Earth exhibit at a local museum. After donning a head-mounted display, she sees Earth as it appears from space. Using a data glove, she zooms in, using higher and higher levels of resolution, to see continents, then regions, countries, cities, and finally individual houses, trees, and other natural and man-made objects. Having found an area of the planet she is interested in exploring, she takes the equivalent of a 'magic carpet ride' through a 3-D visualization of the terrain. (www.isde5.org/al_gore_speech.htm)

Google Earth's technology was far less sophisticated than Gore's Digital Earth, since it could run on almost any home computer and did not require the museum's immersive environment. However, it also fell somewhat short of the full vision, since the speech talks of being able to explore Earth's history, and to visualize Earth futures as the outcomes of simulation models.

The idea of a mirror world, an integrated representation of all that is known about the planet, is much older than Gore's speech, and older than Gore's earlier and much less developed reference in his 1992 book *Earth in the Balance* (Gore, 1992). Humans have a long history of Earth representation, from the earliest drawings on cave walls to the first printed maps and the first images from space. Digital Earth represents the culmination of a long process that includes the advances made in the digital representation of maps in the 1960s, the development of sophisticated three-dimensional graphics, the popularization of the Internet in the 1990s, and the development of massive stores of digital geographic data in the form of data warehouses and geoportals (Maguire and Longley, 2005).

The 2005 launch stimulated a vast number of activities. The release of an application program interface (API) allowed more sophisticated users to add content to the service using KML (Keyhole Markup Language), and Google Earth is now just one of a family of *geobrowsers* that includes Microsoft's Virtual Earth (http://maps.live.com), ESRI's ArcGIS Explorer (www.esri.com/software/arcgis/explorer/index.html) and NASA's open-source World Wind (http://worldwind.arc.nasa.gov/). This chapter explores the significance of geobrowsers for the social sciences, using Google Earth as its example. The next section provides a brief review of the service and the technical challenges that it overcomes. This is followed by a section on the fundamental spatial concepts that underlie the service, and that should be understood by anyone seeking to make effective use of it. The following section addresses issues of the social context of Google Earth that are of interest to academic social scientists, and the final section addresses some of the challenges for the research community that remain to be solved if future generations of geobrowsers are to go further in meeting the needs of the social sciences.

2.2 Major features of Google Earth

2.2.1 Hierarchical tessellation

Paper maps must of necessity flatten the Earth, and over the past five centuries a highly sophisticated technology of map projections has evolved (Grafarend and Krumm, 2006;

Snyder, 1993). Each projection represents some compromise between well-defined objectives. The familiar Mercator projection, for example, is conformal, meaning that it avoids distorting the shapes of small features, but in doing so must distort their areas, making features at high latitudes, such as Greenland, disproportionately large, and failing entirely to portray the Poles. The projection was originally devised by Mercator for navigation, since a course sailed on a constant compass bearing or rhumb line appears straight on the map.

Globes avoided this problem almost entirely, but physical globes were difficult and expensive to make and store. However, by the 1990s the technology of computer graphics had advanced to the point where representation of three-dimensional objects was straightforward. Specialized hardware was needed to display and manipulate three-dimensional objects on screen, but by 2000 some high-end personal computers were being equipped with sufficiently powerful graphics cards, in part in support of video games. This was the environment in which Keyhole developed Earthviewer, initially for intelligence applications, and into which it launched its first publicly available geobrowser in 2000. By 2005, when Google acquired Earthviewer, redesigned its user interface and rebranded it, the necessary graphics cards were part of virtually every personal computer.

The notion that virtual globes avoid the distortions of map projections is somewhat fallacious, of course, since images must still be projected onto the flat screen of the digital display. However, virtual globes use a single map projection to do this, the Perspective Orthographic Projection, which is also the projection inherent in the human visual system. Moreover the user perceives the result as a three-dimensional object, a sense that is reinforced by the ability to manipulate the object in real time. Many users of Google Earth have derived vicarious pleasure from being able to spin the globe in the wrong direction, or to spin it about an arbitrary axis. Nevertheless there are still good reasons for knowing about more traditional map projections. Paper maps will be with us for a while, and many projections allow the whole Earth to be seen at once, which may be important in some domains.

Virtual globes require the ability to zoom rapidly from coarse to fine resolutions, over as much as four or five orders of magnitude. Since the early 1990s there has been much research interest in suitable hierarchical data structures, and in what are known as *discrete global grids*. Many rely on concepts of recursive subdivision, in which a large area of the globe is divided into finer and finer nested elements. Dutton (1999) has written extensively about one of these, which he terms the *Quaternary Triangular Mesh* (QTM), and others have researched aspects of its structure and algorithms (Goodchild and Yang, 1992). In this scheme the Earth is first divided into the eight triangles of an octahedron, with vertices at the Poles and at four points around the Equator (0, 90°W, 90°E and 180°). Each triangle is then subdivided into four, and this process can be continued indefinitely (Figure 2.1). A different scheme is used by ESRI's ArcGlobe, which is based on an earlier product known as GeoFusion. The Earth is first divided into the six faces of a cube, and the faces are then divided recursively into triangles. The use of triangles in both of these schemes is advantageous because they can be rendered efficiently by graphics cards.

Hierarchically nested geographies exist beyond the world of discrete global grids in a variety of forms, from the administrative hierarchy of nation–state–county to the hierarchical embedding of drainage basins and Japanese urban addressing. In QTM and other discrete global grids the hierarchy is geometrically regular, and every element is given a unique code that grows longer as the elements grow smaller and more numerous. Location can be defined

Figure 2.1 Dutton's Quaternary Triangular Mesh, a discrete global grid based on recursive subdivision of the faces of an octahedron

by a single numerical string of variable length, rather than the pairs of strings of latitude and longitude, and spatial resolution is always explicit.

2.2.2 Massive volumes of data

The Earth's surface has approximately 500 million square kilometres. To picture the surface at a spatial resolution of 1 km requires 500 million data elements, and at a spatial resolution of 1 m there are approximately 500×10^{12} data elements. Feeding such volumes of data from the Google Earth server to a personal computer creates truly daunting problems. For example, assuming that each data element consists of a single byte (allowing 256 possible values), it would take 69.4 working years to send the whole Earth at 1 m resolution over a T1 line, the communication speed that is typical of many offices but generally faster than home broadband connections. Over a 56 kbit phone modem the same task would take 12 400 years.

Google Earth manages to create images at these resolutions using a number of clever tricks and strategies. First, there is never any point in downloading data at a resolution finer than that of the computer's screen, which means that only on the order of 1 million data elements at the appropriate spatial resolution need to be present at any time. Second, by requiring the user to download software rather than using a standard browser, Google Earth is able to store data locally, and download only data that are newly needed. Finally, *level-of-detail management* allows lower levels of spatial resolution in the periphery of the field of view and during rapid panning. As a result, it is possible using a fairly modest Internet communication speed to create real-time flights over the Earth's surface.

2.2.3 The API

Google Earth's API allows users not affiliated with Google to create their own applications and extensions. Suppose, for example, that the following text is created:

```
<?xml version="1.0" encoding="UTF-8"?>
<kml xmlns="http://earth.google.com/kml/2.1"> <Placemark>
<Point>
<coordinates>-119.844973,34.415663,0</coordinates>
</Point>
</Placemark> </kml>
```

and stored as a file with the name 'office.kml'. Clicking on the file name will execute Google Earth, pan and zoom to the UC Santa Barbara campus, and add a placemark over the author's office location at latitude 34.415663 north, longitude 119.844973 west (six decimal places corresponds to a precision of roughly 10 cm on the Earth's surface). Similar scripts will paste coloured patches, images, three-dimensional structures and many other kinds of features on the Earth's surface.

The script example above may appear somewhat daunting, and much more elaborate scripts are needed for more complex tasks. A far easier approach is to make use of one of the many tools that have been developed to convert data of various kinds into KML files. For example, Arc2Earth (http://www.arc2earth.com) is a simple-to-use third-party extension to the popular geographic information system (GIS) ArcGIS. It allows the user to exploit the power of Google Earth as a mechanism for publishing the results of GIS analysis.

Google published KML shortly after the release of Google Earth, and since then a vast array of extensions or *mash-ups* have been created. Many are available via Google's own bulletin boards, and many others through simple Web search. An example of interest to social scientists is gecensus.stanford.edu, which creates mash-ups of data from the 2000 US census.

2.3 Fundamental spatial concepts

Google Earth provides a distinct perspective on the Earth, allowing users to view its surface in varying amounts of detail and to display assorted aspects of any location. Underlying this perspective is a number of concepts that form the framework for any examination of the geographic world. In this section several of these are reviewed. Much longer lists exist (see, for example, de Smith, Goodchild and Longley, 2007; www.spatialanalysisonline.com), but while many of the more advanced concepts are relevant to GIS in general, they are of limited interest here given the lack of analytic functionality in Google Earth.

2.3.1 Georeferencing

Specification of location on or near the Earth's surface is clearly central to the operation of Google Earth, which relies primarily on latitude and longitude coordinates. Definitions of latitude and longitude require some assumed mathematical form for the shape of the Earth, which, although approximately spherical, is actually a rather lumpy solid. The geoid is defined as the surface of the oceans, after removing tidal and weather effects, and its imaginary continuation to create a surface of equal gravity under the continents. Global mapping approximates the geoid by an ellipse rotated about the Earth's axis, and since the mid 1980s has settled on the particular parameters known as the World Geodetic System of 1984 (WGS84). However, many countries still maintain their own systems, with the result that maps of adjacent countries may not fit along their borders. GPS receivers will often provide a number of options, including WGS84 and local variants, and the differences can amount to hundreds of metres of displacement in some cases. Thus data exported from a mapping package or GIS may well not fit perfectly over the Google Earth base. Many Google Earth users have commented on the fact that the line of zero longitude misses the Greenwich Observatory by roughly 100 m, a shift that occurred with the adoption of WGS84.

Google Earth also includes various facilities for converting other forms of georeference. Placenames are handled by a gazetteer, which converts recognizable names to coordinates, and queries the user about ambiguous names that might refer to any one of a number of locations. Street addresses require a geocoding service, which is only available for the most developed countries.

2.3.2 Features

The images captured by the human retina or by the imaging sensors of Earth-orbiting satellites are composed of continuous gradations of colour. The human brain and, to a lesser extent, the image processing and pattern recognition systems of today's computers are capable of extracting meaning from such images, primarily by identifying the scene's discrete features. Images are partitioned into trees, vehicles, buildings, mountains, lakes and many other feature types, and perhaps named and otherwise characterized.

In the digital world such features are likely to be represented as points, lines, areas or volumes, with associated characteristics or attributes. When displayed visually, as in Google Earth, points will be converted to symbols, lines will be given some appropriate thickness, and areas or volumes will be coloured or shaded. However, this concept of discrete features is only one of two fundamentally different ways in which the geographic world is represented digitally. The base imagery and topography of Google Earth are both conceptually continuous, and stored and processed as such. Base imagery is rendered more or less as it was obtained from the original acquisition system, though the basic picture elements may have been transformed. Topography is rendered by varying the third spatial dimension, and by computing perspective based on the assumed location of the user's eye.

This duality between what are known as the discrete-object and continuous-field conceptualizations of geographic reality pervades the world of GIS. In Google Earth only the base imagery and topography are conceptualized as fields, although additional imagery can be mashed onto the base. Other layers are almost entirely composed of discrete objects, and it would be difficult to render other field-like phenomena, such as atmospheric temperature or soil moisture content, within the current environment.

2.3.3 Layers

A fundamental concept in GIS is the layer, based on the principle that geographic features can be separated into distinct classes or sets, and that the Earth's surface can be characterized through the overlaying of such layers. Layers are analogous to the separate plates used to print the various colours of topographic maps, since each colour represents a different class of phenomena. In Google Earth, imagery constitutes a patchwork base layer, and all other features are superimposed on it. Linear features such as roads and rivers, and point features such as landmarks, present no problem to this approach. However, when area features such as census tracts are coloured by variables such as population density, the result is a layer that obscures all lower layers, including the base mapping. In such cases it is helpful to make overlaid layers partially transparent.

It is impossible to determine the location of any feature on the surface of the Earth exactly, since any method of measurement has inherent uncertainties. It follows therefore that no two layers will fit precisely, unless one has been derived from the other or has somehow inherited the same positional errors. Misfits of as much as 20 m are common with the layers supplied by Google, such as base imagery and roads (Figure 2.2), and larger errors will tend to occur when data are imported from other sources, perhaps because of issues over the definition of latitude and longitude (see previous section), or because of poorly measured positions. In such cases the normal fix is to zoom out so that misfits are not bothersome, but this may be problematic in very detailed applications.

Figure 2.2 An example of Google Earth misregistration. The roads layer is displaced roughly 20 m to the west with respect to the base image. The small superimposed high-resolution image was georegistered using control points surveyed with GPS, and matches the roads layer but not the Google Earth base image

2.3.4 Spatial context and neighbourhood

One of the great benefits of georeferencing is the access it provides to other information about places. Insights are often to be gained by looking at information in geographic context, since context can often suggest explanations or additional hypotheses. Context can be interpreted in two distinct ways in the context of Google Earth, and indeed any geographic information technology. Horizontal context concerns neighbouring features, and the degree to which they may explain events or conditions. Vertical context concerns the role of features on other layers, and the possibility that events or conditions at some location are attributable to other events or conditions at the same location. Both vertical and horizontal context are readily apparent in Google Earth. The mashup of sites associated with the novels of Jane Austen (http://bbs.keyhole.com/ubb/showflat.php/Cat/0/Number/411188/an/0/page/0) is an excellent example of using Google Earth to provide geographic context to a body of literature and its author.

2.3.5 Scale

Scale has two distinct meanings in a geographic context: the extent of an area of interest, and the degree of detail or resolution of its representation. A large-scale study might refer to one covering a large area of the Earth, or to one that examines some area at coarse resolution. To confuse things further, scale in a mapping context has often meant the ratio of the distance between two points on the map to the distance between the same pair of points on the Earth's surface, known as the representative fraction (RF). In this context a 'large' ratio or scale indicates a map with a fine rather than coarse level of detail.

Google Earth handles scale through its zoom function, allowing the user to slide from coarse to fine and back again. The resolution of the screen defines the fineness or coarseness of the image, and limits the amount of data that must be sent over the Internet to create it. Distance on the Earth can be determined using the measuring tool, and the concept of representative fraction, which is meaningful only for paper maps, is effectively abandoned.

The resolution of base imagery varies markedly over the Earth, from less than a metre in some urban areas to tens of metres elsewhere (Figure 2.3). Resolution is a function of the instrument used to acquire the imagery, which in Google Earth is created by stitching together data from a large number of sources, including Earth-orbiting satellites and aircraft.

2.4 The social perspective

Google Earth presents social scientists both with a powerful tool for displaying the results of research, and with a topic worthy of study in its own right. This section addresses some of the issues that arise in that latter context. Like many other forms of geographic information technology, Google Earth at first sight appears neutral in its impacts, and it is hard to imagine its developers expressing much concern over the subtle ways in which it might influence the world. At a somewhat deeper level, however, such technologies raise interesting and in some cases disturbing questions.

Figure 2.3 Boundary between high-resolution (approximately 1 m) and low-resolution (approximately 20 m) base imagery southeast of Bristol, UK

Many of these questions originate in work on the social significance of mapping that began in the late 1980s, primarily with the research of Brian Harley (2002). Denis Wood's *The Power of Maps* (Wood, 1992) carried this process of deconstruction further, asking whether something as apparently innocuous as a road map might somehow reveal an agenda on the part of its makers. By the early 1990s the growth of interest in GIS, and the lack of concern for its social implications, led to the publication of John Pickles's *Ground Truth* (Pickles, 1995), which raised a host of questions about the social impacts of this burgeoning technology. Those issues that clearly apply to Google Earth are reviewed in the following sections.

2.4.1 A mirror world

Much of Google Earth's appeal and ease of use stems from its apparently faithful replication of the visible features of the planet's surface. The base imagery provides a reasonable rendering of how the surface looks at close to local noon; the topography provides a faithful and unexaggerated relief, and while roads and point features are symbolized, their rendering is close to the familiar one of maps and atlases. It is not too much of a stretch, then, to assume that users feel they are looking at a facsimile of the real world, and some first-time users may even assume that the imagery is current to the minute.

In reality, of course, Google Earth presents a far-from-faithful rendering. The age of imagery varies, and it may be difficult to determine exactly when a given area of imagery was acquired. In March 2007 US Congressman Brad Miller drew attention to the apparent replacement of New Orleans imagery with coverage dating from before 2005's Hurricane Katrina, noting that:

> To use older, pre-Katrina imagery when more recent images are available without some explanation as to why appears to be fundamentally dishonest . . . [and] appears to be doing the victims of Hurricane Katrina a great injustice by airbrushing history. (www.topix.net/us-house/brad-miller/2007/03/subcommittee-criticizes-google-images)

Spatial resolution also varies, and while one can appreciate why areas of Iraq and Darfur should be portrayed at sub-metre resolution, given the current worldwide interest in those areas, it is harder to understand why imagery for some areas of England is no better than 30 m and partially obscured by cloud, while some rural areas of Peru are visible at 1 m.

The view of the world available from Google Earth is essentially that from above. This is also the view of traditional cartography, of course, and has been the subject of extensive critique (see, for example, the feminist critique of Kwan, 2002). Google Earth's has been termed a 'God's-eye view' and 'the view from nowhere'. Where representations of three-dimensional structures are available (mashups of buildings are available for many major cities), it is possible to fly among them, and Microsoft's Virtual Earth goes further by allowing the user to browse images captured at ground level from moving vehicles. Nevertheless the view remains essentially static and distanced.

2.4.2 Privacy

High-resolution images capture information that is often beyond the reach of normal ground-based observation. One can see into backyards and into gated communities, and monitor compliance with regulations regarding brush-clearing and irrigation, as well as possibly illegal activities such as the growing of marijuana. Sub-metre imagery is capable of detecting cars parked in driveways or the destruction of villages in Darfur. The base imagery of Google Earth is in many cases of higher quality than the best mapping publicly available, especially in countries such as India where mapping is regarded as an issue of national security. During the Hurricane Katrina recovery effort, large amounts of information on property damage were available via Google Earth, including photographs taken on the ground. It is clear that Google Earth, and the other geographic information technologies to which it is linked, are capable of an unprecedented level of vision into the daily lives of individuals.

2.5 Research challenges

The first generation of geobrowsers as represented by Google Earth clearly present significant opportunities to the social sciences, both as tools and as subjects for research. Several of the former are reviewed in this final section.

2.5.1 Layer matching

While the misfits between layers are often tolerable, there are inevitably applications that would benefit from their removal. This function is known generally as rubber-sheeting, and is achieved by warping one layer so that it correctly fits an established base. Tools for this purpose exist within the wider GIS field, and could be adapted to the Google Earth environment where ease of use is a primary concern. Misfits also occur in a horizontal sense, when pieces of imagery fail to align, as often occurs where fine-resolution imagery meets comparatively coarse imagery. Again, tools to edge match exist in the wider world and could be adapted.

2.5.2 Rendering the non-visual

Most properties of interest to social scientists are not visible from above, and do not fit the mirror-world metaphor of Google Earth. Cartographers have long recognized the need to map such properties, and have devised techniques such as choropleth mapping for the purpose. However, these are not as appropriate for Google Earth, with its emphasis on the visual and on fine-resolution base imagery. Mashups of census data, for example, look hopelessly crude when made using standard techniques (Figure 2.4). There is a need for fresh thinking on this topic that exploits the rich set of options available in Google Earth.

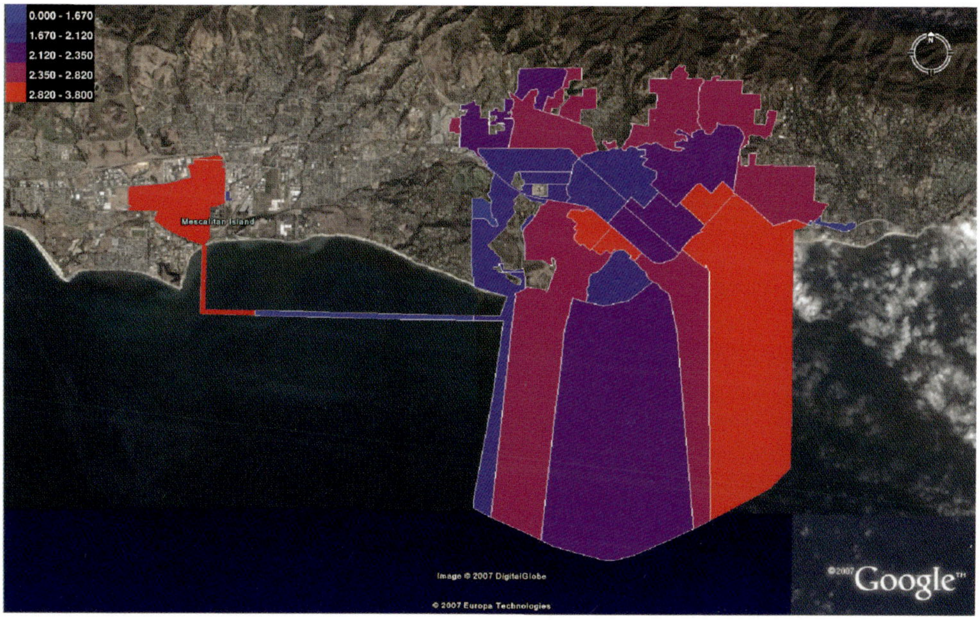

Figure 2.4 Mashup of census summary data by census tract for the city of Santa Barbara. The variable is average household size. It is standard census practice to extend coastal tract boundaries over the ocean. The area to the west includes the airport and lies within the city limits, but California law requires contiguity in city boundaries, hence the thin corridor. Google Earth includes an option to make the census overlay semi-transparent

2.5.3 Analysis and modelling

Google Earth is designed primarily as a tool for visualization, and its functionality for analysis is limited to a few simple measurements of distance and area. GIS users who are familiar with the much greater analytic power of their software have expressed interest in adding to the analytic capabilities of Google Earth, and in possible intermediate products. ESRI's ArcGIS Explorer (www.esri.com/software/arcgis/explorer/index.html) may develop into a leader in this latter category. Analysis and modelling on the curved surface of the Earth is far from straightforward, requiring radically different methods from those devised for a flat two-dimensional world (Raskin, 1994). Yet such capabilities would be invaluable in the analysis of such global phenomena as climate change and trade.

2.5.4 A global data source

The concept of a global map dates from the nineteenth century, when the first proposals surfaced for a uniform coverage of the planet's surface. Today we are close to achieving that goal, given the power of satellite remote sensing and the ability of geobrowsers to integrate data from a patchwork of sources. At the same time many other sources of geographic data are in decline, as a result of cutbacks in funding for national mapping agencies, particularly in less-developed countries. This is especially true of features and properties that cannot be detected in images, such as placenames and social variables. Google Earth has the potential to act as a global source and as a technology for integrating data, but its position in the private sector is problematic. Efforts are under way to adopt standards across competing geobrowsers, and there are increasingly interesting options for both software and data in the public domain. Nevertheless, the goal of a global spatial data infrastructure and a single portal to all that is known about the surface of the Earth remains elusive.

2.6 Conclusions

Google Earth provides at least a first approximation to the vision outlined by Vice-President Gore in 1998. It has captured the imagination of a vast number of people, many of whom had no prior experience in any form of geographic information technology. From the perspective of GIS, it represents a distinct democratization, in which people without any training in the basic concepts of spatial science are able to access at least some of its tools. This includes many social scientists, who for the first time have access to comparatively simple ways of displaying georeferenced data, and gaining the insights that a spatial perspective can provide.

Google Earth and other geobrowsers address what previous generations of developers had seen as insuperable challenges: feeding vast amounts of data through comparatively limited Internet pipes, manipulating three-dimensional images in real time, and zooming through a hierarchical data structure over at least four orders of magnitude of resolution. The software is robust, and designed with a user interface that a child of 10 can learn in 10 minutes. Nevertheless, the concepts that underlie Google Earth are sophisticated, and while most are intuitive, others, such as the basis of georeferencing, are highly technical.

The potential exists for serious misinterpretation and misuse, and there is a growing need for basic education in these and other spatial concepts.

Geobrowsers offer enormous potential to social scientists, both as tools for visualization and as subjects of research. The social implications of fine-resolution imagery and of a proprietary view of the world are profound, and demand attention. Much has been written about the social impacts of GIS over the past two decades, and this work badly needs to be extended to the developing field of geobrowsers.

Finally, Google Earth and its competitors are only a first approximation to the concept of Digital Earth, and could be of much greater value if some obvious limitations could be addressed. From the perspective of the social sciences, the focus on content that is visible from above is problematic, given the abundance of more abstract data sources. A new generation of techniques is needed that can mash such data with the Google Earth base, creating more powerful ways of communicating what social scientists know about the surface of the Earth.

References

de Smith, M. J., Goodchild, M. F. and Longley, P. A. (2007) *Geospatial Analysis: A Comprehensive Guide to Principles, Techniques and Software Tools.* Winchelsea, Winchelsea Press.

Dutton, G. (1999) *A Hierarchical Coordinate System for Geoprocessing and Cartography.* Lecture Notes in Earth Sciences 79. New York, Springer.

Goodchild, M. F. and Yang, S. (1992) A hierarchical spatial data structure for global geographic information systems. *Computer Vision, Graphics and Image Processing: Graphical Models and Image Processing* 54(1): 31–44.

Gore, A. (1992) *Earth in the Balance: Forging a New Common Purpose.* London, Earthscan.

Grafarend, E. W. and Krumm, F. W. (2006) *Map Projections: Cartographic Information Systems.* Berlin, Springer.

Harley, J. B. (2002) *The New Nature of Maps: Essays in the History of Cartography.* Baltimore, MD, Johns Hopkins University Press.

Kwan, M.-P. (2002) Feminist visualization: re-envisioning GIS as a method in feminist geographic research. *Annals of the Association of American Geographers* 92(4): 645–661.

Maguire, D. J. and Longley, P. A. (2005) The emergence of geoportals and their role in spatial data infrastructures. *Computers, Environment and Urban Systems* 29(1): 3–14.

Pickles, J. (ed.) (1995) *Ground Truth: The Social Implications of Geographic Information Systems.* New York, Guilford.

Raskin, R. (1994) *Spatial Analysis on the Sphere.* Technical Report 94-7. Santa Barbara, CA, National Center for Geographic Information and Analysis.

Snyder, J. P. (1993) *Flattening the Earth: Two Thousand Years of Map Projections.* Chicago, IL, University of Chicago Press.

Wood, D. (1992) *The Power of Maps.* New York, Guilford.

3

Coordinated Multiple Views for Exploratory GeoVisualization

Jonathan C. Roberts

School of Computer Science, Bangor University

3.1 Introduction

Over recent years researchers have developed many different visualization forms that display geographical information, from contour plots and choropleth maps to scatter plots of statistical information. Each of these diverse forms allows the user to see their data through different viewpoints. In fact, it is often the case that, when the user sees the information through different views and in different ways, they get a deeper understanding of the information.

Geographical databases often hold a diverse range of different data and are thus complex to understand. For example, spatial data may be held that explains a particular geographical landscape, the land usage or details of buildings found on that land; non-spatial elements may also be stored which detail land ownership, the salary of the land owner and other statistical information about the land owner. The users' goal is to gain knowledge of that information and make sense of a large volume of potentially diverse data with multiple components and different data types.

Thus, to comprehensively understand the information contained within any complex geographical database, the user would need to select some information to display, present it in different forms, manipulate the results, compare objects and artefacts between views, roll back to previous scenarios or previous sessions, understand trends by seeing the data holistically as well as specifically, and finally take measurements of objects and areas in the display. Consequently, developers have created specific exploration environments that utilize coordinated multiple view techniques (CMV), where each of the views are linked together such that any user manipulation in one view is automatically coordinated to that

Geographic Visualization Edited by Martin Dodge, Mary McDerby and Martin Turner

of any other and the environment. These exploratory visualization (EV) environments are highly interactive systems and rely on the premise that 'insight is formed through interaction'.

If we, as developers and researchers, are to provide the best possible environment for the users, we need to take stock of what functionality is being provided by current tools and examine what we are doing well and what we are perhaps not doing so well. This chapter is a systematic analysis of exploratory geovisualization using CMV techniques. We consider the strengths and weaknesses of the area and make a comprehensive review of CMV for geovisualization.

3.1.1 Exploration

The goal of exploration is to search, locate and find out something new. A user starting out on the investigation process may not know anything about the data, let alone the questions to ask. Thus the system should help the user to not only display visual results of the information but also browse and locate pertinent information. Visual exploration, in particular, enables the user to visually investigate the data. In short, the visual exploration enables the user to try out some parameters, instantly view the results of the parameter change, manipulate the data through selection and highlight operations and relate that information to other sources and visualizations. The user may directly select some interesting elements using a bounding box tool, see the selected list of results in both a scatter plot visualization and a textual list of results, and perhaps edit the items on the selection list by adding some more or deleting some in the list. Thus exploration is part of a larger discovery process.

Various researchers have described this discovery process by a variety of models. In geovisualization specifically DiBiase (1990), focusing on the role of visualization in support of earth science research, summarizes this discovery process as four stages. First, exploration reveals the questions and achieves familiarity with the data through testing, experimenting, acquiring the right skills and learning about the underlying model. Second, the user needs to confirm the relationships that exist in the data though comparison operations, relating the information to other explorations and disproving other hypotheses. Third, the results are synthesized; this is achieved by identifying the pertinent features and summarizing the content. Fourth, and finally, the results are presented and they are taught and demonstrated at professional conferences and in scholarly publications.

DiBiase's model emphasizes the uncertainty and goal-seeking nature of the discovery process; it is a conceptual model that encourages the user to personally explore the information, thus to gain a better understanding of the information before presentation to a wider audience. Obviously, these traits are highly important in exploratory geographical visualization and developers need to decide how these features map to individual tools. However, the model de-emphasizes the processes of data preparation and simplifies how the results are gathered and presented to the user.

Sense-making models, on the other hand, are more goal-oriented models and are often discussed in the context of intelligence analysis. These models emphasize the whole process from data preparation to hypothesis presentation (Thomas and Cook, 2005). Specifically, sense-making according to Russell *et al.* (1993) 'is the process of searching for a representation

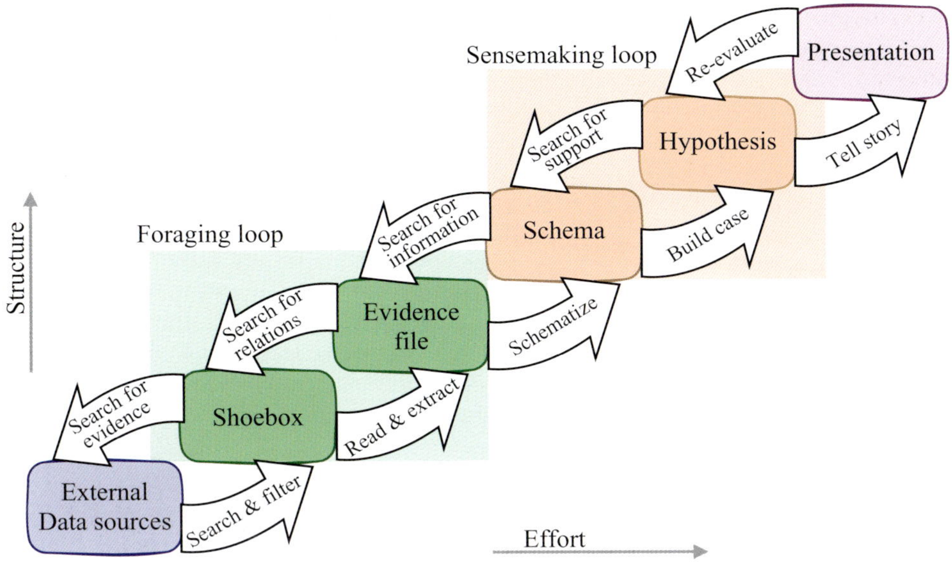

Figure 3.1 Sense-making schema model by Pirolli and Card (2005). Where the data is processed, the user visualizes the information and confirms relationships with the data and then presents their conclusions to the client

and encoding data in that representation to answer task-specific questions'. One of the most comprehensive sense-making models is by Pirolli and Card (2005), Figure 3.1. In their schema model the analyst searches and filters the data for relevant information, which provides a demonstration set that is stored for future reference (known as a *shoebox*). Specific and relevant information or inferences may then be extracted and stored to provide evidence files. Then the data is structured, organized and represented in a schematic way. This is both to highlight interesting facts and to confirm relationships within the data; it matches with the first and second stages of DiBiase's model. This organizational part of the process may be achieved through the representation of the data by an informal diagram or a complex visualization. Then the case is built, and finally presented to the client. This is a process-oriented model, where a developer can clearly map the stages into processes of an exploratory visualization tool.

Exploration itself is a process that tends to generate many possibilities and widens the search space. Techniques of generating multiple views, selecting and highlighting elements, zooming and displaying additional detail all help the user to explore. However, the user still needs to draw conclusions and present that information to colleagues. Thus, for exploration tools to be successful, developers need to provide methods to both widen and narrow the solution space, support the user through an exploration and support them in their presentation of that learned information.

Therefore, exploratory environments require a wide range of features: (1) tools that perform effective *data preparation*; (2) informative *visualization techniques* that display the information comprehensively and clearly and that display the information in different forms; (3) *interaction* techniques that allow the user to manipulate the information; and (4) extensible and easy-to-use *tools and toolkits*. These four parts are extremely important for exploratory visualization. They match well with the models of both DiBiase (1990) and Pirolli and Card (2005); hence, in the next few sections we discuss issues of where we are with exploratory geovisualization under these four headings.

3.2 Data preparation

In our information age data is being produced at a phenomenal rate. In some projects data is being produced at such a rate that is only possible to store it, and with current technologies and techniques it is often impossible to analyse it fully. An ever-increasing amount of data is at high resolution, with companies and agencies creating and processing high-quality digital land use and socio-economic data with spatial attributes. With the ubiquity of phones and global positioning devices, more spatial and time-stamped data is being stored.

Additionally, such geospatial databases are collected by different researchers, created by an assortment of sensors, saved in different formats, held physically at various locations around the world and incorporate many variables and types of data. Hence, the user is faced with several challenges when they wish to integrate the geographical data from multiple databases, across different domains and potentially use the data for a purpose that was not originally envisaged. Each of these challenges impinges upon Exploratory Visualization.

3.2.1 Too much data!

The total quantity of data is often one of the most crucial challenges for a developer to consider. Size certainly impinges upon the processing time. It takes longer to process huge amounts of data, especially because users wish to mine similarities and structure within the information, hence parts of the data need to be correlated with other elements. Friesen and Tarman (2000) write: 'even at tomorrow's gigabit-plus networking bandwidths, sharing these enormous data sets with distributed sites becomes impractical, and we'll need high performance visualization resources to have any hope of near-real-time interaction or observation of this data'. Solutions to processing this huge quantity of data include: parallel and distributed computing or remote computing solutions (Harwick, Coddington and James, 2003), and more recently interest has turned toward service- and grid-based architectures (Aktas *et al.*, 2005; Brook *et al.*, 2007). Spatial datasets certainly bring specific challenges that high-performance computing methodologies can address, but in reality few researchers have looked at solutions to these challenges. Reducing the size of the data obviously would make the exploration more interactive.

Methods such as filtering generate smaller demonstration sets (shoebox datasets) that would be processed more readily, take up less screen space and allow the user to focus on relevant data. Spatial filtering may be achieved through cropping, sampling, averaging (binning), partitioning, clustering and aggregation (Tang and Shneiderman, 2001). These techniques can be achieved through pre-processing the data or closely linking the filtering with the visualization (dynamically visualizing the results). Dynamic queries (Ahlberg and Shneiderman, 1994) permit the user to directly interact with the visualization by adjusting sliders, buttons and menu items to filter the data and instantly view the result in the display. Furthermore, other pre-processing may be achieved through knowledge discovery (KD) and data mining (DM) methods to find patterns in large spatial datasets (MacEachren *et al.*, 1999). In addition, the plain quantity of data often exacerbates the problem. Not only does it slow the interactive process, but saving and loading the data takes time and effort. Hence caching techniques would also speed up the interaction.

Storing and caching the processed (shoebox) data are important not only for interactivity, but they can also be utilized by the investigator as a temporary copy or by other investigators to confirm a hypothesis. Thus, there is a need for both curation of geographical data (especially shoebox information) and storage of information on the provenance of the data and the session. For instance Thomson *et al.* (2005) discuss issues of lineage and provenance in geovisualization and Koua and Kraak (2004a–c) discuss some ideas on tools to support 'knowledge construction throughout the exploration process', but much work is still to be achieved.

While filtering methods pre-process or dynamically select a smaller demonstration set, other processing methods aim to group, aggregate and assign structure to the data. Patterns that are assigned through KD and DM techniques can be visualized. Many of the techniques are non-trivial as it is often not possible to provide brute-force algorithms to find similarities within the data (Miller and Han, 2001). Parameters of these algorithms can be tightly coupled to the interface, such that the results are consequently updated (Maceachren *et al.*, 1999; Koua and Kraak, 2004a–c). For example, by adapting values to (say) the *k*-means algorithm, the user can explore different partitioning configurations (Torun and Duzgun, 2006). Thus, it is possible to combine knowledge discovery, data mining and visualization techniques. In fact, various researchers have called for the closer integration of data mining and visual exploration techniques (Kreuseler and Schumann, 2002; Kocherlakota and Healey, 2005). Classification hierarchies can be used to explore the data (Kemp and Tan, 2005); these can be displayed as graphs and other structured organizations (Rodgers, 2004). Classifications are important, and if, in particular, they are visualized, they allow the user to understand the underlying structure of the data and furthermore can be used for filtering and processing. Diverse categorizations provide different ontologies (Arpinar *et al.*, 2006). These classifications can then be visualized separately and thus provide multiple views on the same information; see Section 3.3. Not only are there challenges with processing the quantity of data, but there are also challenges in displaying it all at once. Venkataraman *et al.* (2006) write: 'LIDAR data can easily exceed 100 Mpixels making it impossible to have detail as well as context on a single desktop display'. Some researchers, particularly from the database community, have looked at pixel-based visualizations (Keim, 2000) and some work has been done in spatial visualization using pixel displays (Keim, *et al.*, 2004), but there is still a limited number of pixels on a screen and hence a limit to the amount of information that can be displayed at once.

The crucial factor is not just the size but the quantity of pixels it can display. High-quality large-format cinema-style video projectors are still very expensive and have huge running costs, typically because of the short life-span of the bulbs. Hence researchers have developed tiled screens. Such multiple display devices have been successfully used for geospatial exploration (Venkataraman *et al.*, 2006). However, these displays have their problems. They need special device drivers to control all the multiple displays, the screens need to be housed in a customized rig, there may be problems with the viewing angles, and there will always be a gap between each screen due to the screen's housing.

3.2.2 Complex and diverse datasets

Not only is it useful to view the information through different forms, and gain better understanding through looking and manipulating the data, but value can be gained by algorithmically fusing information from assorted databases. For instance, a second database may provide information that was missing from the first, errors or discrepancies of the data may be highlighted through the amalgamation of data, and information may be inferred through the integration of the two datasets.

Furthermore, large and complex geographical datasets often contain missing or erroneous information. Missing data obviously affects how the user explores the information. For missing data, the developer needs to work out what to do with it: whether to ignore it by pre-filtering it out, make an assumption and substitute the information, or classify it as missing and treat it specifically. If it is going to be included then the data-processing algorithms may need to be re-written to specifically handle these cases. For instance, what happens when three values are averaged when one value is null? Likewise, erroneous data can be deleted or flagged as being potentially wrong, but again these concepts need to be fully integrated into the system, from data-processing and visualization to interaction and manipulation. If users are to explore the complete dataset, then it is imperative that the developer integrates methods that deal with this information. It is far too easy to develop tools that merely ignore this type of data. Often missing or erroneous information can help the user infer some fact or it can be used to support a hypothesis. Seminal work in this area is Unwin *et al.* (1996) in their MANET (Missings Are Now Equally Treated) system. However, current systems still do little to integrate this uncertain data and much work needs to be done here to fully integrate and allow the user to appropriately manipulate missing information.

3.2.3 Data processing challenges

The sheer size, complexity and diverse nature of geographical datasets definitely have consequences for exploratory analysis. Certainly researchers have developed some clever and useful algorithms to address many of these issues, but the main limitation is that they are not commonplace in most of the general purpose geovisualization tools. For example, techniques such as parallel algorithms or the use of remote high-performance computers have their benefits, but although much work has been done looking at such techniques individually and their application to geographical visualization techniques, they are not commonplace and not included natively in many exploratory geovisualization systems. Data abstraction techniques also have the desired effect of speeding up the processing of large datasets, and again a reasonable amount of research has been achieved. However, although visual filtering methods are in widespread use, there has been little work on tightly integrating data mining techniques with exploratory visualization techniques.

Furthermore, additional research on the curation of geographical data is required. That is, both the storage of temporary datasets from an exploration and also details of the operation history (the commands used in the exploration) should be saved. This would enable provenance of the exploration and permit additional researchers to reproduce and confirm any results.

The use of dense pixel visualization for the representation of abstract data is an area of research that certainly has potential. It permits large amounts of data to be displayed and creates a holistic view of the information. However, such techniques are currently underused in geovisualization.

Finally, there are certainly challenges with incorporating non-existent or erroneous data into a geovisualization system, although there are clear benefits to incorporating such data. Some work has been achieved in visualizing *uncertain information*, but researchers who initially store the data need to save and appropriately mark up the information, and system developers need to integrate uncertainty concepts throughout the whole system, such that uncertainty is a principal component of the system.

3.3 Informative visualizations

The canonical form for geographical display is certainly the map. However, there are many different types of visualization that are used in geovisualizations. In multiview exploratory systems these additional (non-map-based) realizations are often used alongside the map display; all user controls and operations may be associated with any or all of the views to provide a powerful exploratory environment. In this section we discuss various terminologies and implications of multiple views; we review a variety of display forms that are used in geovisualization and consider representation and re-representation methodologies. The forms detailed within this section are not meant to form a comprehensive review, but are intended to give the reader an understanding of the breadth of techniques that are currently available. Indeed, this book and others, such as *Exploring GeoVisualization* (Dykes, MacEachren and Kraak, 2005), provide many more examples of geovisualization forms.

3.3.1 Multiple views

The term *multiple-views* is all-encompassing; it includes any system which allows direct visual comparison of multiple windows, including visualizations from different display parameters. It usually implies that the visualizations are viewed in a desktop windows environment. However, researchers have also used the term to describe various tile-based displays. There are three further implications. The first is that the operations are coordinated and that any operation is linked with any other view, hence the more specific term of *multiple coordinated views* may be used. The second is that the user can instantiate any number of views that they require. The third is that the views are encapsulated in their own window. By having these multiple representations encapsulated in separate windows, the user can easily compare two (or more) representations side-by-side. In fact, there are many examples of developers utilizing *dual-views* (Convertino *et al.*, 2003) to allow the user to compare the information side-by-side.

The term *multiform* on the other hand is specific (Roberts, 2000). It describes how the information can be displayed using different types of representation. For example, one view may show the information as a map, while another window shows the information in a table. In this example, the user can select regions on the map and display the statistical information in the associated table. The environment may also allow the user to investigate another location on the map, and thus the numerical data can be displayed in another table, in a further window. There are other terms in the literature that have specific meaning. For instance, *alternative representations* has been used to explain multiple forms (Koua and Kraak, 2004a–c), but the phrase is also used to promote multiple opinions (Roberts and Ryan, 1997). Users may have a different understanding of the data and hence provide various interpretations of the phenomenon; for example, historic information may be sparse, there may be disagreements between eye-witness reports or interpretations of the data may vary. In archaeology, experts may disagree and their suppositions may change over time as they become influenced by other people's opinions. Alternative representations are also useful in education, in that the learner better understands a process if it is explained or demonstrated in multiple ways.

Multiple views are also used for control. Many of the views utilize direct manipulation, where the user directly controls the information in the window. Consequently, a user may find it easier to manipulate the data in one type of representation in comparison to another; for instance, the user is given two views, one being a three-dimensional network view showing the position of objects found from an archaeological excavation and the other an alphabetical list of the objects found. If the user wished to see where all the pottery vessels were located, then they could select 'pottery vessels' from the textual list to highlight all those in the three-dimensional view. However, if they wanted to see what information is located near a specific location then drawing a bounding box on the plot would be easier. This is also a demonstration of a *master and slave* relationship, where one view directly controls another. Another use of multiple views is to create *focus and context*, where one view depicts a summary of all the information with the other showing detail. In addition, both focus and context information can be displayed in one unified view using distortion techniques.

In a windows environment it is possible to gain information on demand through popup information; these popup views provide specific information dependent on the context in which it was requested (Geisler, 2000). In virtual reality a similar style of master and slave relationship exists. The concept is named *view on a bat* or *worlds in miniature* (Stoakley, Conway and Pausch, 1995). In this case the user has a virtual bat that moves around the virtual world with the user. Different pieces of information can be mapped onto the bat such as a geographical map of the area or some extra information that is relevant to the context of the user at that point in time. Similar concepts occur in other domains. For instance augmented table-top environments use markers to instantiate additional information (Hedley *et al.*, 2002).

Finally, the last type is *small-multiples* (Tufte, 1990). By displaying the data in many different small representations, the user can perceive trends in the data. It is through an overall perception of the texture that the user perceives trends in the information.

3.3.2 Multiple forms

There are many types of geographical visualizations. Inspired by the categorization of Lohse *et al.* (1994), the various geovisualizations can be divided into seven categories: Maps/Cartograms, Networks, Charts/Graphs, Tables, Symbols, Diagrams and Pictures.

Maps communicate spatial meaning and there is a direct association between the physical space and the represented space. There are clearly many different types of maps that are used in geovisualization. Geographical maps aim to represent proportions and the geography of our world and can be two- or three-dimensional. For example, elevation information, such as from LIDAR data, can be mapped to colour or height. Two-dimensional maps are often used as the main reference visualization, with other statistical information being layered on top. Furthermore, different types of data, such as agriculture, transportation, boundaries or population density, can be represented as different layers of the map. Other forms of maps include choropleths, where the land areas are shaded with values proportional to statistical measures of that land, while cartograms distort the map dependent on another statistical measure, e.g. the time it takes to travel from one point to another.

Networks describe relational and associational information, e.g. that a connects to b and then c. Networks include trees, hierarchies, routing diagrams and graph visualizations. A well-known hand-crafted network is that of Harry Beck's London underground map; it is really a network presentation because it represents relational information of the underground rail lines. In fact, aspects of Beck's map have inspired network layout algorithms (Stott and Rodgers, 2004). Network visualizations have been used in geovisualization to represent various types of associated data and have been realized through different layout strategies. Rodgers (2004) provides a good overview of graph techniques for geovisualization. Depictions range from node-edge graphs and treemaps to trails.

Charts display statistical or mathematical information. In this category we include line graphs, histograms, circular histograms, pie charts, surface plots, scatter plots and parallel coordinate plots (Edsall, 2003). Each of these visualizations addresses a specific need; line graphs and histograms visualize continuous data on a two-dimensional plot, surface plots display continuous data in three dimensions, bar charts and pie charts display quantitative information and parallel coordinates display multidimensional data, while scatter plots allow users to see clusters.

Tabular and matrix layouts are popular for displaying statistical quantities and numerical information contained within geographical databases. They are familiar forms and as such can be extremely useful. Spreadsheets enable large amounts of numbers to be shown and manipulated; with table lens views (which utilizes distortion techniques) even more information can be displayed (Rao and Card, 1994), while reorderable matrix techniques map the values into the size attributes of symbols to allow trends and similarities to be viewed (Siirtola and Mäkinen, 2005).

Symbols may be used in two ways. Either they are used to identify individual aspects of the information, such that objects or buildings can be located on a map, or they are used to notify trends. In the latter case, the trend is understood through perceiving the texture that is formed from viewing multiple symbols close together. For instance, Nocke *et al.* (2005) plotted multiple icons onto a map to visualize land usage.

Diagrams realize some process, concept or phenomenon; most are hand-crafted to display a particular phenomenon or result. They are often used in teaching to clearly explain a process. There are many examples in the literature, demonstrating phenomena like how volcanoes function, the migration of various animals or the campaigns of Napoleon's army.

Pictures are often associated with geographical datasets. Aerial photographs or site photographs can be easily associated with ground or land-use data. With the ubiquity of GPS it is easy to take a picture with position information. Applications like Google Maps (maps.google.com; maps.google.co.uk, accessed May 2007) montage multiple aerial photographs to make a detailed view of the Earth, while projects such as 'MyLifeBits' from Microsoft Research utilize cameras that automatically take pictures at regular instances throughout the day and correlate them with positional information from GPS (Bell, Gemmell and Lueder, 2007).

3.3.3 Challenges to create informative visualizations

The geographical and cartographic communities have many years of experience. Developers know how to generate effective and clear map representations. Furthermore, the geovisualization community has borrowed from this long heritage and have created many dynamic and interactive exploratory visualization systems. Systems such as cdv (Dykes, 1997), Mondrian (Theus, 2002) and CommonGIS (Andrienko and Andrienko, 1999) include many different forms, from scatter plots, bar charts and line graphs to parallel coordinate plots. However, there is too much choice. Gahegan (1999) described one of the major challenges with exploratory geovisualization as being the 'vast range of potential approaches and mappings', and the problem still exists today.

A user does not necessarily know which representation is most effective; consequently they may miss out on some important information because of the representation methodology they used. Thus, there is a need to examine which representation form is most effective for a given task, and methods are required to aid the user in their exploration.

3.4 Interaction and manipulation

Both the developer and the user need to make choices, not only how to display the data, but how to interact with the information. There are usually lots of parameters that can be changed to alter the visual appearance of the display. In fact, just the simple act of altering the colourmap can have a significant impact on what is emphasized and what is hidden in the display.

From a developer's standpoint there are some important steps required to generate a representation. There are different models to follow, but the principles remain the same. First, a demonstration dataset is created from the original data. This is achieved through filtering, aggregating and processing the original data. Second, the developer must choose what visual form to use and then decide how to map the data into it. Often the data is first mapped into an intermediate form, which is then rendered and displayed. The intermediate form is a data structure that holds summary or pre-calculated information. For instance, if the visualization is a three-dimensional surface, then triangles, coordinates, colour and other geometric information may be stored. Other data may also be stored, such as cached values to speed up rendering; associated arrays may be calculated to control coordinated views, and a list of the user's selected elements. Third, the user then manipulates the parameters to change how the information is displayed.

One of the initial choices of a developer and user is to decide where the result is displayed, that is, what alters when the parameters are changed. Parameter changes can be made in four areas: first, to control the appearance of the demonstration data; second, to control how the data is mapped to the display; third, to navigate the information such as zooming, clipping or changing the level of abstraction of the information being displayed; and fourth, controlling the environment and other meta-information associated with the system (such as determining where the views are placed on the screen, loading or saving files or managing bookmarks or controlling other exploration management facilities). Finally, interaction can also be *coordinated* and thus affect multiple visualizations simultaneously.

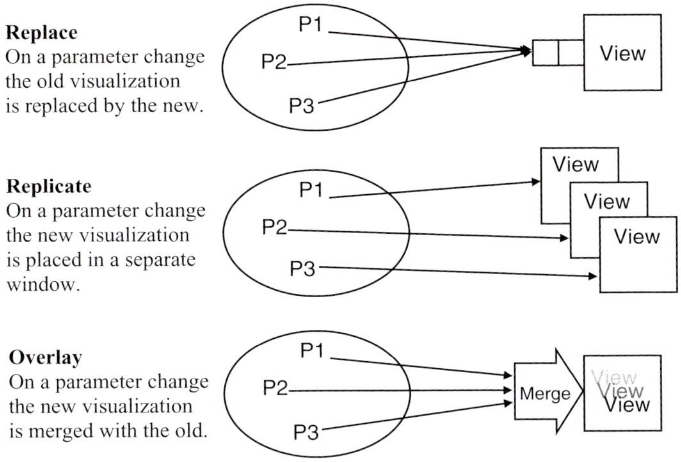

Figure 3.2 On a parameter change the information in the view can be replaced by the new information (replace), displayed in a new window (replicated) or merged with the existing representation (overlay)

3.4.1 Where is the information displayed?

The simple answer to 'where is the result displayed?' is that the information contained within the current view is *replaced* completely by the new information, i.e. when the user changes a parameter a single display is updated. This is often achieved interactively, such that, when the user changes the position of a value slider, the display instantly updates. Such dynamic query interfaces provide pre-attentive visualization of the information. However there are two further models of re-representation: *replicate* and *overlay* (Roberts, 2004), see Figure 3.2. Replication occurs when, on a parameter change, a new window appears with the new information. This is useful when it is imperative to compare information from one parameterization to another; but if unrestricted the screen can get cluttered with many windows and then it is difficult to understand what view represents which parameterization. Overlaying the information is the final strategy. There are several ways to overlay information or join information together from different parameterizations. The information could be stacked on top in another layer, or the new information could be merged together to provide a *difference view* (a difference view explicitly demonstrates the difference between the two parameterizations). Some systems are better at allowing the user to change where the output goes. Module visualization environments (MVE), such as AVS™ (www.avs.com/) and IRIS Explorer™ (www.nag.co.uk/), allow the developer to adapt the flow of information by changing the configuration of the modules. The data flows through different modules (filter, map, render, etc.), concluding in a display module. When the user changes a parameter in a module, the new data flows through the system and the display is updated. However, to create a new view (a replication), the user needs to copy and connect multiple modules in order to perform the operation. The data flow is split and the modules provide a fan-out flow.

Overlaying the information, especially generating difference views, is extremely useful for the user. It shows unambiguously the changes between two parameterizations or between two datasets. However, it is difficult to design intuitive and useful difference modules. Layering is used in geographical visualization; for instance, different land uses can be layered on top of each other and these layers can be separated in space or time such that the user can explore information on individual layers. However, it can be difficult to perceive the information comprehensively on each layer due to the occlusion of one layer to another. It is possible to imagine other strategies to merge the data, such as incorporating different information through various graphical lenses (Tominski *et al.*, 2006), but most general geovisualization systems do not support such operations.

3.4.2 Interactive filtering

Interactive filtering is an important exploratory technique. The user changes a parameter, or the range of values, to reduce the quantity of information that is being displayed. What once was a dense, overpopulated display can clearly display trends and highlight important outliers. Methods range from constraining ranges (Williamson and Shneiderman, 1992) to changing the bin width of a histogram (Theus, 2002). One important aspect of this type of interaction is that the query immediately updates the display.

Dynamic queries originated from work done by Ahlberg, Williamson and Shneiderman (1992). There are many examples where dynamic queries have been used to display geographical information. HomeFinder (Williamson and Shneiderman, 1992) demonstrates how geographical information, in this case houses for sale, can be explored through dynamic queries; IVEE (Ahlberg and Wistrand, 1995), which was later developed into SpotFire™, demonstrates how heavy metals in Sweden can be explored using dynamic queries; Dang, North and Shneiderman (2001) show how choropleth maps can be used with dynamic queries; Burigat and Chittaro (2005) present interactive queries for geographic data on mobile devices. Range sliders can be directly integrated with the visualization; for example, researchers have placed range sliders directly upon parallel coordinate plots (Shimabukuro *et al.*, 2004). They have also been incorporated into many general geovisualization tools (e.g. Andrienko and Andrienko, 2003; Brodbeck and Girardin, 2003; Feldt *et al.*, 2005).

3.4.3 Interactively adapting mapping parameters

Colour is important in geographical visualization. In fact, changing how colour is mapped to the display can considerably alter how the information is perceived. Not only can a colour map affect whether a colour-blind person can perceive the information, but elements may be hidden or emphasized through different colour mappings; colour can permit the user to perceive depth and can be used to delimit areas and borders. It is well known, but infrequently applied, that certain colour maps, such as the rainbow colour map, may mislead the user (Rogowitz and Treinish, 1998). Hence choosing the correct colour map with an appropriate transfer function is important (Brewer *et al.*, 1997; Rogowitz and Kalvin, 2001). Colour itself is one type of *visual variable*, thus it is often be possible to exchange colour with another type (such as grey scale, size or texture). However, it is prudent to exchange the visual variable with one with similar traits, such as exchanging one visual variable that displays continuous data with another that is also perceived continuously (Bertin, 1981).

Direct manipulation techniques allow the user to point to and *pick* interesting elements to select, delete or adapt them. Various researchers have used different names to express the same technique, including *highlighting* (Robinson, 2006), *brushing* (Carr *et al.*, 1986; Becker and Cleveland, 1987) and *painting multiple views* (Buja *et al.*, 1991). However, the techniques remain the same: a user visually selects one or more elements, the elements are stored or processed, and the selected elements are displayed to the user through highlighting.

Commonly the selection operation is coordinated to all windows to allow the user to see how elements in one projection are displayed in others. Linked brushing is important, as the user can brush in one view in one projection and see the results of that operation in other dimensions in other views. Linked brushing can also be used to discover outliers between multiform views (Lawrence *et al.*, 2006).

There are different tools to directly select the required elements. Commonly a freehand lasso is used to delimit the elements (Wills, 1996), but the brush can be a point, line or area. For example, Stuetzle (1987) demonstrates point selection and Becker, Cleveland and Weil (1988) describe a line-brush, while there are many examples of bounding boxes (Becker, Cleveland and Wilks, 1987; Buja *et al.*, 1991; Swayne *et al.*, 2003; Ward, 1994, Piringer *et al.*, 2004; Weaverm, 2006).

Often *highlight* is the only operation that developers implement. However, even 20 years ago, Becker and Cleveland (1987) specified four brush operations: highlight, shadow highlight, delete and label. The *highlight* operation changes the appearance of the selection list in all windows; *shadow highlight* changes the appearance of the selected elements in the view and removes them from all other linked views; *delete* removes the elements from the view; while *label* displays additional information about the selected elements, such as a text label. More recently, researchers have extended these ideas, e.g. *multiple brushes* or *compound brushing*. Multiple brushes allow users to control many brushes, with each new brush in a different colour. Compound brushing combines the results from multiple brush operations. Chen (2004) describes a flexible compound brushing system that permits selections to be combined using various logical operations (e.g. AND, OR, XOR), the operations being controlled through a graph layout, while Wright and Roberts (2005) present a direct methodology named *click and brush*.

Finally, the selected elements need to be visually *highlighted*. Commonly this is through a colour change, but other methods include linked lines, depth of field, transparency, contour lines or shadows surrounding the elements (Robinson, 2006).

3.4.4 Navigation

Navigation and *viewpoint manipulation* can come in various forms. With three-dimensional geovirtual environments the user can walk, fly, clip to remove information, and change the projection transformation (such as from parallel projection to perspective projection). However, users often get lost in their navigation, for instance map users lose orientation information (Gahegan, 1999). Thus, techniques that control and constrain the user's navigation are often useful (Buchholz, Bohnet and Dollner, 2005). Linked master–slave navigation can also allow the user to navigate efficiently in dual view systems (Plumlee and Ware, 2003), see Section 3.3.1. In two-dimensional maps the user can zoom or pan (Harrower and Sheesley, 2005) and scroll, and often the zooming operations are associated with a semantic change (Cecconi and Galanda, 2002). Often multiple views are linked together such that focus and the other provides detail (Convertino *et al.*, 2003).

3.4.5 Interacting with the environment

Users need effective ways to interact with the environment. Often, because of the small screen size, a user needs to move, iconize or delete unwanted windows. However, rather than permitting windows to overlap, some developers have constrained how the windows are displayed. For instance, GeoWizard (Feldt *et al.*, 2005) collages the windows together and allows the user to enlarge or reduce the size of a window by changing the position of the dynamic splitters. In human–computer interaction researchers have proposed various other interfaces, from zoomable user interfaces (Bederson and Meyer, 1998) to elastic windows (Kandogan and Shneiderman, 1997), but these have not been integrated into many exploratory geovisualization systems.

Owing to the fact that exploration is often unguided, a user should be able to interact with the session history and roll back to a previous result. Few geovisualization tools include the rich functionality required to comprehensively manage the session history. Recent research has been achieved in this area, including: vistrails (Callahan *et al.*, 2006) and provenance for visual exploration systems (Streefkerk and Groth, 2006). However, techniques that save the session history, allow users to extend different exploration paths and integrate explorations from different users are missing from most geovisualization systems.

Finally, in a large exploration session it is often difficult to remember which views are linked together and which parameters were used to create a particular view. Techniques to visualize this subsidiary information are named *meta-visualization*. Little work has been done in this area; however MacEachren *et al.* (2001) describe a *watcher window* that presents summary information of which windows are active, who is in control of which windows in a collaborative environment, and which windows are shared, while Weaver (2004) describes a technique that explicitly visualizes which views are linked together through arrows that are annotated on top of the visualizations.

3.4.6 Coordination

The area of coordination is rich and varied and is used in many disciplines (Olson *et al.*, 2001). In fact, any forms of interaction can potentially be linked together: data processing, filtering, mapping, navigation, rendering and interacting with the environment (Boukhelifa, Roberts and Rodgers, 2003). The two most common are *linked brushing* (Buja *et al.*, 1991) and *navigational slaving* (North and Schneiderman, 2000). However, there are many aspects of coordination. For instance, the users' needs may change as they continue on their exploration, thus they may require some views to be initially coordinated and then need to uncoordinate them to follow a new trail.

Roberts (2005) details seven rudiments of coordination based on program variables:

1. coordination *entities* describes what is coordinated, such as data, parameters, process, function or event;

2. entities are *typed*, such that elements of the same type can be connected – some form of translation may be required to connect entities with different types;

3. each entity has temporal and *chronology* information, such that some coordinations persist while others are temporary for a task;

4. each coordination is defined in a *scope* – for example it may be that the entities are coupled for a specific task and then uncoupled when the task has finished;

5. the coordination has a certain *granularity* – for example, some entities may be coordinated point-to-point while for others it may be possible to chain multiple entities together;

6. the coordination of entities may be *initiated* by the user or automatically by the system; and finally,

7. there are different methods to *update* the information when entities are coordinated – for instance, cold linking permits elements to be coupled once, warm linking permits users to decide when the views are updated, and hot linking is when the information is dynamically updated (Unwin, 2001).

There are three principle coordination architectures: (1) constraint-based programming (Mcdonald *et al.*, 1990); (2) a data centric approach taken from the database community, where change to one component of a relational database is tightly coupled to other components (North and Schneiderman, 2000); and (3) the module view controller pattern, where one component observes the model and updates the view when changes are made (Pattison and Phillips, 2001; Boukhelifa *et al.*, 2003). Tools such as Waltz (Roberts and Waltz, 1998) and Improvise (Weaver, 2004) implement many of the ideas of the rudiments. However, few CMV developers incorporate comprehensive coordination capabilities in their systems.

3.4.7 Challenges for interaction and manipulation

Most exploratory systems allow the visualization to be adapted and viewed again from a new parameterization, whether this information replaces the old or is displayed in a new window. However, there are still many challenges. First, not many systems allow the user to overlay the information. One overlay strategy is to generate the difference between the two parameterizations (Suvanaphen and Roberts, 2004). This can be extremely useful as the visualization would clearly show what has changed. However, few systems supply this functionality. Second, it is not clear as to how many views are actually useful. For instance, a user may easily get lost in a explosion of overlapping windows, hence there is certainly a trade-off to using replacement, replication and overlay. Consequently, more research is required to evaluate how many views are useful and to provide guidelines to when each strategy should be used. Third, with large explorations the user can easily forget which view represents what parameterization. Thus methods to support the user in their exploration are required.

Both interactive filtering and interactively adapting mapping parameters are important exploratory techniques and it is clear that much research has been achieved. However, many of these techniques are not implemented in modern exploratory tools. Developers seem to rely on a small subset commands, such as picking and highlighting. Consequently, users are missing out on rich interaction techniques.

Navigation is still tricky. Users get lost in their exploration, they forget where they are and they may start to misinterpret information. Researchers need to create interfaces that clearly demonstrate where a user is, what orientation the data is and what scale it is displayed at. Furthermore, it is important to generate usable and intuitive exploration environments. If the user does not know how to control the environment, or if they get overwhelmed by too many windows, then they will not be effective in their exploration process. Thus, developers need to strike a balance between comprehensive function-rich environments on the one hand, and simple to use interfaces on the other. Developers should continue to research into methods that aid the user in their exploration and store session information.

Finally, coordination is obviously a major part in CMV systems. However, coordination tends to be achieved automatically and often performed silently by a system (such that users do not necessarily know what is connected together). Hence, meta-visualization techniques should be developed to enable the user to see how the system is setup.

3.5 Tools and toolkits

Geographical and spatial data has been long used to demonstrate research ideas in coordination. Felger and Schroeder (1992) in the visualization input pipeline (VIP) describe linked cursors of three-dimensional maps. LinkWinds (Jacobson, Berkin and Orton, 1994) demonstrates coordination using precipitation and ozone depletion; Visage (Roth *et al.*, 1996) demonstrates coordinated manipulation of maps; the tight coupling interface of DEVise (Livny *et al.*, 1997) has been used to demonstrate various examples including looking at product purchases by location; Spotfire (Ahlberg, 1996) uses map visualizations along with other statistical forms.

There are a few different ways to develop CMV systems. Tools such as CommonGIS (Andrienko and Andrienko, 1999), GeoVISTA studio (Takatuska and Gahegan, 2002) and Improvise (Weaver, 2004) contain comprehensive coordination capabilities. However, if developers wish to develop from scratch, then languages such as Java provide a convenient medium for development. For instance, tools like Mondrian (Theus, 2002) have been developed in Java.

In addition, toolkits such as the InfoVis toolkit (Fekete, 2004), Prefuse (http://prefuse.org/) and Piccolo (Bederson, Grosjean and Meyer, 2004) provide developers with the functionality to create complex visualization environments in Java. However, it still takes a lot of careful planning design and time to create effective geovisualization tools, and often developers wish to integrate algorithms and tools from different researchers.

One way to develop prototype tools is to use Flash or scalable vector graphics (SVG), such as those used by Steiner, MacEachren and Guo (2002) and Marsh, Dykes and Attilakou (2006), respectively. These solutions have the added advantage that they are instantly web accessible and available by many remote users. Alternatively, techniques such as the Information Visualization CyberInfrastructure (IVC) software framework, which extends original work on the Information Visualization Repository (Börner and Zhou, 2001), provide a uniform method to interact with a multitude of algorithms by providing a programming interface to algorithms and a user interface to end-users using the Eclipse Rich Client

Platform (RCP). Finally, recently there has been interest in component and bean technologies. For example, ILOG's Jviews (www.ilog.com/products/jviews/) uses Java, while GeoWizard (Feldt *et al.*, 2005) uses Microsoft's .NET to provide geographical visualization components.

3.5.1 Challenges for geovisualization tool developers

There are still many challenges faced by tool developers. First, it takes a lot of time and effort to develop and maintain a tool, but toolkits such as InfoVis toolkit (Fekete, 2004), Prefuse (http://prefuse.org/) and Piccolo (Bederson, Grosjean and Meyer, 2004) certainly aid the developer. Utilizing Flash or SVG to quickly develop and disseminate prototypes is certainly a promising strategy. These also permit the applications to be disseminated on the web, which allows many remote users to operate them and try out the techniques. However, it is still difficult to get the remote users to collaborate in an exploration. Current collaboration techniques are not well integrated with modern CMV systems, and techniques to support efficient and complex collaboration explorations are still naïve.

Finally, interoperability and extensibility still plague developers. Systems such as Börner and Zhou's (2001) IVC certainly provide one method to integrate multiple algorithms and techniques. However, interoperability and extensibility concepts should be included in system development from the start of the project.

3.6 Conclusions

Many geovisualization tool developers seem to ignore or forget the richness of the published research and implement only basic coordination, interaction and navigation effects. Certainly developers and researchers have at their disposal a large heritage of geovisualization techniques. Researchers have proposed many geovisualization forms, a multitude of manipulation strategies and detailed aspects of coordination, brushing and navigation. However, concepts such as handling missing data, visualizing uncertain values, coordination rudiments and interactive manipulation techniques need to be included in the systems and included from the beginning of development. From experience, it is difficult to twist a system to perform techniques that were not originally planned.

Exploration itself is still difficult. It is difficult to know where to go, where users have been and how to control the interface effectively. One challenge is that either systems use a *replacement* strategy and the user fails to understand the history trail through the provided linear history tree or the system provides a *replication* strategy and the user gets lost in an explosion of windows. Thus, researchers should perform more research into how to manage the user's exploration better.

Finally, the web certainly provides a convenient way to disseminate geovisualizations. Consequently, the utilization of Flash and SVG seems promising and should be encouraged. However, there is also a need to allow multiple users to collaborate in these visualization sessions. Although some work has been done in collaboration, it remains a niche technique and the functionality to appropriately manage and merge ideas from multiple participants is still naïve.

References

Ahlberg, C. (1996) Spotfire: an information exploration environment. *SIGMOD Record* 24(4): 25–29.

Ahlberg, C. and Shneiderman, B. (1994) Visual information seeking using the filmfinder. In *CHI '94: Conference Companion on Human Factors in Computing Systems.* ACM Press, pp. 433–434.

Ahlberg, C., and Wistrand, E. (1995) IVEE: an information visualization and exploration environment. Technical Report SSKKII-95.03, Chalmers University of Technology. Available at: www.cs.chalmers.se/SSKKII/ivee.html.

Ahlberg, C., Williamson, C. and Shneiderman, B. (1992) Dynamic queries for information exploration: an implementation and evaluation. In *CHI '92: Proceedings of the SIGCHI Conference on Human Factors in Computing Systems.* New York, ACM Press, pp. 619–626.

Aktas, A., Aydin, G., Donnellan, A., Fox, G., Granat, R., Lyzenga, G., McLeod, D., Pallickara, S., Parker, J., Pierce, M., Rundle, J. and Sayar, A. (2005) Implementing geographical information system grid services to support computational geophysics in a service-oriented environment. In *NASA Earth–Sun System Technology Conference,* University of Maryland, Adelphi, MD, June 2005.

Andrienko, G. and Andrienko, N. (1999) Making a GIS intelligent: CommonGIS project view. In *AGILE'99 Conference on Geographic Information Science,* Rome, April 1999, pp. 19–24.

Andrienko, N. and Andrienko, G. (2003) Coordinated views for informed spatial decision making. In *CMV '03: Proceedings of the conference on Coordinated and Multiple Views in Exploratory Visualization,* Washington, DC, 2003. New York, IEEE Computer Society, p. 44.

Arpinar B., Sheth, A., Ramakrishnan, C., Lynn, U., Azami, M. and Kwan, M. P. (2006) Geospatial ontology development and semantic analytics. *Transactions in GIS* 10(4): 551–575.

Becker, R. and Cleveland, W. (1987) Brushing scatterplots. *Technometrics* 29(2): 127–142.

Becker, R. A., Cleveland, W. S. and Wilks, A. R. (1987) Dynamic graphics for data analysis. *Statistical Science* 2(4): 355–395.

Becker, R. A., Cleveland, W. S. and Weil, G. (1988) *The Use of Brushing and Rotation for Data Analysis.* London, Wadsworth Brooks/Cole, 1988, pp. 247–275.

Bederson, B. and Meyer, J. (1998) Implementing a zooming user interface: experience building pad++. *Software Practical Experience* 28(10): 1101–1135.

Bederson, B., Grosjean, J. and Meyer, J. (2004) Toolkit design for interactive structured graphics. *IEEE Transactions in Software Engineering* 30(8): 535–546.

Bell, G., Gemmell, J. and Lueder, R. (2007) My life bits. http://research.microsoft.com/barc/mediapresence/MyLifeBits.aspx, accessed May 2007.

Bertin, J. (1981) *Graphics and Graphic Information Processing,* translated by Berg, W. J. and Scott, P. New York, Walter de Gruyter.

Börner, K. and Zhou, Y. (2001) A software repository for education and research in information visualization. In*IV '01: Proceedings of the Fifth International Conference on Information Visualisation,* Washington, DC, 2001. IEEE New York, Computer Society, pp. 257–262.

Boukhelifa, N., Roberts, J. C. and Rodgers, P. J. (2003) A coordination model for exploratory multiview visualization. In *CMV '03: Proceedings of the conference on Coordinated and Multiple Views In Exploratory Visualization,* Washington, DC, 2003. New York, IEEE Computer Society, pp. 76–85.

Brewer, C. A., MacEachren, A. M., Pickle, L. W. and Herrmann, D. (1997) Mapping mortality: evaluating color schemes for choropleth maps. *Annals of the Association of American Geographers* 87(3): 411–438.

Brodbeck, D. and Girardin, L. (2003) Design study: using multiple coordinated views to analyse geo-referenced high-dimensional datasets. In *CMV '03: Proceedings of the conference on Coordinated and Multiple Views In Exploratory Visualization*, Washington, DC, 2003. New York, IEEE Computer Society, pp. 104–111.

Brooke, J.M., Marsh, J., Pettifer, S., and Sastry, L.S. The importance of locality in the visualization of large datasets. *Concurrency and computation: practice and experience*, 19: 195–205, 2007.

Buchholz, H., Bohnet, J. and Dollner, J. (2005) Smart and physically-based navigation in 3d geovirtual environments. In *IV '05: Proceedings of the Ninth International Conference on Information Visualisation (IV'05)*, Washington, DC, 2005. New York, IEEE Computer Society, pp. 629–635.

Buja, A., McDonald, J. A., Michalak, J. and Stuetzle, W. (1991) Interactive data visualization using focusing and linking. In *Proceedings Visualization '91*. New York, IEEE Computer Society Press, pp. 156–163.

Burigat, S. and Chittaro, L. (2005) Visualizing the results of interactive queries for geographic data on mobile devices. In *GIS '05: Proceedings of the 13th annual ACM international workshop on Geographic information systems*. New York, ACM Press, pp. 277–284.

Callahan, S. P., Freire, J., Santos, E., Scheidegger, C. E., Silva, C.T. and Vo, H. T. (2006) Vistrails: visualization meets data management. In *SIGMOD '06: Proceedings of the 2006 ACM SIGMOD international conference on Management of data*. New York, ACM Press.

Carr, D. B., Littlefield, R. J. and Nichloson, W. L. (1986) Scatterplot matrix techniques for large *n*. In *Proceedings of the Seventeenth Symposium on the Interface of Computer Sciences and Statistics*, New York, 1986. Elsevier North-Holland, pp. 297–306.

Cecconi, A. and Galanda M. (2002) Adaptive zooming in web cartography. *Computer Graphics Forum* 21: 787–799.

Chen, H. (2004) Compound brushing explained. *Information Visualization* 3(2): 96–108.

Convertino, G., Chen, J., Yost, B., Ryu, Y. S. and North, C. (2003) Exploring context switching and cognition in dual-view coordinated visualizations. In *Coordinated and Mulitiple Views in Exploratory Visualization (CMV03)*, Los Alamitos, CA, 2003. New York, IEEE Computer Society, p. 55.

Dang, G., North, C. and Shneiderman, B. (2001) Dynamic queries and brushing on choropleth maps. In *IV '01: Proceedings of the Fifth International Conference on Information Visualisation*, Washington, DC, 2001. New York, IEEE Computer Society, pp. 757–764.

DiBiase, C. (1990) Visualization in the earth sciences. *Earth and Mineral Sciences, Bulletin of the College of Earth and Mineral Sciences* 59(2): 13–18.

Dykes, J. (1997) cdv: a flexible approach to ESDA with free software connection. In *Proceedings of the British Cartographic Society 34th Annual Symposium*, pp. 100–107.

Dykes, J., MacEachren, A. and Kraak, M. J. (eds) (2005) *Exploring GeoVisualization*. Oxford, Pergamon Press.

Edsall, R. M. (2003) The parallel coordinate plot in action: design and use for geographic visualization. *Computers and Statistical Data Analysis* 43(4): 605–619.

Fekete, J. D. (2004) The infovis toolkit. In *INFOVIS '04: Proceedings of the IEEE Symposium on Information Visualization (INFOVIS'04)*, Washington, DC, 2004. New York, IEEE Computer Society, pp. 167–174.

Feldt, N., Pettersson, H., Johansson, J. and Jern, M. (2005)Tailor-made exploratory visualization for statistics Sweden. In *CMV '05: Proceedings of the Coordinated and Multiple Views in Exploratory Visualization*, Washington, DC, 2005. New York, IEEE Computer Society, pp. 133–142.

Felger, W. and Schroeder, F. (1992) The visualization input pipeline – enabling semantic interaction in scientific visualization. In *Eurographics '92 (Computer Graphics Forum Volume 11 No. 3)*, Kilgour, A. and Kjelldahl, L. (eds). Blackwell, Oxford, pp. 139–151.

Friesen, J. A. and Tarman, T. D. (2000) Remote high-performance visualization and collaboration. *IEEE Computer Graphics Applications*, 20(4): 45–49.

Gahegan, M. (1999) Four barriers to the development of effective exploratory visualisation tools for the geosciences. *International Journal of Geographical Information Science* 13(4): 289–309.

Geisler, G. (2000) Enriched links: a framework for improving web navigation using pop-up views. Technical Report TR-2000-02, School of Information and Library Science, University of North Carolina at Chapel Hill.

Harrower, M. and Sheesley, B. (2005) Designing better map interfaces: a framework for panning and zooming. *Transactions in GIS* 9(2): 77–89.

Hawick, K. A., Coddington, P. D. and James, H. A. (2003) Distributed frameworks and parallel algorithms for processing large-scale geographic data. *Parallel Computing* 29(10): 1297–1333.

Hedley, N. R., Billinghurst, M., Postner, L., May, R. and Kato, H. (2002) Explorations in the use of augmented reality for geographic visualization. Presence: *Teleoperation and Virtual Environments* 11(2): 119–133.

Jacobson, A. S., Berkin, A. L. and Orton, M. N. (1994) LinkWinds: interactive scientific data analysis and visualization. *Communications of the ACM* 37(4): 43–52.

Kandogan, E. and Shneiderman, B. (1997) Elastic windows: evaluation of multi-window operations. In *CHI '97: Proceedings of the SIGCHI conference on Human factors in computing systems*, New York, 1997. New York, ACM Press, pp. 250–257.

Keim, D. A. (2000) Designing pixel-oriented visualization techniques: Theory and applications. *IEEE Transactions on Visualization and Computer Graphics* 6(1): 59–78.

Keim, D. A., North, S. C., Panse, C. and Sips, M. (2004) Visual data mining in large geo-spatial point sets. Special Issue: Visual Analytics. *IEEE Computer Graphics and Applications* 24(5): 36–44.

Kemp, Z. and Tan, L. (2005) Federated geospatial decision support - a marine environment management system. In *Proceedings of GISPLANET 2005*, p. 11.

Kocherlakota, S. M. and Healey, C. G. (2005) Summarization techniques for visualization of large multidimensional datasets. Technical Report TR-2005-35, Department of Computer Science, North Carolina State University.

Koua, E. L. and Kraak, M. J. (2004a) Alternative visualization of large geospatial datasets. *Cartographic Journal* 41(3): 217–228.

Koua, E. L. and Kraak, M. J. (2004b) Geovisualization to support the exploration of large health and demographic survey data. *International Journal of Health Geographics* 3(12).

Koua, E. L. and Kraak, M. J. (2004c) A usability framework for the design and evaluation of an exploratory geovisualization environment. In *IV '04: Proceedings of the Eighth Information Visualisation Conference*, Washington, DC, 2004. New York, IEEE Computer Society, pp. 153–158.

Kreuseler, M. and Schumann, H. (2002) A flexible approach for visual data mining. *IEEE Transactions on Visualization and Computer Graphics* 8(1): 39–51.

Lawrence, M., Lee, E. K., Cook, D., Hofmann, H. and Wurtele, E. (2006) Explorase: exploratory data analysis of systems biology data. In *CMV '06: Proceedings of the Conference on Coordinated and Multiple Views in Exploratory Visualization*, Washington, DC, 2006. New York, IEEE Computer Society, pp. 14–20.

Livny, M., Ramakrishnan, R., Beyer, K., Chen, G., Donjerkovic, D., Lawande, S., Myllymaki, J. and Wenger, K. (1997) Devise: integrated querying and visual exploration of large datasets. In

SIGMOD '97: Proceedings of the 1997 ACM SIGMOD international conference on Management of data. New York, ACM Press, pp. 301–312.

Lohse, G. L., Biolsi, K., Walker, N. and Rueter, H. H. (1994) A classification of visual representations. *Communications of the ACM* 37(12): 36–49.

MacEachren, A. M., Wachowicz, M., Edsall, R., Haug, D. and Masters, R. (1999) Constructing knowledge from multivariate spatiotemporal data: Integrating geographic visualization (gvis) with knowledge discovery in databases (kdd). *Geographic Information Science* 13(4): 311–334.

MacEachren, A., Brewer, I. and Steiner, E. (2001) Geovisualization to mediate collaborative work: tools to support different-place knowledge construction and decision-making. In *Proceedings of International Cartographic Conference*.

Marsh, S. L., Dykes, J. and Attilakou, F. (2006) Evaluating a geovisualization prototype with two approaches: remote instructional vs. face-to-face exploratory. In *IV '06: Proceedings of the Conference on Information Visualization*, Washington, DC. New York, IEEE Computer Society, pp. 310–315.

McDonald, J. A., Stuetzle, W. and Buja, A. (1990) Painting multiple views of complex objects. In *OOPSLA/ECOOP '90: Proceedings of the European Conference on Object-oriented Programming on Object-oriented Programming Systems, Languages, and Applications*. New York, ACM Press, pp. 245–257.

Miller, H. J. and Han, J. (2001) *Geographic Data Mining and Knowledge Discovery*. Boca Raton, FL, CRC Press.

Nocke, T., Schlechtweg, S. and Schumann, H. (2005) Icon-based visualization using mosaic metaphors. In *IV '05: Proceedings of the Ninth International Conference on Information Visualisation*, Washington, DC. New York, IEEE Computer Society, pp. 103–109.

North, C. and Shneiderman, B. (2000) Snap-together visualization: a user interface for coordinating visualizations via relational schemata. In *Proceedings of Advanced Visual Interfaces*, Italy, 2000 , pp. 128–135.

Olson, G., Malone, T. and Smith, J. (2001) *Coordination Theory and Collaboration Technology*. Philadelphia, PA, Lawrence Erlbaum Associates.

Pattison, T. and Phillips, M. (2001) View coordination architecture for information visualisation. In *APVis '01: Proceedings of the 2001 Asia-Pacific symposium on Information visualisation*, Darlinghurst, Australia, 2001. Australian Computer Society, pp. 165–169.

Piringer, H., Kosara, R. and Hauser, H. (2004) Interactive focus+context visualization with linked 2d/3d scatterplots. In *CMV '04: Proceedings of the Second International Conference on Coordinated and Multiple Views in Exploratory Visualization*, Washington, DC, 2004. New York, IEEE Computer Society, pp. 49–60.

Pirolli, P. and Card, S. (2005) The sense-making process and leverage points for analyst technology as identified through cognitive task analysis. In *Proceedings of International Conference on Intelligence Analysis*, McLean, VA, p. 6.

Plumlee, M., and Ware, C. (2003) Integrating multiple 3d views through frame-of-reference interaction. In *CMV '03: Proceedings of the conference on Coordinated and Multiple Views In Exploratory Visualization*, Washington, DC. New York, IEEE Computer Society, pp. 34–43.

Rao, R. and Card, S. K. (1994) The table lens: merging graphical and symbolic representations in an interactive focus + context visualization for tabular information. In *CHI '94: Human Factors in Computing Systems*. New York, ACM Press, pp. 318–322.

Roberts, J. and Waltz, C. (1998) An exploratory visualization tool for volume data, using multiform abstract displays. In *Visual Data Exploration and Analysis V, Proceedings of SPIE*, Erbacher, R. F. and Pang, A. (eds), Vol. 3298. IS&T and SPIE, pp. 112–122.

Roberts, J. C. (2000) Multiple-view and multiform visualization. In *Visual Data Exploration and Analysis VII*, Erbacher, R., Pang, A., Wittenbrink, C. and Roberts, J. (eds), Proceedings of SPIE, Vol. 3960. IS&T and SPIE, pp. 176–185.

Roberts, J. C. (2004) Exploratory visualization using bracketing. In Costabile, M. F. (ed.), *Proceedings of Advanced Visual Interfaces (AVI 2004)*, Gallipoli, May 2004. New York, ACM Press, pp. 188–192.

Roberts, J. C. (2005) Exploratory visualization with multiple linked views. In *Exploring Geovisualization*, Dykes, J., Maceachren, A. and Kraak, K.-J. (eds). Amsterdam, Elsevier, pp. 159–180.

Roberts, J. C. and Ryan, N. (1997) Alternative archaeological representations within virtual worlds. In *Proceedings of the 4th UK Virtual Reality Specialist Interest Group Conference*, Brunel University, Uxbridge, November 1997, Bowden, R. (ed.), pp. 179–188. Available at: www.crg.cs.nott.ac.uk/groups/ukvrsig/.

Robinson, C. A. (2006) Highlighting techniques to support geovisualization. In *Proceedings of the ICA Workshop on Geovisualization and Visual Analytics*, Portland, OR, June 2006.

Rodgers, P. (2004) Graph drawing techniques for geographic visualization. In *Exploring Geovisualization*, MacEachren, A., Kraak, M. J. and Dykes, J. (eds). Oxford, Pergamon Press, pp. 143–158.

Rogowitz, B. E. and Kalvin, A. D. (2001) The 'which Blair project': a quick visual method for evaluating perceptual color maps. In *VIS '01: Proceedings of the Conference on Visualization '01*, Washington, DC. New York, IEEE Computer Society, pp. 183–190.

Rogowitz, B. E. and Treinish, L. A. (1998) *Data Visualization: the end of the Rainbow. Spectrum,IEEE* 35(12): 52–59.

Roth, S., Lucas, P., Senn, J., Gomberg, C., Burks, M., Stroffolino, P., Kolojejchick, J. and Dunmire, C. (1996) Visage: a user interface environment for exploring information. In *Proceedings of Information Visualization*, San Francisco, 1996. New York, IEEE Computer Society, pp. 3–12.

Russell, D. M., Stefik, M. J., Pirolli, P. and Card, S. K. (1993) The cost structure of sensemaking. In CHI '93: *Proceedings of the SIGCHI Conference on Human Factors in Computing Systems*. New York, ACM Press, pp. 269–276.

Shimabukuro, M. H., Flores, E. F., de Oliveira, M. C. F. and Levkowitz, H. (2004) Coordinated views to assist exploration of spatio-temporal data: a case study. In *CMV '04: Proceedings of the Second International Conference on Coordinated and Multiple Views in Exploratory Visualization*, Washington, DC. New York, IEEE Computer Society, pp. 107–117.

Siirtola, H. and Mäkinen, E. (2005) Constructing and reconstructing the reorderable matrix. In *International Conference on Information Visualization*, Vol. 4. New York, IEEE Computer Society, pp. 32–48.

Steiner, E., MacEachren, A. and Guo, D. (2002) Developing and assessing light-weight data-driven exploratory geovisualization tools for the web. In *Advances in Spatial Data Handling, Proceedings of the 10th International Symposium on Spatial Data Handling*, Richardson, D. E. and Oosterom, P. V. (eds). Berlin, Springer, pp. 487–500.

Stoakley, R., Conway, M. J. and Pausch, R. (1995) Virtual reality on a wim: interactive worlds in miniature. In *CHI '95: Proceedings of the SIGCHI Conference on Human Factors in Computing Systems*. New York, ACM Press/Addison-Wesley, pp. 265–272.

Stott, J. and Rodgers, P. (2004) Metro map layout using multicriteria optimization. In *Proceedings 8th International Conference on Information Visualization (IV04)*, July 2004. New York, IEEE Society Press, 355–362.

Streefkerk, K. and Groth, D. P. (2006) Provenance and annotation for visual exploration systems. *IEEE Transactions on Visualization and Computer Graphics* 12(6): 1500–1510.

Stuetzle, W. (1987) *Plot Windows*, Vol. 82. London, Wadsworth Brooks/Cole, pp. 466–475. [Also in *Dynamic Graphics for Statistics*, Cleveland, W. S. and McGill, M. E. (eds). London, Wadsworth Brooks/Cole, 1988, pp. 225–245.]

Suvanaphen, S. and Roberts, J. C. (2004) Textual difference visualization of multiple search results utilizing detail in context. In *Theory and Practice of Computer Graphics*, Lever, P. G. (ed.), Bournemouth, June 2004. EGUK, IEEE Computer Society Press, pp. 2–8.

Swayne, D. F., Buja, A. and Lang, D. T. (2003) Exploratory visual analysis of graphs in ggobi. In *Proceedings of the 3rd International Workshop on Distributed Statistical Computing*, Hornik, K., Leisch, F. and Zeileis, A. (eds), March 2003.

Takatuska, M. and Gahegan, M. (2002) GeoVISTA Studio: a codeless visual programming environment for geoscientific data analysis and visualization. *Computers and Geosciences* 28: 1131–1144.

Tang, L. and Shneiderman, B. (2001) Dynamic aggregation to support pattern discovery: A case study with web logs. In *DS '01: Proceedings of the 4th International Conference on Discovery Science*, London. Berlin, Springer, pp. 464–469.

Theus, M. (2002) Interactive data visualization using Mondrian. *Journal of Statistical Software*, 7(11); available at: www.jstatsoft.org/v07/i11/ (accessed 10 December 2007).

Thomas, J. J. and Cook, K. A. (2005) *Illuminating the Path: The Research and Development Agenda for Visual Analytics*. New York, IEEE Press.

Thomson, J., Hetzler, E., MacEachren, A., Gahegan, M. and Pavel, M. (2005) A typology for visualizing uncertainty. In *Visualization and Data Analysis 2005*, Erbacher, R. F., Roberts, J. C., Grohn, M. T. and Borner, K. (eds), Vol. 5669, pp. 146–157.

Tominski, C., Abello, J., van Ham, F. and Schumann, H. (2006) Fisheye tree views and lenses for graph visualization. *In IV '06: Proceedings of the 10th Conference on Information Visualization*, Washington, DC, 2006. New York, IEEE Computer Society Press, pp. 17–24.

Torun, A. and Duzgun, S. (2006) Using spatial data mining techniques to reveal vulnerability of people and places due to oil transportation and accidents: a case study of istanbul strait. In *ISPRS Technical Commission II Symposium*, Vienna, July 2006, pp. 43–48.

Tufte, E. R. (1990) *Envisioning Information*. Graphics Press.

Unwin, A. (2001) *R* objects, two interfaces! (*R* objects to interfaces?). In *Proceedings of the 2nd International Workshop on Distributed Statistical Computing*, Hornik, K. and Leisch, F. (eds), March 2001.

Unwin, A., Hawkins, G., Hofmann, H. and Siegl, B. (1996) Interactive graphics for data sets with missing values – MANET. *Journal of Computaional and Graphical Statistics* 5(2): 113–122.

Venkataraman, S., Benger, W., Long, A., Jeong, B. and Renambot, L. (2006) Visualizing hurricane katrina: large data management, rendering and display challenges. In *GRAPHITE '06: Proceedings of the 4th international conference on Computer graphics and interactive techniques in Australasia and Southeast Asia*. New York, ACM Press, pp. 209–212.

Ward, M. O. (1994) XmdvTool: integrating multiple methods for visualizing multivariate data. In *Proceedings Visualization '94*, Bergeron, R. D. and Kaufman, A. E. (eds). New York, IEEE Computer Society Press, pp. 326–333.

Weaver, C. (2004) Building highly-coordinated visualizations in improvise. In *Proceedings of the IEEE Symposium on Information Visualization (INFOVIS'04)* Los Alamitos, CA, 2004. New York, IEEE Computer Society, pp. 159–166.

Weaver, C. (2006) Metavisual exploration and analysis of devise coordination in improvise. In *CMV '06: Proceedings of the Fourth International Conference on Coordinated and Multiple Views in Exploratory Visualization*, Washington, DC, 2006. New York, IEEE Computer Society, pp. 79–90.

Williamson, C. and Shneiderman, B. (1992) The dynamic homefinder: evaluating dynamic queries in a real-estate information exploration system. In *SIGIR '92: Proceedings of the 15th Annual International ACM SIGIR Conference on Research and Development in Information Retrieval.* New York, ACM Press, pp. 338–346.

Wills, G.J. (1996) 524,288 ways to say 'this is interesting'. In *Proceedings of the 1996 IEEE Symposium on Information Visualization (INFOVIS '96).* New York, IEEE Computer Society, pp. 54–61.

Wright, M. A. and Roberts, J. C. (2005) Click and brush: a novel way of finding correlations and relationships in visualizations. In *Proceedings Theory and Practice of Computer Graphics,* Lever, L. and McDerby, M. (eds). University of Kent, June 2005, pp. 179–186. Eurographics Association. Available at: www.diglib.eg.org.

4

The Role of Map Animation for Geographic Visualization

Mark Harrower, and Sara Fabrikant

Department of Geography, University of Wisconsin – Madison

Department of Geography, University of Zurich

4.1 Introduction

Many of today's significant research challenges, such as resource management and environmental monitoring, depend upon capturing, analysing and representing dynamic geographic processes. The ability to recognize and track changes in complex physical systems is essential to developing an understanding of how these systems work (Yattaw, 1999). For thousands of years cartographers have been perfecting the representation of dynamic spatio-temporal phenomena with static, spatial representations in the form of two-dimensional maps (Bertin, 1983). As early as the 1930s, cartographers also experimented with adding the time dimension to congruently representing dynamic geographic processes with animated map displays. Dynamic cartographic representations, such as cartographic movies (Tobler, 1970), two- and three-dimensional computer animations (Moellering, 1976, 1980), as well as interactive map animations and simulations, have become increasingly popular since personal computers with growing graphic-processing power have become cheaper, faster and more user-friendly (Peterson, 1995). Even though real-time three-dimensional landscape fly-throughs and interactive map animations of various spatial diffusion processes have become widespread with dissemination through the Internet, it still seems that the cartographic community has only been scratching the surface of dynamic displays (Campbell and Egbert, 1990; Fabrikant and Josselin, 2003) and there is the very real risk that mapping technology is outpacing cartographic theory. This chapter explores the role of animation in geographic visualization and outlines the challenges, both conceptual and technical, in the creation and use of animated maps today.

Geographic Visualization Edited by Martin Dodge, Mary McDerby and Martin Turner

Animation has it roots in the Latin word *animare*, in English, 'bring to life'. Animation is not to be confused with film or video movies. Animations are defined as sequences of static graphic depictions (frames), the graphic content of which, when shown in rapid succession (typically 24–30 frames per second), begins moving in a fluid motion. The use of animated maps spans the spectrum of disseminating spatial knowledge to a wide audience (e.g. animated weather map loops on television) to data exploration for knowledge discovery by experts (e.g. highly interactive exploratory spatial data analysis tools for scientists). Animated maps have become increasingly popular in recent years because they congruently represent the passage of time with changing graphic displays (or maps). With Google Earth or similar virtual-globe interfaces that can render rich, immersive three-dimensional maps in real-time, the dynamic change of one's point of view or geometric perspective on a three-dimensional landscape has become as easy as one mouse click. However, this technology goes beyond just virtually exploring the surface of the Earth: expert users of animated maps can now explore 'attribute space' using the same kind of techniques developed to explore 'geographic space', for example, when cycling through different themes of a geographic dataset, or by moving along a timeline that represents ordered data values of a certain variable of interest.

An example of this is Peterson's (1995, pp. 50–51) non-temporal animation[1] depicting the percentage of births to mothers under the age of 20 for the United States. The first frame in the animation is a two-class map and the last frame is a seven-class map. The legend is represented as a histogram with bars indicating the number of observations in each category.

Regardless of the map-use goal, unlike static maps that do not change, the individual frames of an animated map are on-screen briefly and there is little time to examine fine details (Monmonier, 1994; Harrower, 2003). In other words, there are obvious cognitive and perceptual limits that must be understood and used to inform map design. We believe that exceeding these limits – which is easy to do with today's massive and complex dataset coupled with powerful computer graphic cards – is likely to leave the user frustrated or unsure of what they have seen. Since our visual memory has limits (Marr, 1982) we will simply not see the finer details in the animation, and only the most general patterns and clusters will be noticed (Dorling, 1992). Basic map-reading tasks, such as comparing colours on a map with those on a legend, become significantly more difficult when the map is constantly changing and thus 'the compression of time as well as space in dynamic cartography poses new problems requiring the recasting, if not rethinking, or the principles of map generalization' (Monmonier, 1996, p. 96).

Their fleeting nature combined with the problem of split attention suggests, then, that the primary utility of animated maps is to not to emphasize specific rates for specific places, but rather to highlight the net effect of the frames when run rapidly in sequence and to gain an overall perspective of the data (Dorling, 1992). According to Ogao and Kraak (2002, p. 23), 'animations enable one to deal with real world processes as a whole rather than as instances of time. This ability, therefore, makes them intuitively effective in conveying dynamic environment phenomena'. Unlike static maps, animated maps seem especially suited to emphasizing the change between moments (Peterson, 1995) and static maps are often insufficient for the task of representing time because they do not directly represent – or foster hypotheses about – geographic behaviours or processes. Because animated maps can

[1] Examples of non-temporal map animations are available on the web at: http://maps.unomaha.edu/books/IACart/book.html#ANI

explicitly incorporate time they are (potentially) better suited to this task and cartographers have long-sought to exploit the potential of animated maps over the past 50 years (Campbell and Egbert, 1990; Harrower, 2004).

In the foreword to *Interactive and Animated Cartography*, Mark Monmonier extols: 'In rescuing both makers and users of maps from the straitjacket of ossified single-map solutions, interactive mapping promises a cartographic revolution as sweeping in its effects as the replacement of copyists by printers in the late fifteenth and early sixteenth centuries' (Peterson, 1995, p. ix). Although the 'promises' of this revolution have been at times slow to materialize, there is justifiable enthusiasm for animated and interactive mapping systems. Moreover, the concurrent and rapid maturation in the last 15 years of (i) animated mapping software, (ii) widespread and powerful computers, (iii) fast Internet connections and (iv) an explosion of rich temporal-spatio data – and tools for viewing those data – has allowed animated mapping to flourish (e.g. NASA World Wind, Google Earth). Unfortunately, academic theories and field-validated best practice have not kept pace with these technological changes, and as a research community we are still learning how to get the most out of our animated maps and, quite simply, know when or how to best use them. Animated maps do not replace static maps, nor are they are not intrinsically better or worse than static maps; they are simply different. Like any form of representation (words, images, numerical formulas), animated maps are better suited to some knowledge construction tasks than others. Understanding what those tasks are is one of the key research challenges for geovisualization (MacEachren and Kraak, 2001; Slocum *et al.*, 2001).

Research into the cognitive aspects of map animations would help to shed light on how users understand and process information from these dynamic representations. The construction of sound theoretical foundations for the effective representation of spatio-temporal phenomena and the adequate depiction of fundamental spatial concepts and relationships, including people's understanding thereof, is not new to GIScience and geovisualization, but due to wider usage of GIS tools outside of geography, it has gained new importance in the past few years.

Preliminary research has shown that animation can reveal subtle space–time patterns that are not evident in static representations, even to expert users who are highly familiar with the data (MacEachren *et al.*, 1998). A good example of the power of animated maps to stimulate new knowledge is provided by Dorling and Openshaw (1992). In their investigation of leukaemia rates in northern England, previously unrecognized hotspots (localized in both space and time) emerged from an animated map of these data. The cancer hotspot in a specific area lasted only a few years and had been missed in previous (i.e. static) analysis because the temporal component of the data had been collapsed, thus 'grossly diluting situations such as these by amalgamating years of low incidence with a pocket of activity' (Dorling and Openshaw, 1992, p. 647). The animation also revealed a second unexpected process, which they described as a 'peculiar oscillation' between leukemia cases in Manchester and Newcastle with an approximately five-year periodicity. Fresh insights such as these provide a useful starting point for more formal spatial analysis.

Wood (1992) chastises cartographers for trying to distil time out of the map and states that 'time remains the hidden dimension' in cartography.

> But the map *does* encode time, and *to the same degree* that it encodes space; and it invokes a temporal code that empowers it to signify in the temporal dimension (Wood, 1992, p. 126).

Timeless maps are problematic because they portray the world in an 'eternal present' and eliminate the concept of process. Langran (1992, p. 22) suggests that historically cartographers actively avoided dealing with time by mapping mostly static things with static maps 'thereby shifting the burden of dealing with the temporal phenomena to the map user'. Being constrained by static display technology (e.g. paper), the majority of Western maps traditionally have emphasized space over time and, thus, are more often used to represent states rather than processes. However, with the rise of geographic visualization (for experts) and highly interactive web maps (for the public), representing time has become routine in cartography. An active group of researchers have called for – and begun to develop – empirically validated theories and practices upon which to create a robust understanding of 'temporal cartography'.

4.2 Types of time

One of the primary reasons for making animated maps is to show spatial processes. As Dorling and Openshaw (1992, p. 643) note:

> It is self-evident that two-dimensional still images are a very good way (if not the only way) of showing two-dimensional still information. However, when the underlying patterns (and processes) start to change dynamically, these images rapidly begin to fail to show the changes taking place.

Time and change, however, are broad concepts and, just as there are different types of attribute data (e.g. categorical data, numerical data), there are different types of time. More importantly, much as there are different graphic 'rules' for different kinds of spatial data (e.g. sequential colour schemes match with numerical data series, qualitative colour schemes with categorical data), there is reason to believe that different types of time require different graphic approaches, although further work is needed in this area to elucidate what these might be.

One of the first geographers to move beyond simple linear concepts of time was Isard (1970), who characterized four types of time: universe time which is absolute and linear, cyclic time such as diurnal patterns, ordinal time which records the relative ordering of events, and time as distance in which the spatial dimension is used to represent time. Frank (1994) identifies three basic kinds of time: linear, cyclic and branching. Linear time depends upon the scale of measurement. Linear time measured at the ordinal scale is the sequence of events, whereas linear time measured numerically is duration. Duration can be continuous or discontinuous. Cyclic time expresses the idea of repetition or recurrence in the sequence of events that can occur in either a regular or irregular manner. Branching time is used to describe future possibilities. The further into the future one goes, the greater the number of possible temporal paths. In complementary work, Haggett (1990) describes four types of temporal change in geography: constants, trends, cycles and shifts. Constants (i.e. no change) and trends (i.e. linear change) are long-term changes. Cycles describe recurring patterns and shifts describe sudden changes (not necessarily recurring).

While there is no agreed upon list of the 'basic' kinds of geographic time, there are commonalities within these frameworks: Isard, Frank and Haggert all include the concepts of *linear time* and *cyclic time* in their respective typologies. These are perhaps the 'core' ideas of time in geography and, as we discuss below, are reflected in widespread use of both cyclic and linear temporal legends (Edsall *et al.*, 1997; Esdall and Sidney, 2005) and ontologies of space–time (Hornsby and Egenhofer, 2000).

4.3 The nature of animated maps

In cartography, two basic animations types are known: temporal animation and non-temporal animation (Dransch, 1997). Temporal animation deals with the depiction of dynamic events in chronological order and depicts actual passage of time in the world. In a temporal animation, 'world time' (e.g. days, centuries) is typically proportionally scaled to 'animation time' (e.g. typically seconds). For example, a population growth animation based on decennial census data is mapped such that 10 years of world time represent a constant time unit in the animation (temporal scale). Examples of temporal animations are population growth, diffusion processes of diseases, commodities and the like, or wild fire spreads and glacier movements.

Non-temporal animations use animation time to show attribute changes of a dynamic phenomenon. The morphing technique is a good example of a non-temporal animation. For instance, animation time is utilized to show a phenomenon's transformation from an orthographic two-dimensional map depiction (e.g. 'god's eye' view) into a perspective three-dimensional view. Other very popular non-temporal animation examples are fly-bys or fly-throughs of three-dimensional terrain, where the viewer's perspective changes over time (e.g. animation of camera motion).

Figure 4.1 demonstrates the flexibility and variety of animated maps that can be found today. The photo-realistic fly-over map (upper left) takes viewers on a high-speed flight

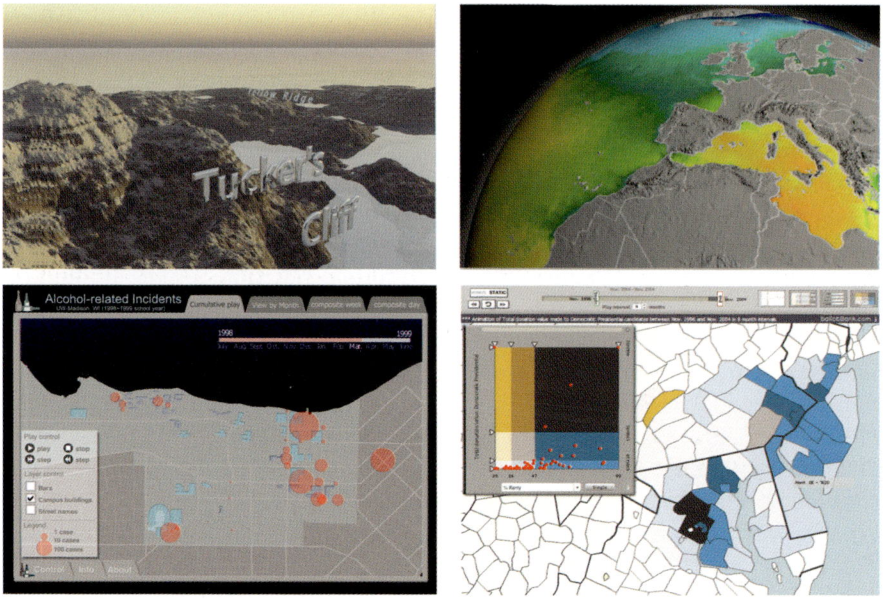

Figure 4.1 Four types of animated maps (clockwise from top left): 'Fly-over' animated map (Harrower and Sheesley, 2005), animation of sea surface temperatures (NASA/Goddard Space Flight Center Scientific Visualization Studio, 2006), Ballotbank.com (Heyman, 2007) and UW-Madison alcohol-related incidents (Liu and Qi, 2003)

over a fractal-generated landscape (which mimics natural forms), annotated with three-dimensional text placed directly on the landscape. Unlike static two-dimensional maps, few rules have been established in how to most effectively label three-dimensional immersive maps (Maass, Jobst and Döllner, 2007), including issues such as text size, placement, label density and text behaviour (e.g. should the three-dimensional text 'track' the user, always presenting an orthogonal face?). The MODIS sea surface temperature animation (upper right) highlights the importance of the Gulf Stream in the North Atlantic by using four years of high-resolution space–time data that captures subtle behaviours not easily seen in static snapshots. This animation demonstrates that a well-designed map can successfully present gigabytes of raw satellite data without overwhelming the reader, in part because individual data points (pixels) merge to form a coherent larger picture (i.e. distinct ocean currents, seasonality and overall north–south gradient). Ballotbank.com (lower right) can create animated bivariate choropleth maps that allow for both temporal and attribute focusing (both forms of information filtering) as well as on-the-fly temporal aggregation, allowing uses to adjust the 'temporal granularity' of the animation: often what one sees in an animation is a product of the ways in which the data are presented, especially the temporal and spatial resolution of the data, and the look of the animation can vary dramatically by changing either (Harrower, 2002). Thus, allowing the user to make these kinds of adjustments to the map themselves is both useful and ethical. The map of alcohol-related incidents (lower left) employs *temporal re-expression* controls, allowing users to animate through the data by composite hour of day, day of the week and month to see how patterns of drinking behaviour change as the basic unit of time changes.

4.3.1 Characteristics of animated maps

Animated maps, sometimes called movie maps or change maps, are primarily used to depict geographic change and processes. Static maps present all of their information simultaneously; animated maps present information over time. Thus, animated maps have an additional representational dimension that can be used to display information. Increasing the running time of an animation increases the total amount of data that can be represented, but at a cost to the user. As the length of the animation increases so too does the difficulty of remembering each frame of the animation. Put another way, although the amount of data that can be represented within an animation is virtually unlimited, there is a finite amount of information the user can distil from the animation and store in their short-term visual memory. As a result, animated maps are typically less than a minute in duration. They are more analogous to television commercials than feature-length films. One practical reason for this is the limitations of visual working memory (Sweller, 1988). Another reason is that animated maps are temporal abstractions. As condensed forms of knowledge, animated maps are intentionally scaled-down representations of the world.

Just as static maps have a spatial scale, temporal animated maps have a temporal scale. This can be expressed as the ratio between real-world time and movie time. For example, five years of data shown in a 10 s animation would have a temporal scale of 1:157 million. Although it is possible to build animated maps that vary their temporal scale as they play – to focus on important moments, or blur-out others – most animated maps keep a constant temporal scale. Additional aspects of the map include its temporal granularity/resolution

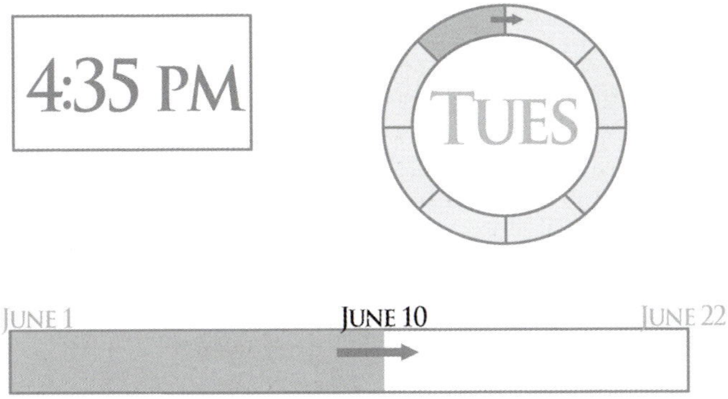

Figure 4.2 The three most common kinds of temporal legend – digital clock, cyclical and bar. The graphical cyclic and bar legend can communicate at a glance both the specific instance and the relation of that moment to the whole (source: M. Harrower)

(the finest temporal unit resolvable) and pace (the amount of change per unit time). Pace should not to be confused with frames-per-second (fps): an animation can have a high frame rate (30 fps) and display little or no change (slow pace). Perceptually, true animation occurs when the individual frames of the map/movie no longer are discernable as discrete images. This occurs above roughly 24 fps – the standard frame rate of celluloid film – although frames rates as low as five fps can generate a passable animation effect. Higher frame rates yield smoother looking animations, although modest computers will have trouble playing movies at high frame rates, especially if the map is a large raster file.

The passage of time or the temporal scale is typically visualized along side the map animation through a 'temporal legend'. Figure 4.2 shows three different kinds of temporal legend: digital clock, cyclical time wheel and linear bar. The advantage of graphical temporal legends, such as the time wheel and linear bar, is that they can communicate at a glance both the current moment (e.g. 4:35 p.m.) and the relation of that moment to the entire dataset (e.g. halfway through the animation). Because it is often assumed that these kinds of legends support different map-reading tasks, designers often include more than one on a single map. Cyclic legends, for example, can foster understanding of repeated cycles (e.g. diurnal or seasonal), while linear bars may emphasize overall change from beginning to end.

By making a temporal legend interactive, the reader can directly manipulate the playback direction and pace of the movie, or jump to a new moment in the animation (known as non-linear navigation). This has become a common interface action in digital music and video players and many map readers now expect to be able to directly interact with temporal legends to control the map. One unresolved problem with legend design is split attention: because animated maps, by their very nature, change constantly, the moment the reader must focus on the temporal legend they are no longer focusing on the map and may miss important cues or information. If they pause the animation to look at the legend, they lose the animation effect. The more the reader must shift their attention between the map and the legend, the greater the potential for disorientation or misunderstanding. Proposed but untested solutions to the problem of split attention include (1) audio temporal legends and (2) embedded temporal legends that are visually superimposed onto the map itself.

In their suggestions for improving animated maps, Johnson and Nelson (1998) suggested giving the subjects the freedom to have direct control over the animated sequence. Moreover, Andrienko, Andrienko and Gatalsky (2000) claim that the greatest understanding may be achieved when the animation is under user control and the geospatial data can be explored in a variety of ways.

Temporal tense is an issue explored by Wood (1992, p. 126): '[t]ense is the direction in which the map points, the direction of its reference in time'. Building on basic linguistic concepts of present, past and future, Wood demonstrates that *all* maps possess 'temporal codes'. These temporal codes allow us to construct a temporal topology for geographic data. Like its spatial counterpart, the concept of temporal topology allows the relationship between temporal entities to be understood and encoded. The development of temporal analytical capabilities in GIS such as temporal queries requires basic topological structures in both time and space (e.g. an object is both *to the left of* and *older than* another object; Peuquet, 1994).

4.4 Potential pitfalls of map animation

While there are some representational tasks for which animation seems especially well-suited (e.g. showing motion), equally there are some representational tasks for which animation is poorly suited. For example, animating changes in property ownership of a neighbourhood over the past 10 years is unwise because these changes are discrete events that can be conceptualized as happening without time. Although buying property is a complex human process, the actual cadastral change is applied at an instant of time (e.g. noon on 1 January 2007). Creating a linear temporal animation of this event might be ineffective (Goldsberry, 2004), and certainly would be boring since the animation would depict long periods of no change, punctuated by periods of instantaneous change that might easily be missed (*Nothing. Nothing. Nothing. Something. Nothing.*). Put another way, important changes often occur over very short time intervals, and thus a static map with ownership names and dates, or even a simple data table, is likely to be a better choice for the tasking of retrieving dates.

Choropleth maps are seldom created with more than 10 data classes, seven data classes being the often-cited upper limit for good map design (Slocum *et al.*, 2005). These limits derive from psychological studies performed a half century ago (Miller, 1956) that revealed that most individuals can process seven (plus or minus 2) 'chunks' of information at once. Probably class limits are even lower for animated maps considering the increased human memory load required to remember earlier map frames when looking at later ones in the animation sequence (Goldsberry, Fabrikant and Kyriakidis, 2004; Harrower, 2003). Does this mean that animated maps should contain no more than seven frames? Clearly not, as people are capable of working with and understanding animations composed of thousands of individual frames, but the question remains: what are the cognitive limits to the complexity of animated maps? In other words, at what point do animated maps become too data-rich for the user? What forms the basic mental chunks of an animated map? How can the size of these chunks be increased? Although answers to these questions are few, we suspect it is driven in part by the length of the animation (i.e. running time), the complexity of the spatial patterns depicted (i.e. spatial heterogeneity) and the complexity of the patterns of change (i.e. temporal heterogeneity).

4.4.1 Cognitively adequate animation design

We still know very little about how effective novel interactive graphical data depictions and geovisualization tools are for knowledge discovery, learning and sense-making of dynamic, multidimensional processes. Cognitive scientists have attempted to tackle the fundamental research question of how externalised visual representations (e.g. statistical graphs, organizational charts, maps, animations etc.) interact with people's internal visualization capabilities (Tversky, 1981; Hegarty, 1992). Experimental research in psychology suggests that static graphics can facilitate comprehension, learning, memorization, communication of complex phenomena and inference from the depiction of dynamic processes (Hegarty, 1992; Hegarty, Kriz and Cate, 2003). However, in a series of publications surveying the cognitive literature on research on animated graphics (that did not include map animations), Tversky and colleagues claim they failed to find benefits to animation for conveying complex processes (Bétrancourt, Morrison and Tversky, 2000; Bétrancourt and Tversky, 2000; Morrison, Tversky and Betrancourt, 2000; Tversky, Morrison and Bétrancourt, 2002). In the cartographic literature, results on animation experiments are not conclusive in part because they depend on how 'success' is measured. For example, in think-aloud experiments comparing passive, interactive and inference-based animations for knowledge discovery (e.g. differing in their interactivity levels), Ogao (2002) found that animations did play a crucial role in facilitating the visualization of geospatial data. Different animation types are used at specific levels of the exploratory process, with passive animation being useful at earlier observatory stages of the exploration process, and inference-based animation playing a crucial role at later stages of discovery, such as in the interpretation and explanation of the phenomenon under study. Similarly, participants in a study by Slocum *et al.* (2004) suggested in a qualitative assessment that map animations and small multiples are best used for different tasks, the former being more useful for inspecting the overall trend in a time series dataset, the latter for comparisons of various stages at different time steps.

In studies reviewed by Tversky and colleagues, typical 'success' measures are either response time, also known as *completion time* (a measure of efficiency) or *accuracy of response* (a measure of quality). In some experiments comparing map animations with static small multiple displays participants answer more quickly, but not more accurately, with animations (Koussoulakou and Kraak, 1992); in other experiments they take longer, and answer fewer questions more accurately (Cutler, 1998), or the time it takes to answer the question does not matter for accuracy at all (Griffin *et al.*, 2006).

There is a fundamental problem with these kinds of comparative studies. To precisely identify differences in the measures of interest, the designs of the animation and the small multiples to be compared require tight experimental control to the extent that it might make a comparison meaningless. Animations are inherently different from small multiples. Making an animation equivalent in information content to a small multiple display to achieve good experimental control may actually mean degrading its potential power. Animations are not simply a sequence of small multiples. Good animations are specifically designed to achieve more than just the sum of their display pieces.

As mentioned earlier, there are many design issues to consider for the construction of potentially useful map animations. Dynamic geographic phenomena may not only change in position or behaviour over time, but also in their visual properties (e.g. attributes). Moreover, the observer or camera location may change in position, distance and angle in relationship to the observed event. Lighting conditions that illuminate the scene and dynamic events may also change over time (e.g. light type, position, distance, angles, colour attributes). The speed with which an animation is viewed is crucial (Griffin *et al.*, 2006) and whether the viewing order of the frames in an animation is predefined (i.e. without any interaction mechanisms) (Hegarty, Kriz and Cate, 2003). The effectiveness of a small multiple displays on the other hand depends mostly on the appropriate number and choice of the small multiples, that is, how many and which of the key events (macro steps) are selected to discretely represent the continuous, dynamic process. Well-designed small multiple displays depict the most thematically relevant (pre-selected) key events, and unlike non-interactive animations allow viewers to inspect the display in their own pace and viewing order.

Figure 4.3 shows a test participant's eye movement patterns overlaid on an identical small multiple map display, but during two different data exploration tasks. The graduated circles show eye fixation durations (the larger the circle, the longer the fixation) and the connecting lines represent rapid eye movements between fixations, called *saccades.* In Figure 4.3(a) the task is to gain an overall impression of the small multiple display, and to verbally describe the patterns that are discovered during its visual exploration. When animating the gaze tracks, one can see that the viewer is not exploring the display in the implied sequence of the small multiple arrangement, but going back and forth between the maps several times or jumping between different rows of maps.

In Figure 4.3(b) the task was to specifically compare two maps within the display. The significantly different viewing behaviour from Figure 4.3(a) suggests that small multiple displays will never be informationally equivalent to non-interactive animations as Tversky, Morrison and Bétrancourt (2002) are implying. When map use contexts require a user to compare items in a time-series (across time, space or attribute) the non-interactive animations (locking a viewer into a pre-defined sequence) will always add cognitive load, as the viewer will have to wait and remember the relevant information until the respective comparative displays comes into view (see Figure 4.4).

The small multiple map display on the other hand allows the user to freely interact with the data in the viewing sequence they deem necessary for the task. In other words, the experimental data (Figure 4.3) suggests that non-interactive animations should be made interactive to be informationally equivalent to small multiple map displays.

Two animation types depicted in Figure 4.4 dynamically depict the same information shown in the static small multiple map display (see Figure 4.3). Figure 4.4(a) represents a frame of a non-interactive animation, only containing a start button, while the interactive animation depicted in Figure 4.4(b) allows a viewer to start and stop the animation at any time and change its display speed and movement direction.

The great power of carefully designed animations, however, are their ability to display micro steps in complex systems that might be missed in small multiple displays (Slocum *et al.*, 2001; Jones *et al.*, 2004). The apprehension of micro steps is directly related to the perception of apparent motion problem. As Griffin *et al.* (2006) note, this is still an unsolved issue in animation research. The perception of apparent motion (i.e. being able to visually interpolate fluid motion from discrete jumps in position between images) depends on (1) the relationship between the duration of a frame in the animation being displayed, (2) the frame rate in the animation (how many frames are displayed per unit time) and (3) the distance of an object moving across the screen. Rensink, O'Regan and Clark (1997) have demonstrated that observers have great difficulty noticing even large changes between two successive scenes in an animation when blank images are shown in between scenes (e.g. simulating a flicker). This *change-blindness* effect operates even when viewers know that

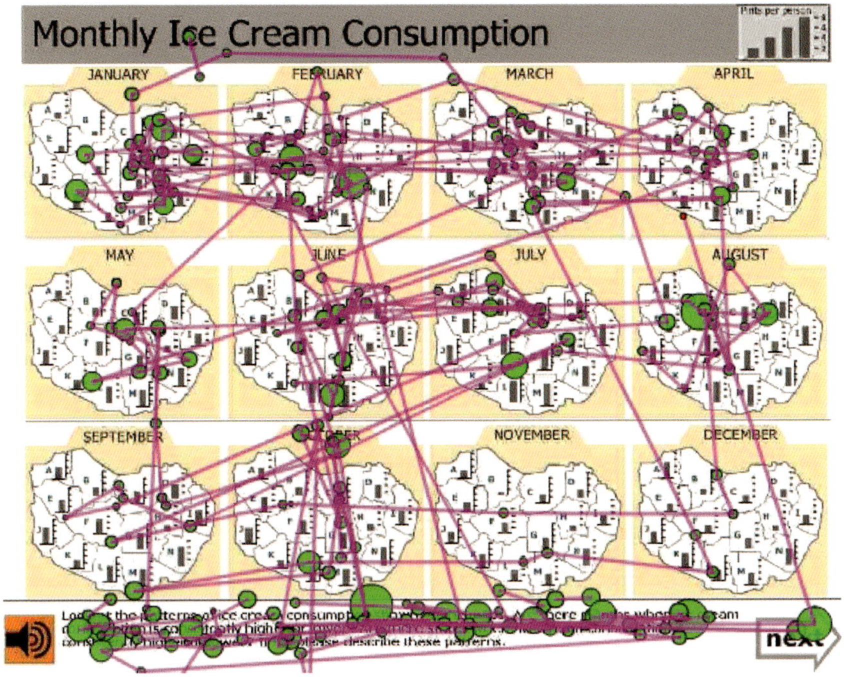

(a)

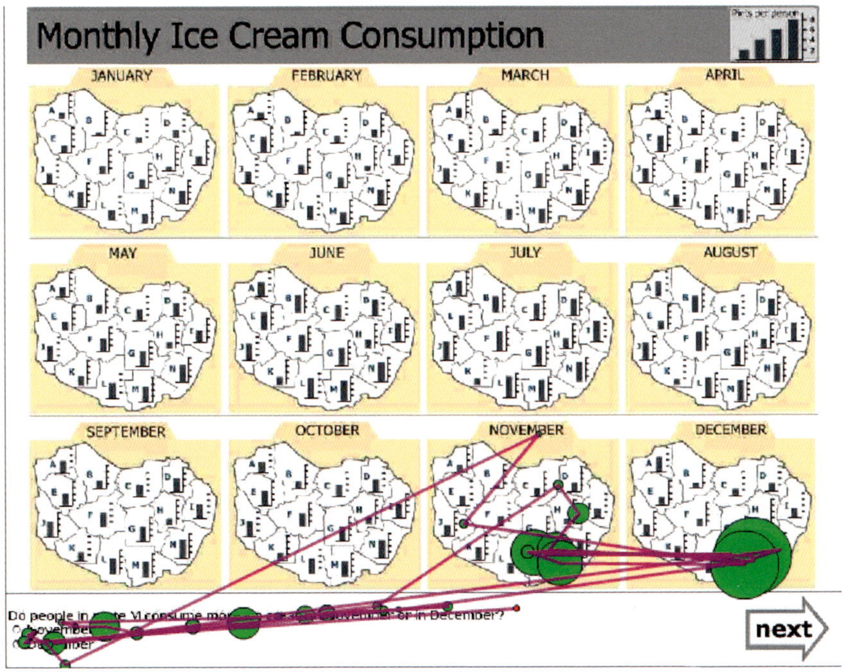

(b)

Figure 4.3 Task-dependent viewing behaviour of two identical small multiple map display stimuli (source: S. I. Fabrikant)

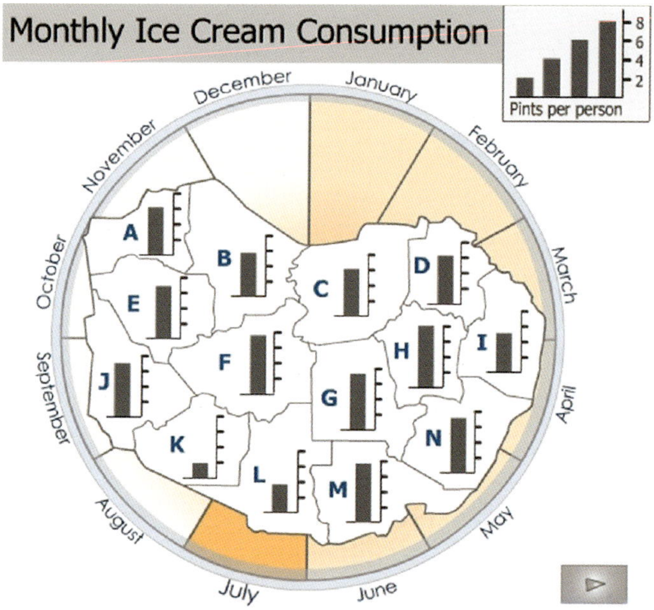

(a)

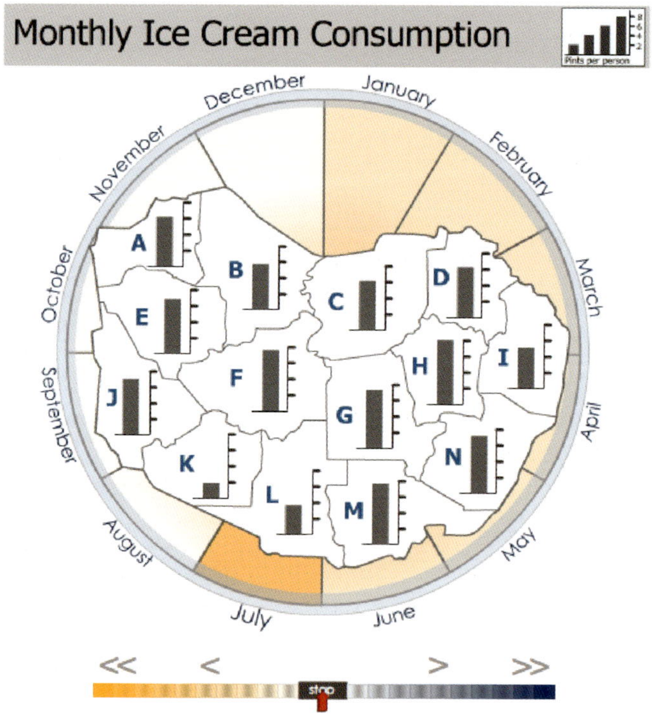

(b)

Figure 4.4 Frames from a non-interactive (a) and (b) interactive animation (source: S. I. Fabrikant) they will occur. Rensink (2002) suggests that only about four to five items can be attended to at any moment in time in animated scenes. Visual transitions in dynamic scenes can be controlled through dynamic design variables, as for example proposed by Lasseter (1987) and DiBiase *et al.* (1992), and these variables could potentially mitigate the change blindness effect. These key design issues need to be carefully considered such that thematically relevant items are made perceptually salient in the animation (Fabrikant and Goldsberry, 2005).

Animation sceptics like to cite Lowe's (1999, 2003, 2004b) experiments on complex interactive weather map animations as examples of animation failures. Lowe (1999) found that participants tended to extract information based on perceptual salience rather than thematic relevance. He additionally suggests potential reasons why animation failed, such as the participant's lack of relevant domain expertise, the complexity of the depicted system and more importantly the manner in which the system was depicted. This begs the question of whether better graphic design principles for animation may make relevant thematic information more perceptually salient, and thus overcome the suggested drawbacks of animation? The level of user manipulation or human–display interaction is one such element of display design.

However, Lowe (2004a) suggests that simply providing user control (e.g. interaction capabilities) to an animation does not always result in the desired learning improvements. He found that participants neglected the animations' dynamic aspects, perhaps because of available user controls. Most of the time participants either stopped the animation and investigated still frames, or the animation was being viewed in a stepwise fashion, one frame at a time. This raises additional research questions such as what kinds of interaction controls are needed for dynamic map displays, and how these controls should be designed such that they are more efficiently used.

4.5 Conclusions

As mentioned earlier, we do not believe it to be a question of whether animations are superior to static maps or not, but as geovisualization designers we are rather interested in finding out how animations work, identifying when they are successful, and why (Fabrikant, 2005). Although Tversky and collaborators emphasize that sound graphic principles must be employed to construct successful static graphics, they do not elaborate which ones. The same authors further suggest that research on static graphics has shown that only carefully designed and appropriate graphics prove to be beneficial for conveying complex phenomena, but they do not offer specific design guidelines. As geovisualization designers we assert that design issues must be carefully considered not only for static graphics, but also especially for dynamic graphics, considering that animated graphics add an additional information dimension (e.g. that of time or change). Ad-hoc design decisions of graphic test stimuli, due to lack in graphic design training, may be an additional reason why animations might have failed in past research. Research in cartography suggests that traditional graphic design principles may only partially transfer into the dynamic realm, and therefore design for animation needs special attention (Harrower, 2003). To systematically evaluate the effectiveness of interactive and dynamic geographic visualization displays for knowledge discovery, learning and knowledge construction, Fabrikant (2005) proposes a research agenda and sketches a series of empirical experiments aimed at developing cognitively adequate dynamic map displays (see sample test stimuli shown in Figures 4.3 and 4.4). The proposed empirical studies adhere to experimental design standards in cognitive science, but are additionally grounded on a solid dynamic design framework borrowed from cartography, computer graphics and cinematography. The goal is to investigate how dynamic visual variables (DiBiase *et al.*, 1992) and levels of interactivity affect people's knowledge construction processes from dynamic displays as compared with static displays.

In order to realize the full potential of animated maps within GIScience – and more broadly, within society – we need to better understand for what kinds of representational tasks they are well suited (and just as importantly not well suited) and how variations in the design of animated maps impact our ability to communicate and learn. By doing so we will broaden the cartographic toolkit available to GIScientists.

In a broader context, better understanding of the human cognitive processes involved in making inferences and extracting knowledge from highly interactive graphic displays is fundamental for facilitating sense-making of multidimensional dynamic geographic phenomena. Better understanding will lead to greater efficiency in the complex decision-making required to solve pressing environmental problems and societal needs.

References

Andrienko, N., Andrienko, G. and Gatalsky, P. (2000) Supporting visual exploration of object movement. *Proceedings of the Working Conference on Advanced Visual Interfaces AVI 2000*, Palermo, 23–26 May 2000. New York, ACM Press, pp. 217–220.

Bertin, J. (1983) *Semiology of Graphics*. Madison, WI, University of Wisconsin Press.

Bétrancourt, M. and Tversky, B. (2000) Effect of computer animation on users' performance: a review. *Le Travail Humain* 63(4), 311–330.

Bétrancourt, M., Morrison, J. B. and Tversky, B. (2000) Les animations sont-elles vraiment plus efficaces? *Revue D'Intelligence Artificielle* 14: 149–166.

Campbell, C. S. and Egbert, S. L. (1990) Animated cartography: thirty years of scratching the surface. *Cartographica* 27(2): 24–46.

Cutler, M. E. (1998) The Effects of Prior Knowledge on Children's Ability to Read Static and Animated Maps. Unpublished MSc Thesis. Department of Geography, University of South Carolina, Columbia, SC.

DiBiase, D., MacEachren, A. M., Krygier, J. B. and Reeves, C. (1992) Animation and the role of map design in scientific visualization. *Cartography and Geographic Information Systems* 19(4): 210–214.

Dorling, D. (1992) Stretching space and splicing time: from cartographic animation to interactive visualization. *Cartography and Geographic Information Systems* 19(4): 215–227.

Dorling, D. and Openshaw, S. (1992) Using computer animation to visualize space–time patterns. *Environment and Planning B: Planning and Design* 19: 639–650.

Dransch, D. (1997) *Computer-Animation in der Kartographie: Theorie und Praxis*. Heidelberg, Springer.

Edsall, R. M., Kraak, M.-J., MacEachren, A. M. and Peuquet, D. J. (1997) Assessing the effectiveness of temporal legends in environmental visualization. *Proceedings of GIS/LIS '97*, Cincinnati, OH, 28–30 October, pp. 677–685.

Edsall, R. M. and Sidney, L. R. (2005) Applications of a cognitively informed framework for the design of interactive spatiotemporal representations. In *Exploring Geovisualization*, MacEachren, A. M., Kraak, M.-J. and Dykes, J. (eds). New York, Guilford Press, 577–589.

Fabrikant, S. I. (2005) Towards an understanding of geovisualization with dynamic displays: issues and prospects. *Proceedings of the American Association for Artificial Intelligence (AAAI) 2005 Spring Symposium Series: Reasoning with Mental and External Diagrams: Computational Modeling and Spatial Assistance*, Stanford University, Stanford, CA, 21–23 March 2005, pp. 6–11.

Fabrikant, S. I. and Goldsberry, K. (2005) Thematic relevance and perceptual salience of dynamic geovisualization displays. *Proceedings of the 22th ICA/ACI International Cartographic Conference*, A Coruña, 9–16 July (CDROM).

Fabrikant, S. I. and Josselin, D. (2003) La 'cartactive', cartographie en mouvement: un nouveau domaine de recherche pluridisciplinaire ou un plan de la géomatique? In *Revue Internationale de Géomatique, Numéro Special 'Cartographie Animée et Interactive'*, Josselin, D. and Fabrikant, S. I. (eds). Paris, Hermès, Lavoisier, pp. 7–14.

Frank, A. (1994) Different types of 'times' in GIS. In *Spatial and Temporal Reasoning in Geographic Information Systems*, Egenhofer, M. and Golledge, R. (eds). New York, Oxford Press, pp. 40–62.

Goldsberry, K. (2004) Stabilizing Rate of Change in Thematic Map Animations. Unpublished Masters Thesis, Department of Geography, University of California Santa Barbara, CA.

Goldsberry, K., Fabrikant, S. I. and Kyriakidis, P. (2004) The influence of classification choice on animated choropleth maps. *Proceedings, Annual Meeting of the North American Cartographic Society*, Portland, ME, 6–9 October 2004.

Griffin, A. L., MacEachren, A. M., Hardisty, F., Steiner, E. and Li, B. (2006) A comparison of animated maps with static small-multiple maps for visually identifying space-time clusters. *Annals of the Association of American Geographers* 96(4): 740–753.

Haggett, P. (1990) *The Geographer's Art.* Cambridge, MA, Basil Blackwell.

Harrower, M. (2002) Visualizing change: using cartographic animation to explore remotely-sensed data. *Cartographic Perspectives* 39: 30–42.

Harrower, M. (2003) Designing effective animated maps. *Cartographic Perspectives* 44: 63–65.

Harrower, M. (2004) A look at the history and future of animated maps. *Cartographica* 39(3): 33–42.

Hegarty, M. (1992) Mental animation: Inferring motion from static displays of mechanical systems. *Journal of Experimental Psychology: Learning,Memory,and Cognition* 18(5): 1084–1102.

Hegarty, M., Kriz, S. and Cate, C. (2003) The roles of mental animations and external animations in understanding mechanical systems. *Cognition and Instruction* 27(4): 325–360.

Heyman, D. (2007) Ballotbank: Directed Geographic Visualization for Elections and Political Science. Unpublished MSc Thesis, Department of Geography, University of Wisconsin – Madison.

Hornsby, K. and Egenhofer, M. (2000) Identity-based change: a foundation for spatio-temporal knowledge representation. *International Journal of Geographical Information Science* 14(3): 207–224.

Isard, W. (1970) On notions of models of time. *Papers of the Regional Science Association* 25: 7–32.

Johnson, H. and Nelson, E. S. (1998) Using flow maps to visualize time-series data: comparing the effectiveness of a paper map series, a computer map series, and animation. *Cartographic Perspectives* 30: 47–64.

Jones, A., Blake, C., Davies, C. and Scanlon, E. (2004) Digital maps for learning: a review and prospects. *Computers and Education* 43: 91–107.

Koussoulakou, A. and Kraak, M. J. (1992). Spatio-temporal maps and cartographic communication. *The Cartographic Journal* 29: 101–108.

Langran, G. (1992) *Time in Geographic Information Systems.* London, Taylor and Francis.

Lasseter, J. (1987) Principles of traditional animation applied to three-dimensional computer animation. *ACM Computer Graphics* 21(4): 35–44.

Liu, J. and Qi, F. (2003) *Alcohol-related incidents at UW – Madison*. Student class project. Available at: http://www.geography.wisc.edu/~harrower/Geog575/finalProjects03/drinking.swf

Lowe, R. K. (1999) Extracting information from an animation during complex visual learning. *European Journal of Psychology of Education* 14(2): 225–244.

Lowe, R. K. (2003) Animation and learning: Selective processing of information in dynamic graphics. *Learning and Instruction* 13: 157–176.

Lowe, R. K. (2004a). User-controllable animated diagrams: the solution for learning dynamic content. In *Diagrams 2004: Diagrammatic Representation and Inference,Lecture Notes in Artificial Intelligence 2980*, Shimojima, A. (ed.). Berlin, Springer, pp. 355–359.

Lowe, R. K. (2004b) Interrogation of a dynamic visualization during learning. *Learning and Instruction* 14: 257–274.

Maass, S., Jobst, M. and Döllner, J. (2007) Depth cue of occlusion information as criterion for the quality of annotation placement in perspective views. In: *The European Information Society: Leading the Way with Geo-information*, Fabrikant, S. I. and Wachowicz, M. (eds). Lecture Notes in Geoinformation and Cartography. Berlin, Springer, pp. 473–486.

MacEachren, A. M. (1994) Time as a cartographic variable. In *Visualization in Geographic Information Systems*, Hearnshaw, H. and Unwin, D. (eds). New York, Wiley, pp. 115–129.

MacEachren, A. M. (1995) *How Maps Work: Representation, Visualization and Design*. New York, Guilford Press.

MacEachren, A. M. and Kraak, M.-J. (2001) Research challenges in geovisualization. *Cartography and Geographic Information Science* 28(1): 1–11.

MacEachren, A. M., Boscoe, F. P., Haug, D. and Pickle, L. W. (1998) Geographic visualization: designing manipulable maps for exploring temporally varying georeferenced statistics. *Proceedings of the IEEE Information Visualization Symposium*, Research Triangle Park, NC, 19–20 October, pp. 87–94.

Marr, D. (1982) *Vision: a Computational Investigation into the Human Representation and Processing of Visual Information*. San Francisco, CA, W. H. Freeman.

Miller, G. A. (1956) The magical number seven, plus or minus two: some limit on our capacity for processing information. *Psychological Review* 63: 81–97.

Moellering, H. M. (1976) The potential uses of a computer animated film in the analysis of geographic patterns of traffic crashes. *Accident Analysis and Prevention* 8(4): 215–227.

Moellering, H. (1980) The real-time animation of three-dimensional maps. *The American Cartographer* 7(1): 67–75.

Monmonier, M. (1994) Minimum-change categories for dynamic temporal choropleth maps. *Journal of the Pennsylvania Academy of Science* 68(1): 42–47.

Monmonier, M. (1996) Temporal generalization for dynamic maps. *Cartography and Geographic Information Systems* 23(2): 96–98.

Morrison, J. B., Tversky, B. and Betrancourt, M. (2000) Animation: does it facilitate learning? *Proceedings of the Smart Graphics 2000: AAAI Spring Symposium*, Stanford, CA.

NASA/Goddard Space Flight Center Scientific Visualization Studio (2006) *MODIS Sea Surface Temperature Highlighting the Gulf Stream (2002 to 2006)*; http://svs.gsfc.nasa.gov/vis/a000000/a003300/a003389/index.html

Ogao, P. J. (2002) Exploratory Visualization of Temporal Geospatial Data Using Animation. Unpublished Ph.D. Dissertation. University of Utrecht, Department of Geography, Utrecht.

Ogao, P. J. and Kraak, M.-J. (2002) Defining visualization operations for temporal cartographic animation design. *International Journal of Applied Earth Observation and Geoinformation* 4(1): 23–31.

Peuquet, D. J. (1994) It's about time: a conceptual framework for the representation of temporal dynamics in geographic information systems. *Annals of the Association of American Geographers* 84(3): 441–461.

Peterson, M. P. (1995) *Interactive and Animated Cartography*. Englewood Cliffs, NJ, Prentice Hall.

Rensink, R. A. (2002) Internal vs. external information in visual perception. *Proceedings of the Second International Symposium on Smart Graphics*, Hawthorne, NY, 11–13 June 2002, pp. 63–70.

Rensink, R. A., O'Regan, J. K. and Clark, J. J. (1997) To see or not to see: the need for attention to perceive changes in scenes. *Psychological Science* 8: 368–373.

Slocum, T. A., Blok, C., Jiang, B., Koussoulakou, A., Montello, D. R., Fuhrmann, S., and Hedley, N. R. (2001) Cognitive and usability issues in visualization. *Cartography and Geographic Information Science* 28(1): 61–75.

Slocum, T. A., Sluter, R. S., Kessler, F. C. and Yoder, S. C. (2004) A qualitative evaluation of maptime, a program for exploring spatiotemporal point data. *Cartographica* 39(3): 43–68.

Slocum, T. A., McMaster, R. B., Kessler, F. C. and Howard, H. H. (2005) *Thematic Cartography and Geographic Visualization*, 2nd edn. Englewood Cliffs, NJ, Prentice Hall.

Sweller, J. (1988) Cognitive load during problem solving: effects on learning. *Cognitive Science* 12: 257–285.

Tobler, W. R. (1970) A computer movie simulating urban growth in the Detroit region. *Economic Geography* 46(2): 234–240.

Tversky, B. (1981) Distortion in memory for maps. *Cognitive Psychology* 13: 407–433.

Tverksy, B., Morrison, J. B. and Bétrancourt, M. (2002) Animation: can it facilitate? *International Journal of Human–Computer Studies*, 57: 247–262.

Wood, D. (1992) *The Power of Maps*. London: Guilford Press.

Yattaw, N. (1999) Conceptualizing space and time: a classification of geographic movement. *Cartography and Geographic Information Systems* 26(2): 85–98.

5

Telling an Old Story with New Maps

Anna Barford and Danny Dorling

Department of Geography, University of Sheffield

5.1 Introduction: re-visualizing our world

This is an unusual chapter – it presents a story through maps. The story comes from a project that made 366 world maps of different variables in the year 2006. The project was undertaken by the Social and Spatial Inequalities group based in the Department of Geography at the University of Sheffield in collaboration with Mark Newman, a physicist working at the University of Michigan. All the maps that are shown here are available for free on our website: www.worldmapper.org.

Maps of the world are powerful tools in shaping how we think of the world; the word 'imagine' is primarily concerned with images and pictures. When we imagine the world it is likely that a map comes to mind, or maybe the earthrise photograph taken on the Apollo 8 mission in 1968. Visualizing the world according to the size of land masses is currently a popular image. Yet there are an infinite number of subjects, relating to how people live in the world, which could be represented by a world map. We (all of us that is, not just the authors) are lucky enough to now have data available to map some of these. Almost none of the data shown here were available worldwide in, say, 1968.

Below is a description of how, using world data, one can change the sizes of territories to show distributions of variables. Using this technique with colleagues, we have created hundreds of apparently distorted world maps. To try to explain why we have done this, the remainder of this chapter is made up of a story that not only concerns the connections between people shown on the same map, but also the connections between what is shown in different maps.

Geographic Visualization Edited by Martin Dodge, Mary McDerby and Martin Turner

5.2 Method and content

The maps shown in Figure 5.1 are cartograms. Cartograms are not a new idea in work that seeks to visualize data. Areas on maps (sizes) can be changed to effectively show a range of numbers that describe the way that people live at the world, at national and local scales (see for example Kidron and Segal's *The State of the World Atlas*, 1984, or Dorling and Thomas's cartograms of the UK, 2004). What is new here is that these maps are a step closer to retaining the physical shape of territorial boundaries as they appear on a land area map, whilst managing to expand, shrink or maintain the area within these new maps in proportion to almost any variable (compare the territory shapes in the two maps in Figure 5.1). This preservation of boundary shape is achieved using Mark Newman's new version of a cartogram-generating computer algorithm (based on Gastner and Newman, 2004), which is itself based upon a diffusion equation from the physics of heat transfer and molecular mixing.

To explain further what you are looking at, take the example of a population map – the space on the map available for territories is divided between territories according to the number of people living in each territory, thus the same area in any territory represents the same number of people. This enables a more democratic representation of the population of the world, where people are treated as equals regardless of where they live. The size of the territory indicates the proportion of the world population that lives there. The sea presents a problem for such density-equalizing projections, because people do not live there. However, not mapping the sea would distort world maps so much that the benefit of mapping (a legible visualization of data) would be negated. Therefore the sea is given a 'neutral buoyancy' or fixed area, despite the lack of people recorded as living there. Antarctica has been treated in the same way as the sea. Another key to interpreting these visualizations is that they only ever show and compare absolute values, never rates.

In this chapter we aim to illustrate how datasets can be mapped and used to describe the way people live, ranging from the spread of adult literacy to the volume of car production to the proliferation of nuclear weapons. These maps allow for comparison between countries, and facilitate a greater understanding of how the world fits together. Mapping these subjects provides an original way of conceptualizing the world in terms of what really affects our daily lives: who has access to clean drinking water, where are most of the cars in the world driven, what proportion of all malaria cases are in each country? When imagining the human world, it is meaningful to think in terms of these subjects, not just in terms of land area.

Until recently, there was a lack of worldwide territory-level data on many variables, which made a project such as this impossible. The United Nations Millennium Development

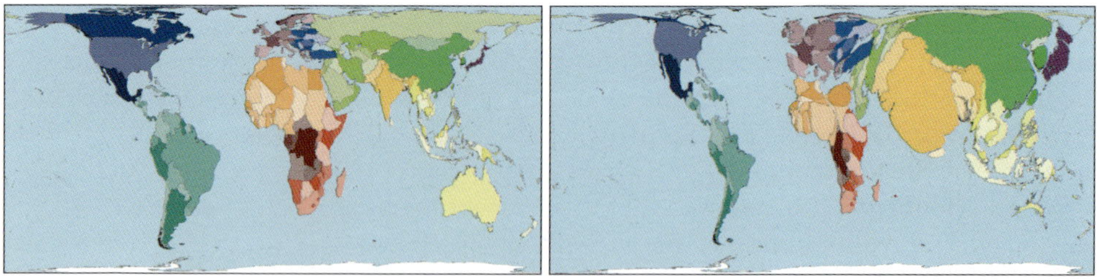

(a) (b)

Figure 5.1 (a) Land area (WORLMAPPER MAP 1); (b) population (Worldmapper Map 2). These maps are available at: www.worldmapper.org

Goals gave an impetus for much more state-level data to be recorded, for preliminary datasets to be corrected, and for more areas to be counted. Many of the datasets used in the Worldmapper website are sourced from agencies of the United Nations, such as the United Nations Development Programme, the World Health Organization and the United Nations Statistics Division. The availability of this data has made world mapping possible for the 366 subjects that we have mapped; and as we write, more databases are becoming available and more maps are becoming feasible.

The aim of this project is to communicate this data. We consider these maps to be useful, effective and interesting vehicles to achieve this aim, but read on to see if you agree. It is important to communicate world data because it is about us, and it can greatly increase our understandings of the world and our positions within it. Each of us has some right and some responsibility to understand how our life relates to those of others. Beyond this talk of rights and responsibilities, we hope that these pictures are also interesting and thought-provoking. This project is politically timely, given the recent demands to 'Make Poverty History', because our maps illustrate the extent of worldwide inequalities that result in poverty. Of course maps themselves do not change anything; they are just – if well drawn – accessible, legible descriptions of many facets of the world. Yet, as Rosa Luxemberg (1871–1919) stated: 'the most revolutionary thing one can do is always to proclaim loudly what is happening'.

Text Box 5.1: Promises

The UK Government is committed to tackling the problem of child poverty. In March 1999, the Prime Minister Tony Blair set out a commitment to end child poverty forever:

> And I will set out our historic aim that ours is the first generation to end child poverty forever, and it will take a generation. It is a 20-year mission but I believe it can be done.

Compare with:

> within a decade no child will go to bed hungry, . . . no family will fear for its next days bread and . . . no human being's future and well being will be stunted by malnutrition.
>
> (Henry Kissinger, *First World Food Conference*, Rome 1974)

Source: Gordon (2004).

World human geography and history are not well known. If you are reading this, it is doubtful that you know who Rosa Luxemberg was and how she was tortured and died for proclaiming loudly what was happening. However, you can type Rosa's name into an Internet search engine and find out more about where, when and how she lived. You could not do that in 1968, or 1978 or even 1988. The history and geography of those whose individual stories are not recorded is much harder to grasp. In the remainder of the century since Rosa's death over 130 million people – mostly adults – have been killed worldwide in genocides (more than in all the wars of that century). At the end of the twentieth century, due almost entirely to avoidable poverty, over 100 million children were dying in their first five years worldwide per decade. A further 30 million were born dead – per decade. Almost all of these child deaths are in 'developing countries'. Every 10 years the largely avoidable deaths of infants and the still-born match the estimate of all adults killed in genocides in the last century. Two simple bar charts in Figure 5.2 sanitize this information on how children still suffer the most, when they are still too young to have taken their chance to change things. And this is despite many pledges made to protect children from forms of poverty (two of which are shown below).

The remainder of this chapter tells a story with maps and just a few words. This is a story of where we are now, told through the births of the poorest, and the deaths of the richest. The maps and text boxes are all taken from a sample of less than a tenth of those available on the Worldmapper website. All that is original here are the few new words introducing, linking and concluding this story, and the way in which these pictures have been arranged to tell this story.

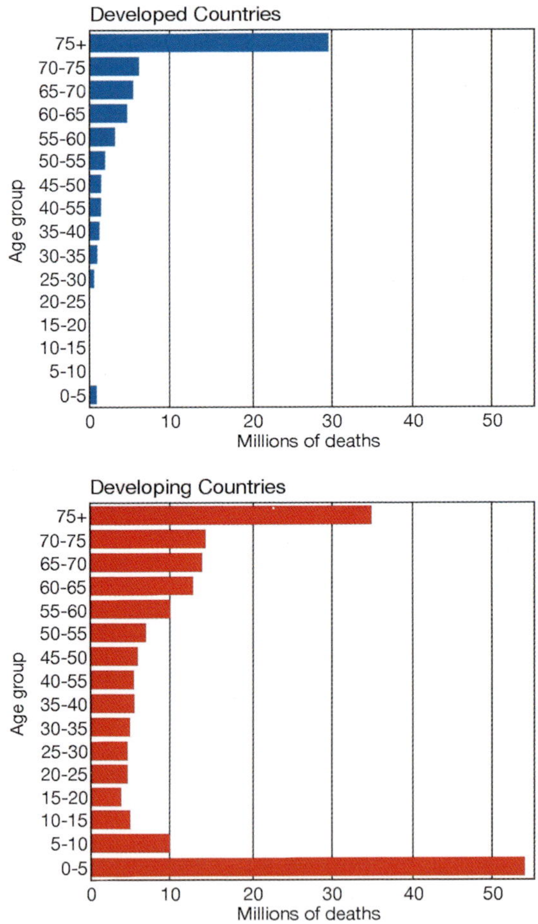

Figure 5.2 Deaths by age group, worldwide 1990–1995. From 'the rhetoric and reality of child poverty' [Adapted from: *The State of the World Population* (1998) reported in Smith and Bræin (2003) courtesy of Earthscan Publications]

5.2.1 Stillbirths

In many places it is common for death to occur at or before birth: 3 million are born dead every year. This is almost as many deaths as those that occur in the first week of life. To those 3 million annual neonatal deaths add another 4 million who die in the remainder of their first year; and another 3 million who die every year in the next 4 years of life. In sum this is 10 million that are born alive under 5 in total dying each year, over 50 million dead in 5 years – and that is not counting the 15 million born dead every 5 years. Figure 5.3 shows where most are stillborn and Text Box 5.2 provides a little context. This is the map of those who never get to live.

Text Box 5.2: Stillbirths

Stillborn means dead at birth. In 2000 there were 3.3 million stillbirths worldwide; more than a third of these occurred in Southern Asia.

The number of stillbirths for every 1000 births varies between regions; where parents live greatly affects the chance of a stillbirth. In Central Africa, there are 42 stillbirths for every 1000 births; the Western European rate is a tenth of this. Mauritania has the highest rate of stillbirths at 6.3%. Generally, the richer the territory, the fewer stillbirths occur there.

> He was beautiful; perfect in every way, except he never took his first breath, and we never heard him cry
>
> (Bereaved Parent, undated)

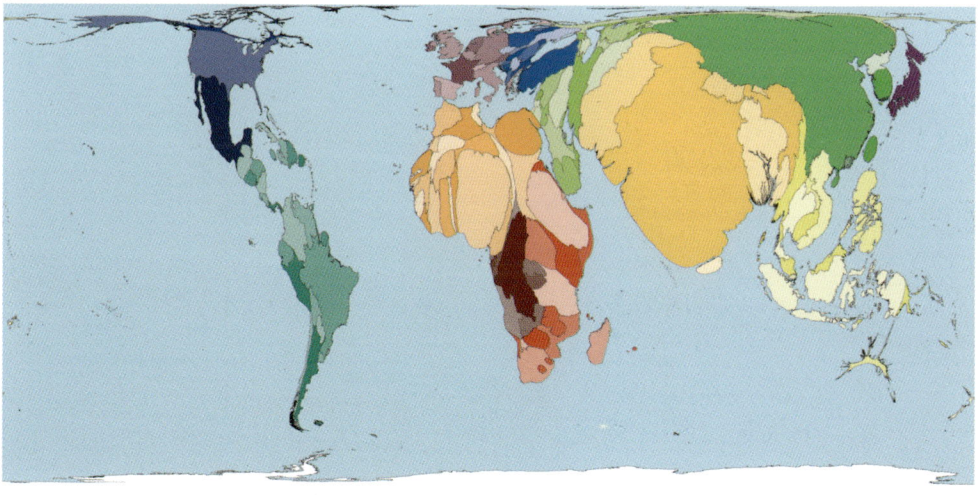

Figure 5.3 Territory size shows the proportion of all stillbirths over 28 weeks of gestation worldwide that occurred there in 2000 (Worldmapper Map 259)

5.2.2 Births

Thankfully, for every 3 million stillbirths worldwide, 133 million babies are born alive and 123 million of those will live to see their fifth birthday. Most people in the world now live in a country where fewer babies are born than would be required to maintain the population at its current size were there no migration. In the terminology used by demographers, fertility rates are now below 'replacement level' in most of the world. Thankfully (again), immigration from abroad can maintain the populations of territories where less than one baby is born per adult life. Figure 5.4 shows where most babies are born alive each year and Text Box 5.3 gives more detail of the variations between places.

Text Box 5.3: Births

A total of 133,121,000 babies were born in the year 2000. In territories with the fewest births per person, more people are dying than are being born. As with all population statistics, even for this vital one, figures are rough estimates.

More children are born each year in Africa than are born in the Americas, all of Europe and Japan put together. Worldwide, more than a third of a million new people will be born on your birthday this year.

The birth of a baby is an occasion for weaving hopeful dreams about the future
(Aung San Suu Kyi, 1997)

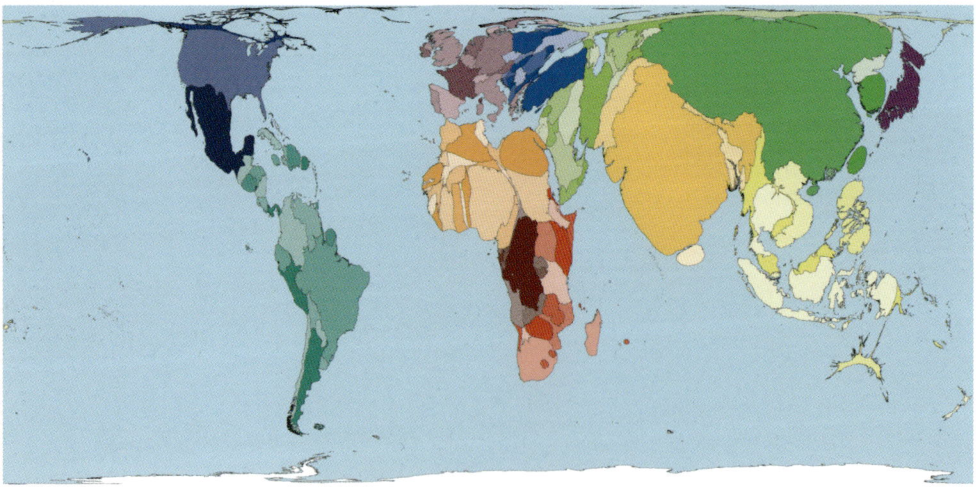

Figure 5.4 This map shows the proportion of the world's total births for each territory (Worldmapper Map 3)

5.2.3 Births attended

A majority of us now enter the world with the help of skilled personnel – mostly midwives. Worldwide, the death of women during childbirth is still the biggest killer of women at child-bearing ages and in most cases those deaths are where the births are not attended by skilled health personnel. In the United States and Canada there are few midwives and most births are attended by physicians, which is far more costly. Infant mortality in the United States is much higher than in most of Europe and Japan. However, as Figure 5.5 shows, the large majority of attended births now occur outside of the affluent triad of North America, Western Europe and Japan; as this is where the majority of all births occir, it would be pitiful if this were not the case.

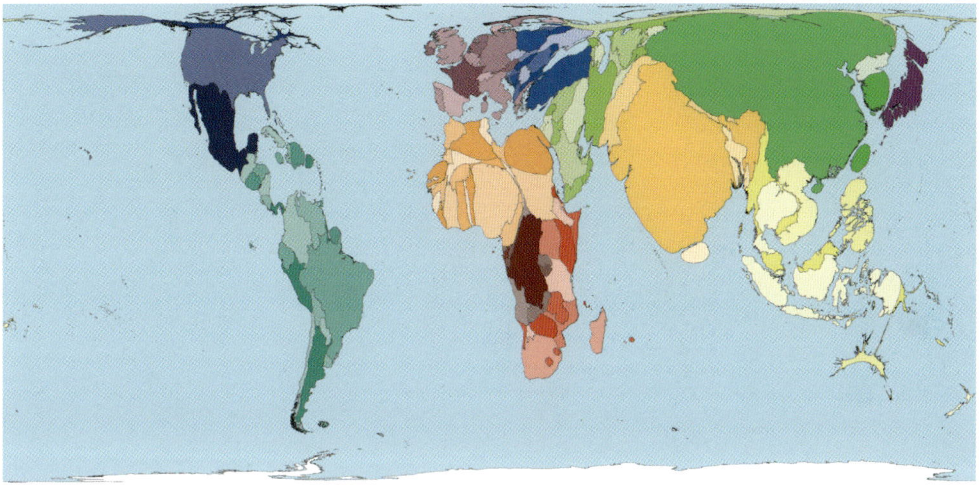

Figure 5.5 This map shows the worldwide distribution of all attended births (Worldmapper Map 4)

Text Box 5.4: Births Attended

Worldwide, 62% of births are attended by skilled health personnel. This ranges from 6% in Ethiopia, to practically 100% birth attendance in Japan.

The total number of births attended, shown on the map, depends partly on how many women there are and how many babies they have.

If you were recently born in Cambodia, there is twice the chance your birth would have been attended, than if you were born in Chad.

> how many of us realize that, in much of the world, the act of giving life to a child is still the biggest killer of women of child-bearing age?
>
> (Liya Kibede, 2005)

5.2.4 Childhood diarrhoea

However, even the majority of those whose journey into life is assisted are not entering a safe, clean or especially caring world:

> Seven out of 10 childhood deaths in developing countries can be attributed to just five main causes – or a combination of them: pneumonia, diarrhoea, measles, malaria and malnutrition. Around the world, three out of four children seen by health services are suffering from at least one of these conditions. (World Health Organization, 1996, 1998; quoted by Gordon, 2004)

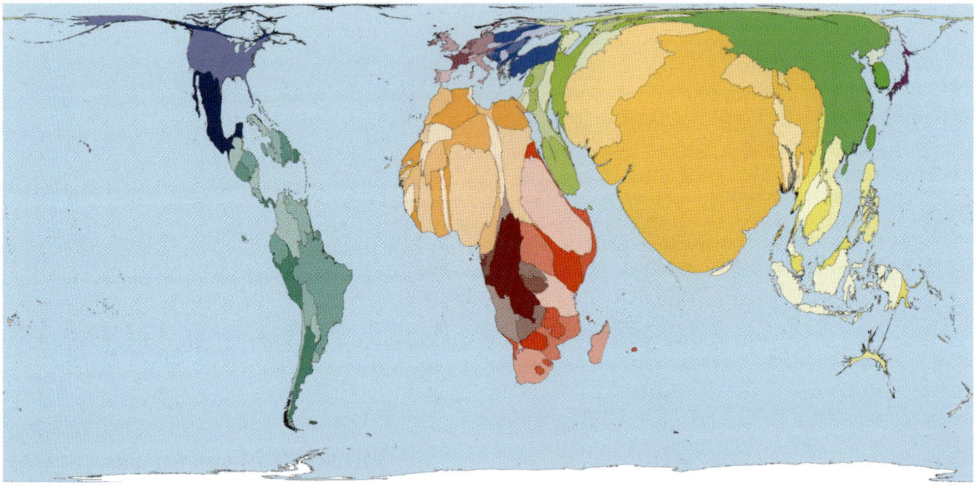

Figure 5.6 Territory size shows the proportion of worldwide cases of diarrhoea found in children aged 0–4 living there (Worldmapper Map 233)

Text Box 5.5: Childhood Diarrhoea

Diarrhoea is common amongst children. In an average two week period, an estimated 82 million children aged 0–5 years old have diarrhoea. Diarrhoea varies in severity – some children recover quickly, a small proportion but a large number die. Access to clean water and rehydration salts can reduce prevalence and minimize the impact.

The highest prevalence of diarrhoea amongst children was recorded in Niger, where 4 in every 10 children had diarrhoea in a typical two-week period. Most children in Niger will have many episodes a year, causing chronic debility.

> I now know that how critical it is to wash hands with soap before eating so as to prevent germs from entering my body. This will protect me from infections such as diarrhoea
>
> (Manoj Patel, 2005)

5.2.5 Child labour

For the majority that survive to their fifth birthday, 123 million little humans per year (give or take a dozen million for how poorly we record our children worldwide), life is not suddenly safe and easy. Now, for many, they are a resource. Some 67 million children aged 10–14 work. The numbers who work aged 5–9 are not known, but it is likely that their distribution is similar to that of the labours of their older brothers and sisters shown here, below. Much child labour helps to produce goods consumed in the rich triad (that is North America, Western Europe and Japan).

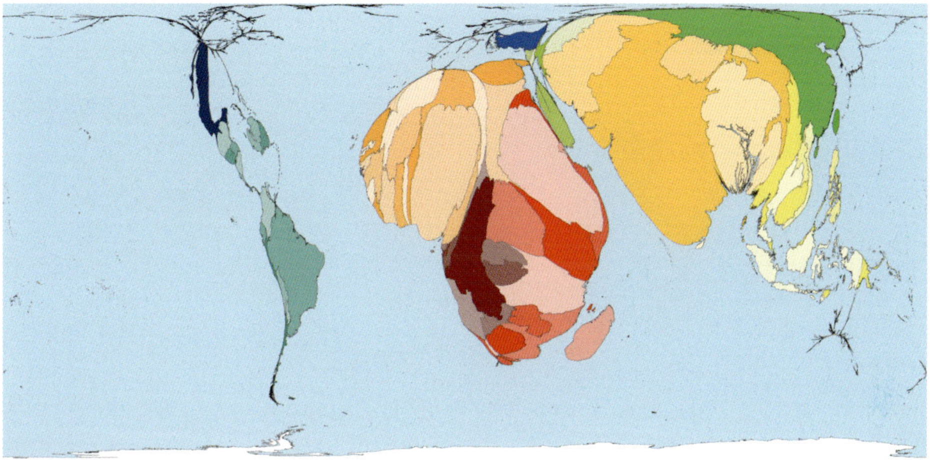

Figure 5.7 Territory size shows the proportion of the worldwide child work force (aged 10–14) that lives there (Worldmapper Map 135)

Text Box 5.6: Child Labour

Nine of the 10 territories with the highest proportions of child labourers are in Africa. The anomaly is Bhutan. At the other extreme, Italy, which has the lowest proportion of children living there, also has a very low percentage of children who work.

The map shows that most child labour occurs in African and Southern Asian territories. India has the highest number of child labourers, twice as many as China, where the second highest population of child labourers lives.

No children work in Japan; in Western Europe there are 13,000 child labourers. These are the regions with the smallest workforces of children.

> At our homes we had done a lot of ploughing, planting, weeding and harvesting; we had hewn wood and drawn water; we had tended sheep, goats and cattle; we had done one hundred and one odd jobs.
>
> (Ndabaninga Sithole, 1959)

5.2.6 Youth literacy

'Everyone has the right to education', according to the Universal Declaration of Human Rights (1948). The second Millennium Development Goal is to achieve universal primary education. In 2002, five out of six eligible children were enrolled in primary education

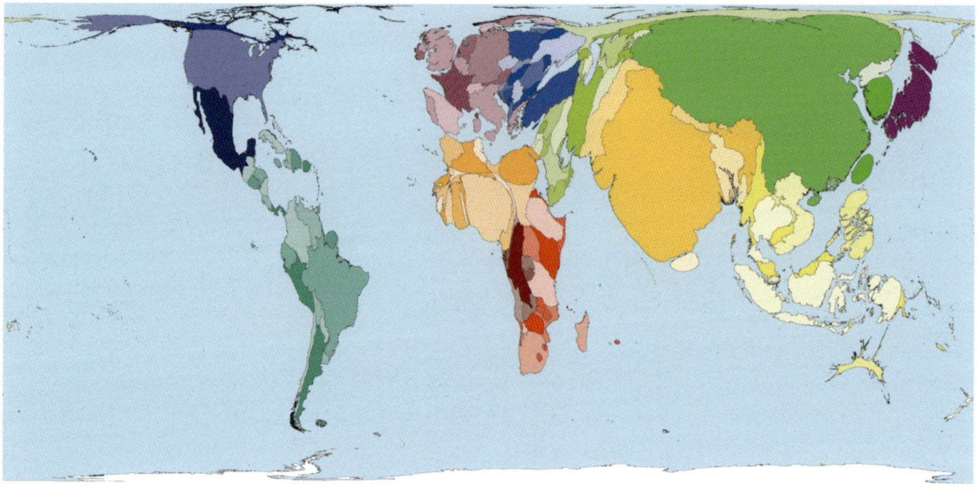

Figure 5.8 Territory size shows the proportion of all 15–24 years old who are literate that lives there (Worldmapper Map 195)

> ***Text Box 5.7: Youth Literacy***
>
> A minimal definition of literacy is being able to read, write and understand a short and simple statement about one's everyday life. Of all 15–24 year olds living in the world, 88% are literate. Over half of this age group live in Asia.
>
> Most of the young people living in most territories can read and write. Only five territories have lower than 50% youth literacy rates. Four of these five are in Northern Africa. Japan recorded the highest rate of youth literacy for a single territory.
>
> The highest number of literate youth live in Eastern Asia, where the youth literacy rate is 98.9%. Of the 12 regions, eight have youth literacy rates of over 95%.
>
> The freedom promised by literacy is both freedom from – from ignorance, oppression, poverty – and freedom to – to do new things, to make choices, to learn.
>
> (Koïchiro Matsuura, 2001)

worldwide (see Worldmapper Map 199 to know where). However, enrolment does not guarantee attendance, or completion. Despite this, 88 per cent, over five out of six of children and young adults aged 15–24, have at least minimal literacy. This is the first generation ever to see a majority able to read and write. Their children may be the first generation ever to see a majority that has access to information over the Internet – but only if it proliferates as mobile phones have done (see Worldmapper Map 334). What will these billions whose grandparents could not read make of their world?

5.2.7 Female youth unemployment

Only the most literate of youth are unemployed. Unemployment is mainly a luxury of the youth that live in that same rich triad that was referred to earlier. Unemployment rates are near zero where there are near zero unemployment benefits, where adults (and children) scour rubbish tips and in our largest cities where thousands beg to live. Unemployment is a mark of economic success worldwide – being able to pay folk not to have to beg. Data on unemployment are collected by the Organization for Economic Cooperation and Development for their member countries, which collect data because they provide benefits. Nevertheless, it would be far better if no one was unemployed. Workers often strike to try to prevent redundancies (Worldmapper Map 358). A universal citizens' wage would eliminate unemployment. Here we show the world drawn with territories in proportion to the 5 million youngest women and girls who are out of work and claiming benefits.

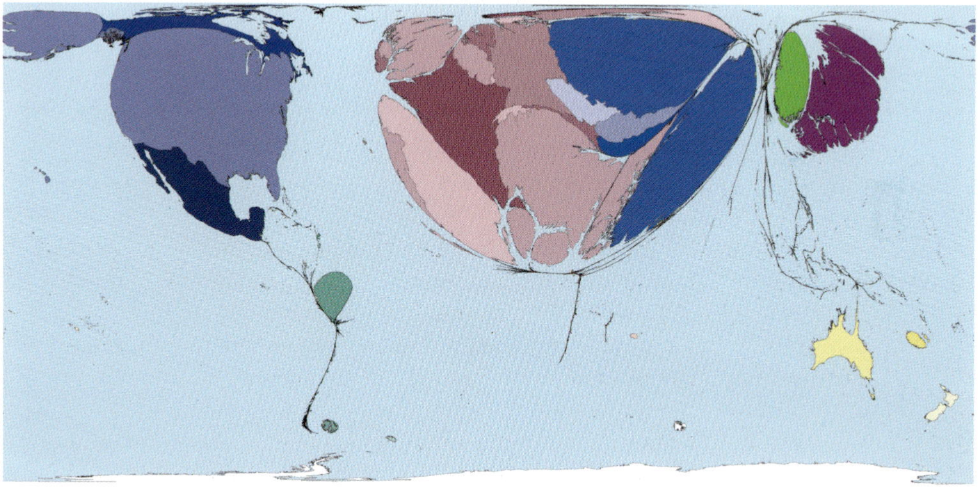

Figure 5.9 Territory size shows the proportion of all young (aged 15 or 16–24) unemployed women in territories that form the Organization for Economic Cooperation and Development that lives there (Worldmapper Map 145)

Text Box 5.8: Female Youth Unemployed

This map only shows the unemployed young women that live in the territories of the Organization for Economic Cooperation and Development (OECD). Most of these women, who are aged between 15 or 16 and 24, live in Western Europe. Within this region the highest rates of young unemployed women were in Greece, Italy, Spain and France, in 2002. Unemployment is high both where there are few jobs and where people are supported and so are not compelled to take any work just to survive.

Only a small proportion of territories are shown here because the majority of territories are not part of the OECD.

> In a majority of... Europe... there is a higher youth unemployment rate for women than for men, but in many countries the differences are not large'
> (Economic Commission for Europe, 2003)

5.2.8 Teenage mothers

Very few of the 26 million teenage mothers in the world live in those territories where many of their peers are unemployed. Teenage motherhood is normal in much of the poor majority of the world. That is partly why it is not seen as normal in most of the triad. These teenagers are mothers to over a fifth of all the children born into the world. They are disproportionately likely to see their children die before them – not because of their age when giving birth – but because of the circumstances in which they are living. Most of the fifth of young people in the world who are the children of teenagers are born into poverty. Where education rates are high, and especially where female literacy is high, and particularly where there is female unemployment – few children have children themselves.

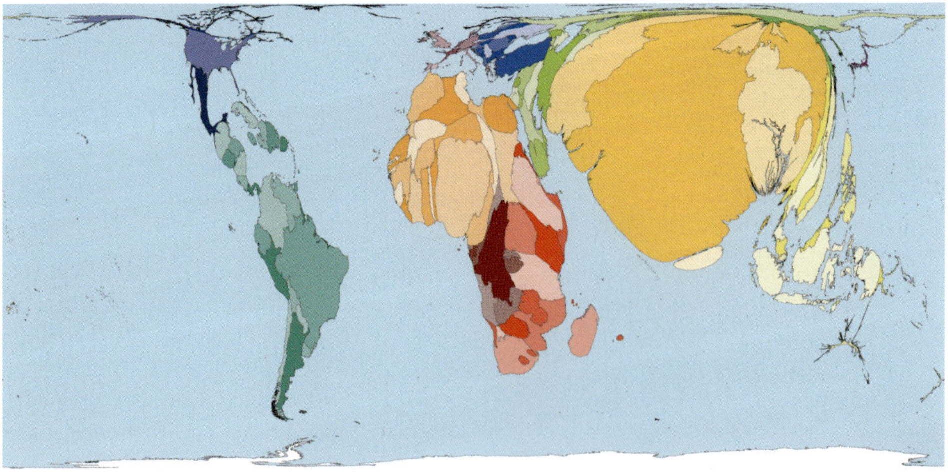

Figure 5.10 Territory size shows the proportion of all teenage (15–19 year old) mothers that lives there (Worldmapper Map 136)

Text Box 5.9: Teenage Mothers

A teenage mother, as mapped here, is a girl/woman aged between 15 and 19 years old who has at least one child. A map of teenage fathers might look similar to this.

A third of all the teenage mothers in the world live in India. There are three times more teenage mothers living in Southern Asia than live in any other region. The fewest recorded teenage mothers live in Japan.

Being a young parent is normal in some parts of the world, whilst in other parts teenage parenthood is a rarity. Where it is a rarity mothers tend to be considerably older – but internationally comparable data on older mothers does not yet exist.

> Attitudes towards young mothers … shift in relation to prevailing moral values, and also to some extent reflect economic conditions
>
> (Debbie A. Lawlor, 2002)

5.2.9 Condom use by men

It was the contraceptive pill, and its use by women, that brought fertility levels to below replacement level for most of the world. However the pill offers no protection from sexually transmitted infections (chlamydia, gonorrhoea, HIV and various others), which are more prevalent than ever (because, in Britain at least, there is still too narrow-minded a concern about pregnancy alone). The pill is also hard to come by in much of the world. Condoms are a little easier to find, although again in the poorest parts of the world they can be too expensive to use. It is estimated that in the 1990s the cost of one year's supply of condoms (however many that might be!) was less than 1 per cent of the Gross National Product (GNP)

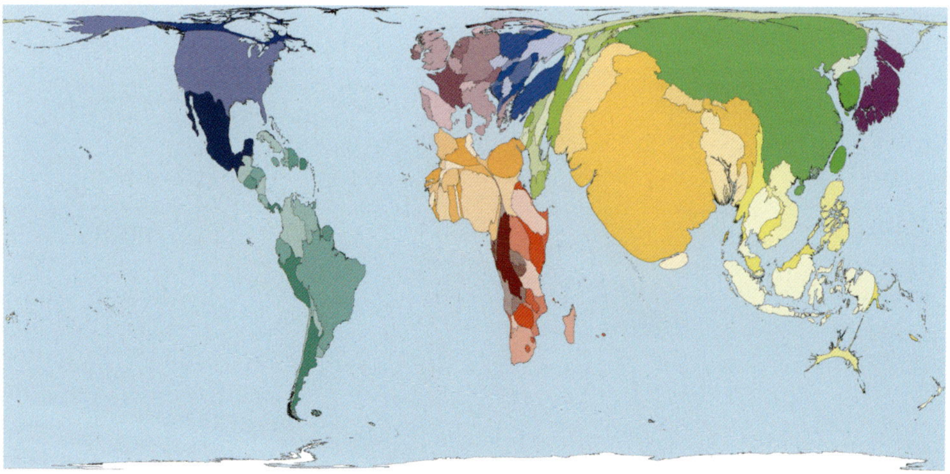

Figure 5.11 Territory size shows the proportion of all boys and men aged 15–24 years old worldwide who used a condom last time they had high-risk sex. 'High-risk sex' is defined as sex with someone who these men are not normally living with or married to (Worldmapper Map 226)

per person in France, Japan, Singapore, the UK and the United States; 23–27 per cent of GNP per person in the Central African Republic, Madagascar and Mali; and a massive 42–45 per cent of GNP per person in Myanmar (then Burma) and Burundi; data from Smith and Bræin (2004). There are also some who do not want the poor of the world to have access to these little pieces of polyurethane or latex.

Text Box 5.10: Condom Use by Men

Of all the older boys and younger men in the world, 47% report having used a condom at the last time of high-risk sex. The region with the highest condom use is North America, at 63.5%. The lowest condom use is in the Middle East at 36.8%. Male and female condom use is not the same for several reasons. Homosexual sex disconnects figures for men and women. Also, many men might sleep with the same women, whilst other women do not have 'high-risk' sex. This is particularly the case because female prostitution is more common than male prostitution. There are also reporting biases.

When condoms are expensive, the costs of sexual freedom can be very high
(Dan Smith, 2003)

5.2.10 Condom use by women

The map of condom use by young women is not quite the same shape as that of use by young men. This despite the fact that usually one condom is shared between a woman and a man! Other times condoms will just be used between men, and sometimes shared just by women (these uses being solely concerned with protecting sexual health). Where education is poor, especially the education of young women, contraception use tends to be poorer. This is due to both lack of access to contraceptives and not being empowered enough to

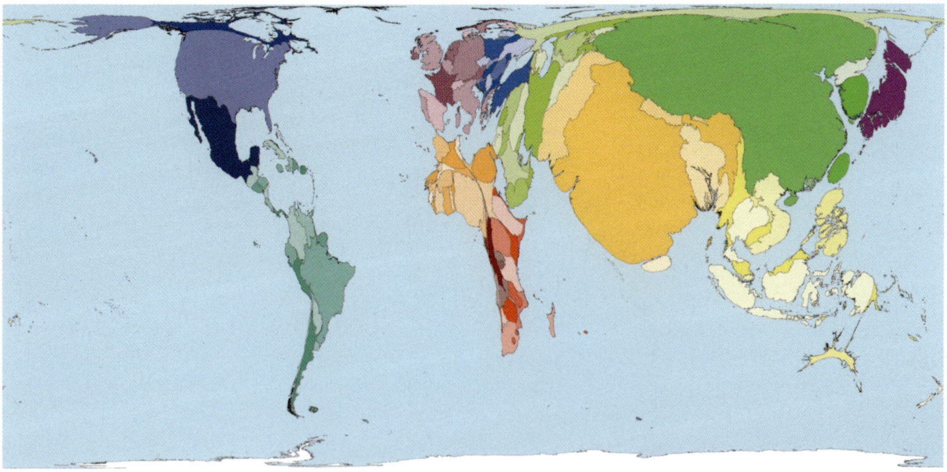

Figure 5.12 Territory size shows the proportion of all girls and women aged 15–24 years old worldwide who used a condom last time they had high-risk sex (Worldmapper Map 225)

ensure their use. The maps here only show the world re-shaped so that area is in proportion to the number of times young women use a condom when having sex with someone they are not living with or are not married to. It is a map of protected sex outside marriage or co-habitation (this includes infidelity). Its inverse – soon to appear on the website – will be of the world shaped by sex that is unprotected by polyurethane or fidelity.

Text Box 5.11: Condom Use by Women

Condoms (including female condoms) are the only contraceptive that also provides the users with protection from sexually transmitted infections (STI). It is sensible for someone to use a condom if they have sex with a person who they are not certain does not have any infections. Often people are themselves unaware of whether they have an STI.

This map shows the distribution of young women who did use a condom last time they had 'high-risk' sex. That is sex with someone they are not married to or do not live with. Not everyone has access to condoms, can demand their use or wants to use them. Only 40% of young women around the world used condoms the last time they had risky sex.

> In many societies, it is taboo even at home to speak about sexual matters, sexual choices and sexual diseases
>
> (Clive Wing, 2005)

5.2.11 Agricultural women

Some 282 million women work as farmers worldwide. The world shaped by the sum of their labours sees the Americas thin whilst Southern and Eastern Asia bulge. It is a world where there are few tractors (Worldmapper Map 121), let alone combine-harvesters,

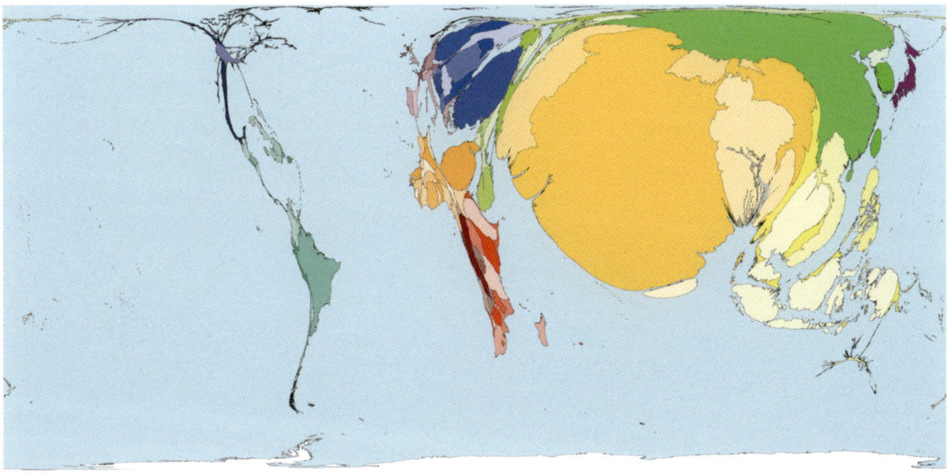

Figure 5.13 Territory size shows the proportion of worldwide female agricultural workers living there (Worldmapper Map 127)

computer-operated irrigation systems or health and safety regulations. It is what the world of the majority was before the industrial revolution, a world that includes much subsistence still, and in which out of every seven farmers and agricultural labours, three are female.

Text Box 5.12: Agricultural Women

Of all the female agricultural workers in the world, 39% live and work in India. The arable agricultural produce of India includes jute, rice, oilseed, cotton, sugar cane, wheat, tea and potatoes. Indian pastoral agriculture includes farming cattle, water buffalo, goats, poultry, sheep and fish. Over half of the female agricultural workers in the world are in Southern Asia.

The United Arab Emirates has the lowest percentage of the population who are female agricultural workers.

This map of female agricultural workers (where Asian territories are large) is almost an inversion of the per person distribution of working tractors (where the most are found in richer regions).

> whoever could make two ears of corn or two blades of grass grow on a patch of land where only one grew before, does a greater service to mankind . . . than the whole gang of politicians put together
>
> (King of Brobingdag, 1754)

5.2.12 Agricultural men

There are subtle differences to the world map made up of men working in agriculture as compared with that of women farmers. An additional 115 million farmers are involved, the Americas are not so thin, but in general the distribution of men working in agriculture is

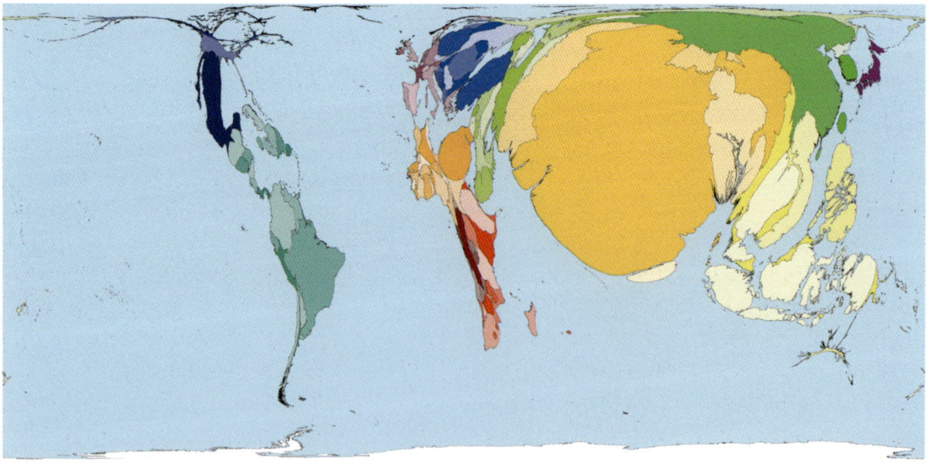

Figure 5.14 Territory size shows the proportion of male agricultural workers worldwide living there (Worldmapper Map 128)

similar to that shown above for women. In rich countries where the proportion of farmers is less than a single percentage of all those in work, it is easy to forget that worldwide one in five workers work on the land. Some of the lowest incomes are recorded for these groups – often there is no formal or regular income at all. In contrast, a few of the agricultural men recorded in this map living in parts of the triad are extremely rich farmers – who have not touched actual soil for many years.

Text Box 5.13: Agricultural Men

This map of where the men who are farmers live and work shows a broadly similar distribution to the equivalent map for women. However a larger proportion of the male agricultural workers of the world are found in Brazil, Colombia and Mexico, as compared with female. In total there are 1.4 male agricultural workers to every female agricultural worker in the world.

The territory in which female agricultural workers made up the smallest percentage of the population, also had the second highest proportion of male agricultural workers. That territory is the United Arab Emirates. The populations of Singapore and Argentina have very low proportions of agricultural workers – both male and female.

> If I could live life again I would study. I would also continue to work in the fields. If you only study, then you forget that someone needs to grow things to eat and you can't eat money.
>
> (Edgar, 2006)

5.2.13 Industrial men

In 2002 there were 519 million men working in industry around the world. When these male industry workers are combined with their female counterparts, they make up almost a quarter of all workers in the world. The work of these people is what supplies the other half of the world's workforce, those who do not work, and of course the workers themselves, with shoes, cars, clothes, ready meals and tea bags (amongst most other things). Yet as the Worldmapper trade maps show, it is not only luxury items such as toys and valuables that find their way mainly to the lands of the rich. More mundane items such as cars, clothing, electronics and computers also arrive disproportionately where richer people live. This is despite the men and women who produce all of these items together having a similar distribution to that of total population (see Worldmapper Map 2 in Figure 5.1) – note that particular industries have particular distributions: the highest net exports of vehicles are from Japan and South Korea; high net machine exports are from Germany, Italy and Japan; there are high net computer exports from Japan, Singapore, the Philippines, Taiwan and South Korea. Of course, high net earnings from exports are not necessarily related to the work that goes into production.

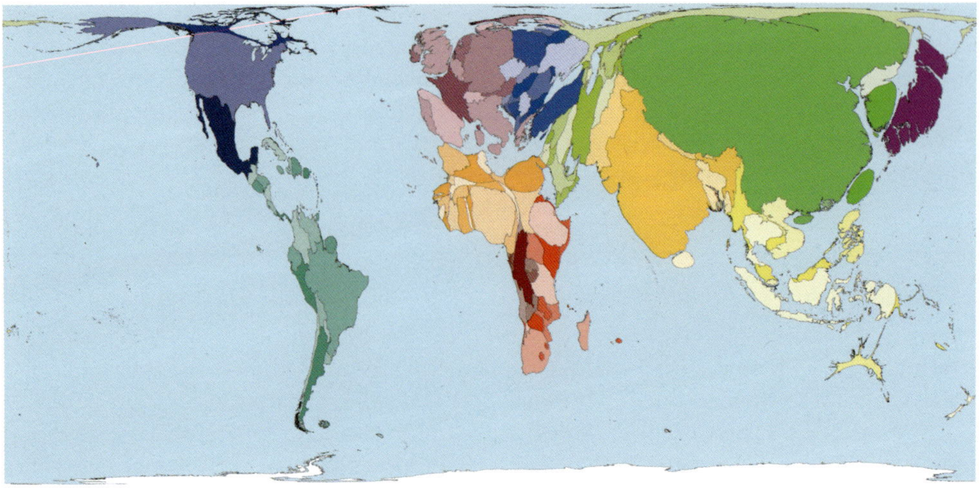

Figure 5.15 Territory size shows the proportion of male industrial workers worldwide living there (Worldmapper Map 130)

Text Box 5.14: Industrial Men

In total 519 million men work in industry. Industry here means manufacturing, the production of tangible goods. A third of the men that do industrial jobs live in Eastern Asia, the majority being in China. As a proportion of the population, the most industrial men live in the Eastern European territories of the Czech Republic and Slovakia. The territory with the smallest proportion of the population being men that work in industry is Bangladesh. Most of the territories with very low proportions of the population that are men working in industry are located either in parts of Southern Asia or in the Middle East.

> They earn about $2,400 a year – nearly five times the average per capita income... Rajesh Kumar Raghavji Santoki, 28, was earning more than $500 a month, and owned a house, a motorcycle and van
>
> (Amy Waldman, 2005)

5.2.14 Industrial women

There are fewer women than men who work in industry; in 2002, 188 million women worked in this sector. Women working in industry made up 3 per cent of the entire world population – if we assume that half the population is male, then 6 per cent of all women were working in industry. Six per cent might sound small, until put into the context of the many other possibilities that people have throughout their lives: studying (at school or university), agricultural work, services work, child care, not working due to old age, life before school, unemployment, and so on. By now you may not be surprised to know that maps of these other occupational possibilities are also available at www.worldmapper.org.

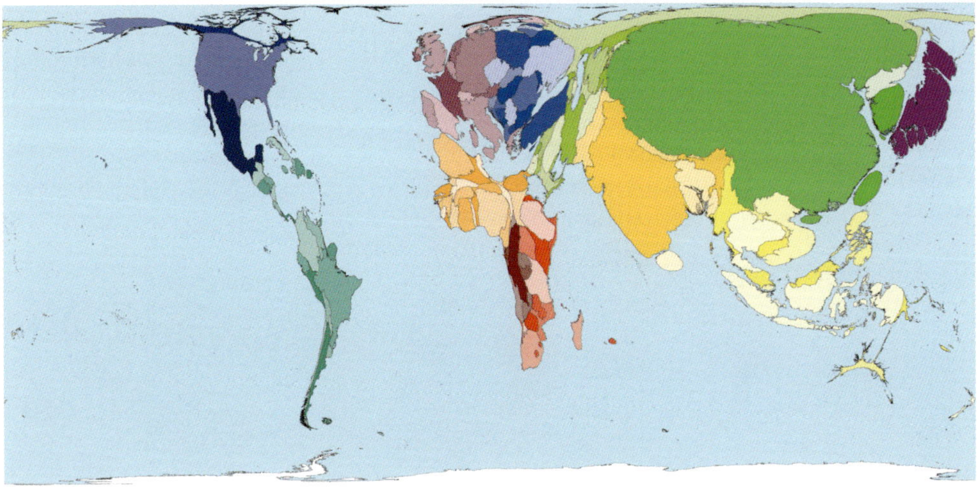

Figure 5.16 Territory size shows the proportion of worldwide female industrial workers living there (Worldmapper Map 129)

> ### *Text Box 5.15: Industrial Women*
>
> Over a third of all the women who work in industry live in China. The main industries of China include the production of iron, steel, machines, armaments, textiles, chemical fertilizers, processed food, toys, automobiles and electronics. Despite China's high total of female workers in industry, it is Eastern European territories where the highest proportions of the population are female industrial workers.
>
> Southern Asia and Asia Pacific also have relatively high numbers of women working in industry. It is the very richest and the very poorest of regions where the fewest of the world's industrial women live.
>
> > In terms of the distribution of women within the industrial sector, women are highly concentrated in garments and textiles industries
> >
> > (United Nations Development Programme China, 2003)

5.2.15 Services men

The largest occupation group is services; this is true whether considering all people working in services or both men and women when counted separately. A total of 879 million men work in services. Men working in services (which is more than women) combined with women in services are 54 per cent of all workers worldwide. No longer are agriculture and industry the main employers – what does this suggest about the stage of our current world civilization? It suggests that we have gone beyond the majority working to produce food and technology for survival (though of course industrial workers spend little of their time producing items that are essential for survival in the territory where they are consumed). Now more than half the global workforce spends their working hours not producing, but servicing. Services, like industry and agriculture, are not necessarily tied to supplying people in or near to where they live. The international scope of services is facilitated by the massive spread of communication technologies (see Figures 5.31 and 5.32), in the same way that transport links (see Worldmapper Maps 33–40) within and between territories allow for trade in raw agricultural materials and manufactured products.

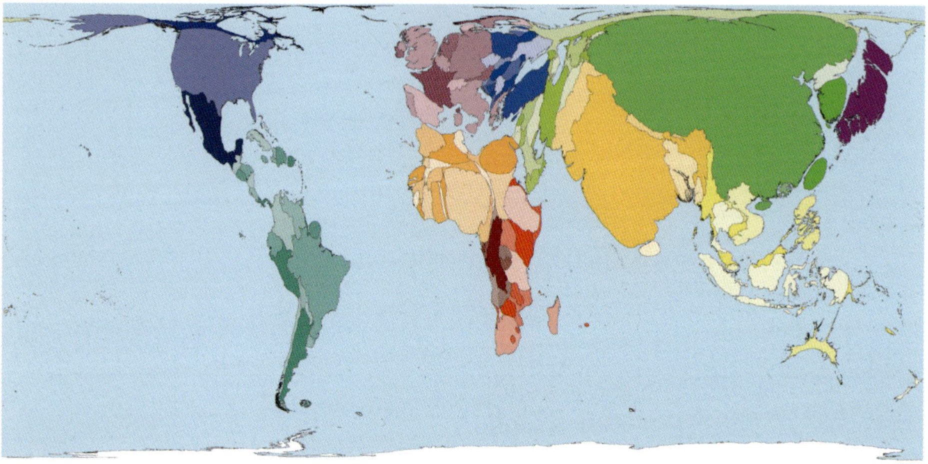

Figure 5.17 Territory size shows the proportion of worldwide male services workers living there (Worldmapper Map 132)

Text Box 5.16: Services Men

Service work does not produce a material object. Services include tasks such as call centre work, hospitality, armed forces and transportation. More men than women work in services.

The most services men work in China; then India; then the United States. Some 14% of the world's population are men that work in the service sector. The lowest percentage of men working in services is 5.5%, in Haiti. Services workers live in every territory in the world, as many services must be performed *in situ*. Some services, such as call centres and data entry, could occur anywhere, so long as there are good channels of communication.

> These guys know accountancy, have computer skills, speak English and they are ready and willing – and that combination is a killer
>
> (Kiran Karnik, 2003)

5.2.16 Services women

More than a third of the world's 753 million female services workers live (and work) in China. Given that a fifth of the world's population lives in China it is not a great surprise that a high proportion of such workers might live there. Most of these services are for the people who live in this country, some for people who live elsewhere. The difference between the relative sizes of China and India is most interesting here – given how similarly sized they are when shaped just by population alone. Just as women living in these two most populous of territories have very different opportunities depending on the country they were born into – so the economic futures of these two countries are likely to differ markedly by what work women are allowed and expected to carry out in each.

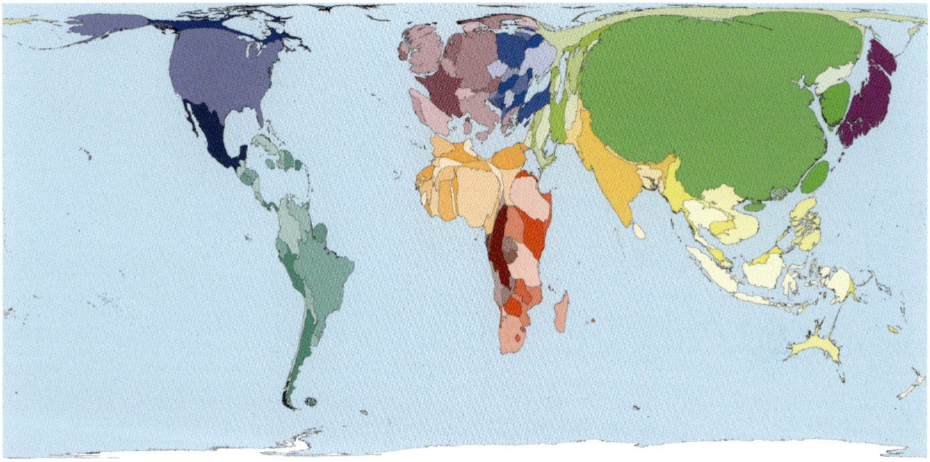

Figure 5.18 Territory size shows the proportion of worldwide female services workers living there (Worldmapper Map 131)

Text Box 5.17: Services Women

The populations with the highest proportions of women that work in the service industries live in Western European, North American and Eastern Asian territories. The populations with the lowest proportions of women working in services live in territories that are located in Southern Asia and the Middle East.

In Sweden 23% of the population is made up of women working in services. Therefore a majority of economically active women in Sweden work in the services sector.

Worldwide most women working in services live in China, making a female services workforce of 260 million that is 35% of all female services workers.

> Girls are asking, 'Do we get overtime? What are the benefits?' Guangdong needs workers. Zhejiang and Shanghai need workers. They have more choices. So it's difficult to find workers.
>
> (Kathy Deng, 2006)

5.2.17 Living on $1 a day

Living on a dollar a day is a measure intended to define a very severe level of poverty. This map (Figure 5.19) has been entitled 'The Wretched Dollar' on our website because, as a measure of poverty, a dollar a day is a wretched definition. To live on US$1 a day or less is below the sustainability rate; it is not a success if people are living on only just over a dollar a day. A more reasonable definition of the absolute poverty line is that set by Seebohm Rowntree

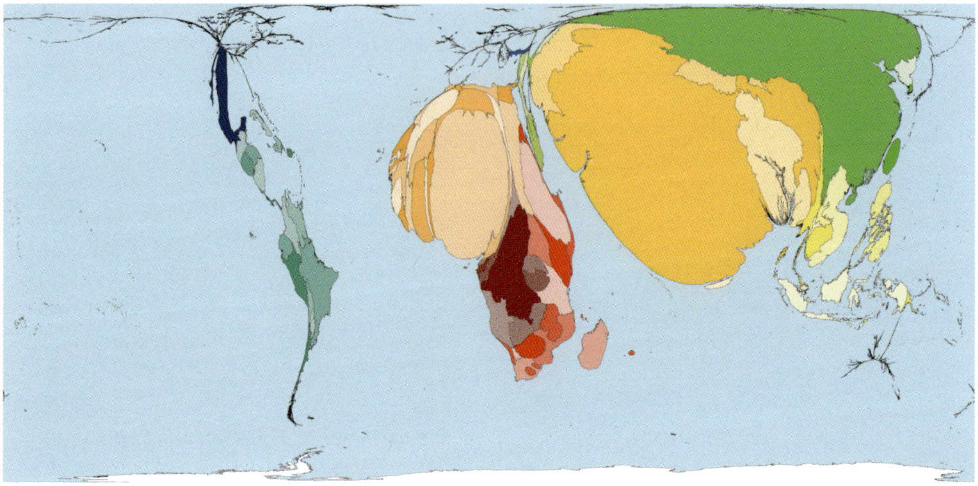

Figure 5.19 Territory size shows the proportion of all people living on less than or equal to US$1 in purchasing power parity a day (Worldmapper Map 179)

Text Box 5.18: The Wretched Dollar (up to $1 a day)

The first Millennium Development Goal is to halve, between 1990 and 2015, the proportion of people who live on the equivalent of US$1 a day, or less. In 2002, an estimated 17% of the world population lived on this amount. They lived on less than or equal to what, to be precise, US$1.08 would have bought in the United States in 1993. In over 20 territories more than a third of the population lives on less than US$1 a day. All but two of these territories are in Africa.

The largest population living on US$1 a day is in Southern Asia, most of whom live in India.

> The mass of the people struggle against the same poverty, flounder about making the same gestures... It is an underdeveloped world, a world inhuman in its poverty
>
> (Franz Fanon, 1961)

almost 100 years ago, which in current prices means living on around roughly US$2 a day. Being a wretched definition of poverty due to setting such a low line, those people living on such a small amount may also be considered to have lives that are wretched. The term echoes the title of Franz Fanon's book *The Wretched of the Earth*, from which the accompanying quote is sourced. Another reason for using the term 'wretched' is because of the inaccuracy inherent in this definition of poverty. The number used when the World Bank counts is not actually a dollar anyway, it is US$1.08 according to 1993 purchasing power parity.

Nevertheless, a map of those people who really do not have enough income to live on represents approximately 17 per cent of the people of the world, that was just over a billion people in 2002. Many of those living on more than US$1 a day are also living in abject poverty.

5.2.18 Living on up to $2 a day

Compared with the map of those who survive on just one wretched dollar a day, this map of those living on $2 a day has a very similar shape. Territories that have swollen most to accommodate on the map many more people who live in absolute poverty at this basic subsistence level include India, China, Nigeria, Ethiopia, Egypt and Bangladesh. Turkey and Indonesia have a larger proportion of all people living on $2 a day than on $1 a day, and so appear to grow most in size when these two maps are contrasted. Reading a map of people earning under a dollar a day misses the 1.625 billion people who earn between $1 and 2 a day. In total there were 2.698 billion people living on less than or equal to $2 a day when these estimates were made at the start of the current millennium.

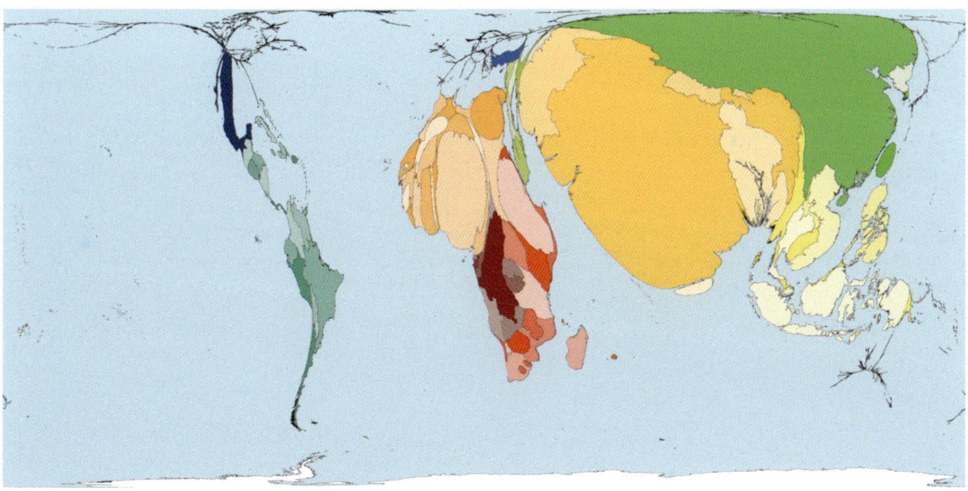

Figure 5.20 Territory size shows the proportion of all people living on less than or equal to US$2 in purchasing power parity a day (Worldmapper Map 180)

Text Box 5.19: Absolute Poverty (up to $2 a day)

Absolute poverty is defined as living on the equivalent of US$2 a day or less. In 2002, 43% of the world population lived on this little. This money has to cover the basics of food, shelter and water. Medicines, new clothing, and school books would not be on the priority list.

When almost an entire population lives on this little, it is unsurprising if undernourishment is high, education levels are low and life expectancy short. In both Nigeria and Mali, nine out of every 10 people survive on less than US$2 a day.

South America has a relatively small poor population, yet 39 million people have less than US$2 a day in Brazil.

> Trickle-down theory – the less than elegant metaphor that if one feeds the horse enough oats, some will pass through to the road for the sparrows
>
> (John Kenneth Galbraith, undated)

5.2.19 Living on up to $10 a day

As this income series increases the dollar value cut-off, new territories become visible on the map. Now at least seven Eastern European territories are visible; more Middle Eastern territories appear and they are bigger than before; South America has started to expand; North America (one-third of the so-called 'rich triad') also reports many people living on very low incomes. This map shows a total of 3.499 billion people – which was over half the

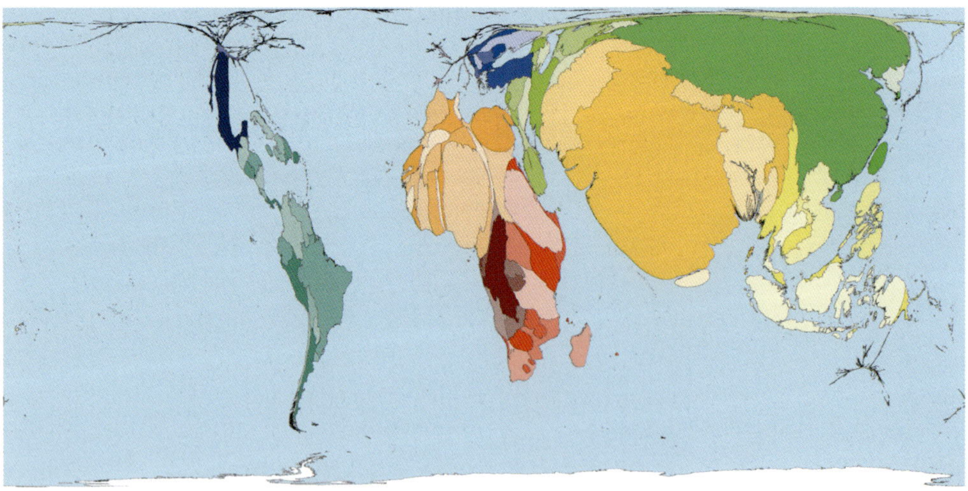

Figure 5.21 Territory size shows the proportion of all people living on US$10 purchasing power parity or less a day worldwide that lives there (Worldmapper Map 153)

world population at the time the data were collected (and still is the majority as we write, given current trends). The map shows everyone from the previous map, plus the extra 801 million who earn between $2 and 10 a day. The definition here is now shifting from absolute poverty to relative poverty, as we slide up the scale of how much people have to live on. Here are the homes of the poor of the earth, including those who are wretched and those living in absolute poverty.

Text Box 5.20: Living on up to $10 a Day

In Indonesia US$10 buys more than it does in the United States, so comparing earning in US$ alone does not allow for the cost of living changing between places. The map shows purchasing power parity (PPP) – someone earning PPP US$10 in Indonesia can buy the equivalent of what PPP US$10 would buy in the United States. As such, more practical assessments of individuals' earnings can be made.

In seven out of the 12 regions more than half of the population live in households where the people live on below PPP US$10 a day. In Central Africa 95% of households have workers earning this little; in Western Europe and Japan less than 1% of the population does.

> There is no work here, and when you do find a job, you earn pathetically low wages. I'm a factory watchman, and I earn the equivalent of eight dollars for a 12-hour day.
>
> (Pirana, 2005)

5.2.20 Living on $10–20 a day

As the range in earnings we consider doubles from $0–10 to $10–20 a day, the shape of the world shifts again to reflect the distribution of these 1.115 billion middle-income people. Already Central African territories have begun to shrink themselves off the income map, due to the tiny proportion of people earning this much/little. The map series suggests that whether earning $10–20 a day is understood to be good or bad depends on where you live. In Western Europe, where territories are now just appearing on the map, the people shown are those in the lowest earning brackets there (most of those whose income is so low will not actually be working). The opposite is true of those Central African territories, where some of the higher earners from that territory are shown. If we are honest with ourselves, this amount of money is still not enough anywhere to be included in society locally in Europe or global society in the case of African territories.

All these income maps show dollars as measured having adjusted for purchasing power parity, so maps show earnings according to what can be bought in the place in which they are earned. If what can be bought in Western Europe for $15 (measured in purchasing power parity) is not enough, then of course it is not enough in Central African territories either. If it is enough to live on, then all of the people about to be represented in the following maps could afford to give up some of their spare money so the people on preceding maps can live more meaningful lives.

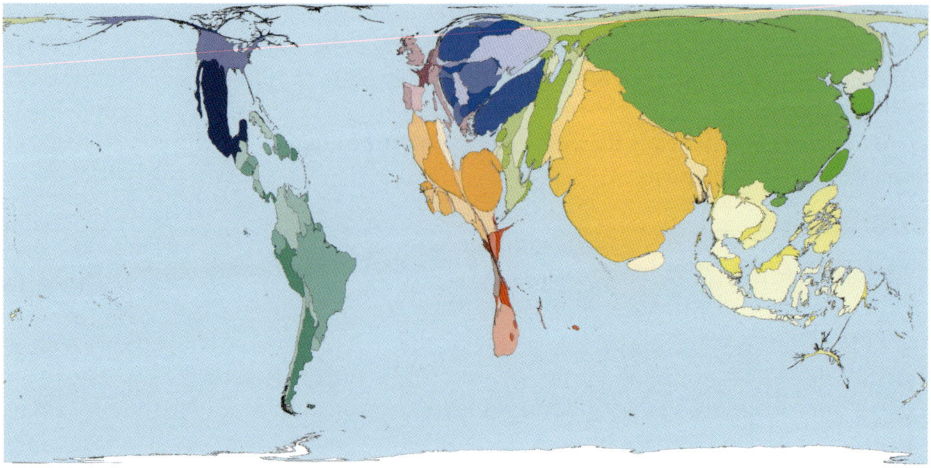

Figure 5.22 Territory size shows the proportion of all people living on PPP US$10–20 a day worldwide that lives there (Worldmapper Map 154)

Text Box 5.21: Living on $10–20 a Day

The territory where the most people live on US$10–20 a day, when measured in purchasing power parity (PPP), is China. In China 26% of the population live on this much – that is 332 million people.

The regions of Eastern Europe, the Middle East and Eastern Asia have the highest percentages of people living on PPP US$10–20 a day. That is 36% in Eastern Europe, and 25% in both the Middle East and Eastern Asia. In both poor and rich regions low percentages of the population live on PPP US$10–20, for opposite reasons. In poor territories the wages needed to achieve this are unaffordable, in rich territories unacceptable.

> If you want to get rid of poverty, we need to empower the poor. Not to treat them like beggars
>
> (Hugo Chavez, 2005)

5.2.21 Living on $20–50 a day

What are the average earnings worldwide? One of every seven people lives on between $20 and 50 a day. Five in every seven live on less than the people shown in this map. One person in every seven lives on more than these people (at a guess you are one of this last category, or soon will be). Of course this story does not concern just seven people, it is about the lives of more than 6 billion people; 884 million of them are shown below. The new addition to the map in comparison to those shown above is Japan, because almost no one in Japan lives on less than $20 a day. South Korea has also expanded, whilst India has shrunk considerably. African territories are becoming increasingly difficult to distinguish – from North to South we can identify Morocco, Algeria, Tunisia, Libyan Arab Jamahiriya, Somalia, South Africa,

Swaziland and Lesotho from Figure 5.23. Can you identify any others? Given that these maps show 54 territories in Africa, this list is nasty, brutish and short (after Hobbes, 1651).[1]

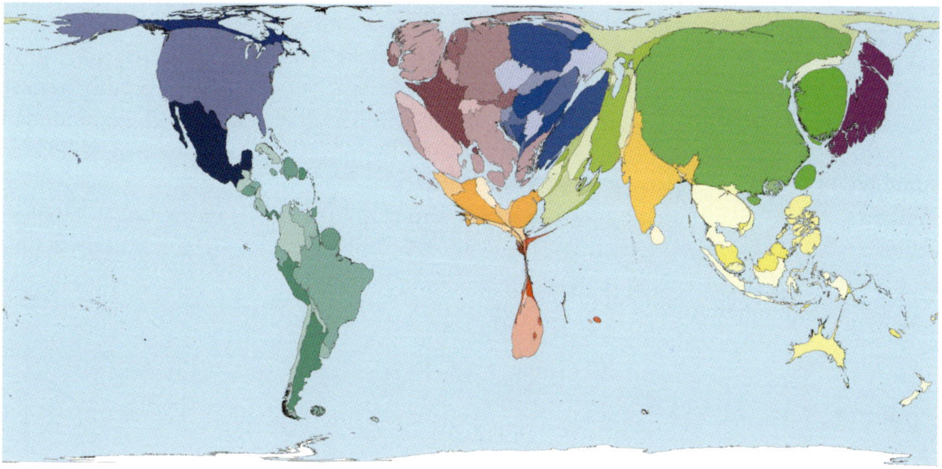

Figure 5.23 Territory size shows the proportion of all people living on PPP US$20–50 a day worldwide that lives there (Worldmapper Map 155)

> ### *Text Box 5.22: Living on $20–50 a Day*
>
> The largest population living in households where people rely on US$20–50 a day, in purchasing power parity (PPP), is China (192 million people). The second largest population in this range, less than half the Chinese total, lives in the United States (80 million). Russia has the third biggest population (41 million) living on this daily amount, which is half the number of its equivalent in the United States.
>
> The highest regional percentages of the population living on PPP US$20–50 a day are found in Europe. However an even higher percentage of Eastern Europeans rely on PPP US$10–20 a day, whilst a higher percentage of Western Europeans earn PPP US$50–100 a day.
>
> > English language speaking guides can earn about US$25 per day; French, German, Chinese and Thai language speakers, US$30 per day; Japanese speakers, US$35 and Italian language speakers upwards of US$40 per day
> >
> > (Myanmar Times, 2006)

[1] Hobbes (1651) argued in *Leviathan* that the lives of masterless men living without laws and subjection to coercive power would be 'solitary, poor, nasty, brutish, and short'. See the *Stanford Encyclopedia of Philosophy*, accessed on 12 January 2007: http://plato.stanford.edu/entries/hobbes-moral/: 'Lives may now be nasty, brutish and short not because we lack "civilisation", but our civilisation is lacking. Laws and coercive power often maintain inequalities more effectively than they challenge them'.

5.2.22 Living on $50–100 a day

The number of people being shown on each of these maps is falling as the money to live on increases, despite the range increasing first from $1 ($0–1 a day) to $10 ($0–10 and $10–20), then to $30 ($20–50), now $50 ($50–100), next to $100 ($100–200), then to near infinity ($100–). From the maps showing a $10 range the numbers have decreased steadily with the increased range (see Figure 5.24). The information in this graph is needed to understand how these maps fit together – not only do most poor people live in India, China, Asia Pacific and African territories; also most people are poor.

Figure 5.24 can also be read with the *x*-axis showing how much money people have to live on. At the range of $50–100 a day, there are 488 million people, which is equivalent to

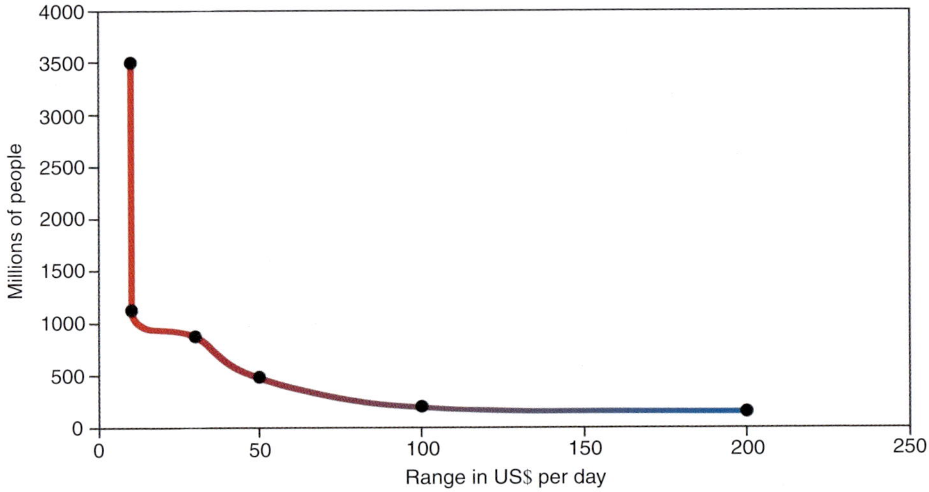

Figure 5.24 Income per day (*x*-axis in dollars) and people (*y*-axis in millions)

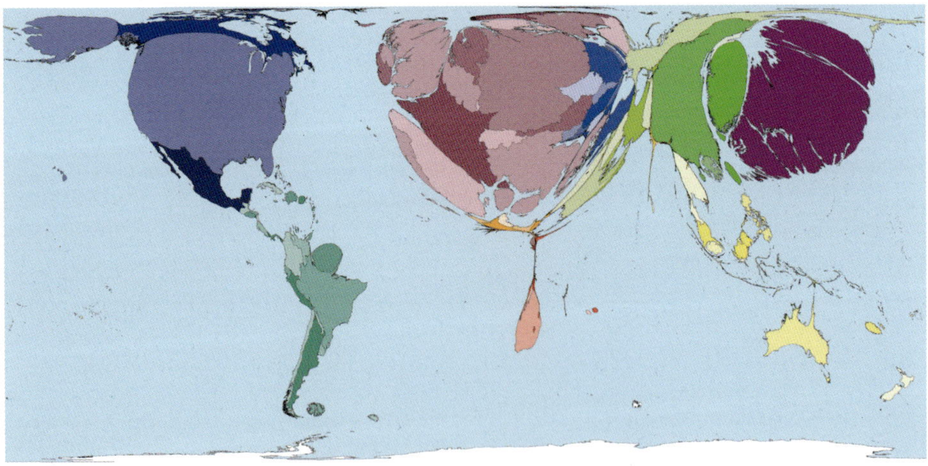

Figure 5.25 Territory size shows the proportion of all people that rely on US$50–100 purchasing power parity a day worldwide that lives there (Worldmapper Map 156)

one in every 13 people. Most of these people live in the rich triad, but the map in Figure 5.25 also shows that there are also a notable proportion of people with these earnings living in South America, Australia, China, Russia, South Africa, Middle Eastern and Eastern European territories.

Text Box 5.23: Living on $50–100 a Day

The majority (61%) of the population of Japan live in households which rely on between US$50 and US$100 in purchasing power parity (PPP). North America, Western Europe and Japan are particularly large on this map because large numbers of people live on PPP US$50–100 a day.

Southern Asia and Central Africa have almost completely disappeared. Indonesia, Vietnam, Cambodia, Laos People's Democratic Republic, Myanmar, Timor-Leste and much of Southeastern Africa are not visible. This is because very few people there live on PPP US$50–100 a day.

> It's as though the people of India have been rounded up and loaded onto two convoys of trucks . . . The tiny convoy is on its way to a glittering destination . . . The other convoy just melts into the darkness and disappears
>
> (Arundhati Roy, 2002)

5.2.23 Living on $100–200 a day

Only one person in every 32 has between $100 and 200 to live on, every day. That makes a total of 203 million people shown on this map (Figure 5.26). This map is very similar to that of people living on $50–100, in that most of the same territories have enough area that they are visible. Yet there have been more subtle alterations in the distribution of this very rich group of people. The United States has expanded, whilst China, South Korea, Tunisia, Algeria, the Czech Republic, Poland, Turkey and Hungary have shrunk. The socio-economic comparisons between these maps are interesting, but the relatively subtle nature of the difference between these maps also demonstrates how the algorithm behind the maps stretches territories so their relative positions and shape behave elastically. As China shrinks, it pulls Russia inwards with it. The expansion of the United States pushes Alaska past the artificial divide this map shows as slicing through the Pacific Ocean, off the far left of the map and onto the far right. Simultaneously, South America seems to have been compressed so its shape becomes slightly wider, and Western Europe moves Eastwards to make space for the United States and the sea that it displaces. As the total map area remains constant, territories between Europe and Asia have lost area, which was necessary for the United States to be able to compress Eurasia from both sides.

Figure 5.26 Territory size shows the proportion of all people living on PPP US$100–200 a day worldwide that lives there (Worldmapper Map 157)

Text Box 5.24: Living on $100–200 a Day

In all regions except for North America, Western Europe and Japan, less than 2.1% of the population live in households which live on US%100–200 purchasing power parity (PPP) a day each. Within North America, Western Europe and Japan, 16–19% of the population live on this amount.

In 75 territories less than one in 1000 people live on this much, despite using purchasing power parity where a higher value is given to the currency in territories where it is cheaper to live. As the measure of purchasing power parity takes into account the cost of living in each territory, earning PPP US%150 in Ethiopia means that the same goods and services could be bought as with US$150 in Germany.

> Every man is rich or poor according to the degree in which he can afford to enjoy the necessaries, conveniences, and amusements of human life
>
> (Adam Smith, 1776)

5.2.24 Living on over $200 a day

This is a map of extremes. Figure 5.27 shows a tiny group of people making a huge amount of money. Only one in 121 people in the world earns this much, that is just 53 million people in total. These people live primarily in the United States. Other than this dominance by one territory, the other change from the previous map to this one is that almost all territories in Eurasia located East of Austria have marked decreases in size. Japan is included in this trend – generally an affluent and economically successful territory, Japan has neither many super rich, nor many extremely poor people. When incomes are pooled over households, as is done here, Japan is the most internally equitable third of the triad areas. Having more than $200 a day at your disposal would create an awkward situation – what would you spend your money on? Would there really be a reason to have quite so much? Does having this amount of money confirm your success in life? It has been argued by Tim Kasser (2002) that materialistic values, which are likely if you appear as a speck of colour on this map, actually decrease your happiness and well-being. He states that, after being able to satisfy your basic needs, increases in income have little effect on happiness. Thus, whether we take a moral or self-interested stance, it is not necessary to have this much! To take this series from the abstract to the real, consider which map you are represented on, which maps your friends or family might be shown in, where might your acquaintances appear?

Figure 5.27 Territory size shows the proportion of all people living on over PPP US$200 a day worldwide that lives there (Worldmapper Map 158)

Text Box 5.25: Living on over $200 a Day

In 2002, 53 million people in the world lived in households in receipt of US$200 purchasing power parity (PPP) per day. Of these high earners, 58% lived in the United States.

Western Europe and South America are also home to quite large populations of high earners. Within Western Europe the most very high earners live in the UK, Italy and France. The highest earners of South America live primarily in Brazil and Argentina. Few very high earners live in Southern Asia, Northern Africa, Eastern Europe and Central Africa.

> I still don't understand how a man can justify awarding himself a 40% pay rise when he is already on a huge salary, the like of which those of us in the public sector will never see, especially with a 3% annual pay rise
>
> (Geraldine, 2001)

5.2.25 Tourist expenditure

Having a lot of money facilitates travel, which could act as a vehicle for the redistribution of money. However, most money spent on international tourism is spent in Western Europe, Japan, and North America. There are two reasons for this – firstly we spend more where prices are higher, which is the case in richer territories. Secondly, whilst 'third world tourism' is popular amongst certain age and interest groups, most international tourist trips are made

Figure 5.28 Territory size shows the proportion of world international tourist spending by residents of each territory (Worldmapper Map 24)

> ***Text Box 5.26: Tourist Expenditure***
>
> Territory size shows the spending of residents (in US$) when they make tourist visits abroad. The four biggest tourist spenders are the United States, Germany, the UK and Japan.
>
> The average tourist spending in 2003 was US$92 per person in the world. However, this is unevenly distributed. At a territorial level per capita spending ranges from US$6005 to 4 US cents. The highest per capita tourist spenders are the Luxembourgeois, Kuwaitis and Austrians. Afghanistanis, the Burmese/Myanmars and Ethiopians spend the least per capita as tourists.
>
> > Carrying cash from one developed country to another is simply not necessary anymore. Using Visa payment cards is easier and safer.
> >
> > (Kamran Siddiqi, 2005)

to European territories (most border crossing tourists are Europeans; see Worldmapper Maps 19 and 20). This raises questions of levels of disposable income, interest in the world beyond the borders within which you live, but also the importance of borders to this map. This map shows movement over territorial borders. The high number of borders to land area within Europe, contrasted with India, China and the United States, affects the shape of this map. Travelling a distance over which you could easily remain within the same territory, were you in a large territory, in Europe could mean travelling to one, or several, territories.

5.2.26 Aircraft passengers

When we travel, how do we travel? This map shows the number of people flying with airlines registered in that place. In 2000 there were almost 2 billion aircraft passengers. These were not 2 billion different people. A small group of people who fly frequently make up the bulk of air passengers and most of them fly with carriers based in the United States and Europe, and often within the United States and Europe (and to a lesser extent Japan – where lower income inequality makes flying more expensive for the rich). Thus, outside of Japan, most of these people are flying where the train lines are best provided. Where there are the most roads already built. Where electronic means of communication are best established and where the infrastructure is most securely established to communicate in ways that do not require travel. However, they have the resources, the money, they can, and so they do travel most where the necessity may be least. Note also the (compared with their populations) very large size of New Zealand and relatively large size of Australia.

Figure 5.29 Territory size shows the proportion of worldwide aircraft passengers flying on aircraft registered there (Worldmapper Map 29)

Text Box 5.27: Aircraft Passengers

Of all the air passengers in the world, 40% fly on aeroplanes registered in the United States. These flights are both domestic and international. In the year 2000 there were 1.6 billion aircraft passengers. In these statistics, every time a person takes a flight, they are counted as an aircraft passenger. Some people are passengers many times in a year so far fewer than 1.6 billion individual people fly in a year.

Most of the world's cities are now within 36 hours of each other

(Peter Haggett, 2001)

5.2.27 Valuable net imports

The movement of valuables around the world in 2002 was responsible for trade worth a total value of US$76 billion. What do we like so much that we collectively spend just under the annual Gross Domestic Product of the Philippines on it over the same 12 month period every 12 months? The answer is pearls, precious and semi-precious stones (0.9 per cent of international trade; in the following parentheses we give per cent of international trade made up by this item), silver, platinum and similar metals (0.2 per cent), developed cinema film (less than 0.05 per cent), watches and clocks (0.3 per cent), printed matter (0.5 per cent), works of art (0.2 per cent), gold and silver ware, and jewellery (0.4 per cent), musical instruments and parts (0.5 per cent), mail (less than 0.05 per cent), special transactions (1.8 per cent), old coins that are non-gold (less than 0.05 per cent) and other, non-monetary form of gold (0.4 per cent). Territories where people spend more than they earn from trade in valuables are shown in Figure 5.30 (the map of net valuables exports is available online, as are paired import/export maps of almost all world trade).

Figure 5.30 Territory size shows the proportion of worldwide net imports of valuables (in US$) that are received there. Net imports are imports minus exports. When exports are larger than imports the territory is not shown (Worldmapper Map 70)

Text Box 5.28: Net Valuable Imports

The United States alone imports 59% of all valuables, net. The second biggest net importer, Italy, imports about a tenth of the US$ value of imports to the United States. The third biggest importer of valuables, the United Arab Emirates, imports only two-thirds of the valuables (US$ net) that Italy imports.

The United Arab Emirates has the highest per person spending on net valuable imports (US$). This is US$1270 per person. One reason that this territory can afford high net imports of valuables is because of its 'liquid gold' – petroleum. Despite exporting diamonds and platinum, South Africa is a high net importer of valuables.

> Will the people in the cheaper seats clap your hands? All the rest of you, if you'll just rattle your jewellery
>
> (John Lennon, 1963)

5.2.28 Royalties and licence fees net exports

One part of trade that we may not consider when we try to comprehensively list items traded, is that in royalties and licence fees. The reason for this omission is copyright's immaterial nature – it gives someone the right to use an idea or creation and in return the seller signs a contract then reaps the financial rewards. Thus years after the creative moment, someone can still be earning from it. Those populations where there is a net income from this trade are the United States, the UK, France, Sweden, Paraguay, and a few others that can barely be seen.

Figure 5.31 Territory size shows the proportion of worldwide net exports of royalties and licence fees (in US$) that come from there. Net exports are exports minus imports. When imports are larger than exports the territory is not shown (Worldmapper Map 99)

Text Box 5.29: Net Royalty and License Fee Exports

Only 18 (out of 200) territories are net exporters of licence fees and royalties. This means that a few people living in less than a tenth of the territories in the world between them receive the US$30 billion of net export earnings for these services.

The International Monetary Fund explains that royalties and licence fees include 'international payments and receipts for the authorised use of intangible, non-produced, non-financial assets and proprietary rights . . . and with the use, through licensing agreements, of produced originals or prototypes . . . '. Thus these export earnings are payments for past ideas.

> Ideas shape our world. They are the raw materials on which our future prosperity and heritage depend.
>
> (Kamil Idris, 2006)

5.2.29 Internet use 1990

In 1990 internet access was the preserve of a very few privileged folk, and often not the richest but the most well educated (or those working in education). This is a map of the head-start people, companies and corporations had in what is now the World Wide Web marketplace. Compare this map with that shown above and the similarities are uncanny, but there are also key differences in the distribution of new opportunities as compared with that distribution of old rights. Compare this map with that below – of how the Internet has spread out in just a dozen years – and the speed of infrastructure development in new technologies becomes clear. Spreading the web will not be like building railroads or canals. However, whether extending the scope of these means of communication extends opportunities and

Figure 5.32 Territory size shows the proportion of worldwide Internet users who lived there in 1990 (Worldmapper Map 335)

information, or always operates to make people ever more unequal is yet to be seen. The vast majority of the royalties secured by sales made on the internet result in money flowing back to the triad, so the story so far is that more equitable access to the net is leading to a yet more inequitable global distribution of wealth.

Text Box 5.30: Internet Users 1990

In 1990 the World Wide Web was a new idea, and only 3 million people used it worldwide; 77% of these people were living in North America – most lived in the United States.

Other Internet users of 1990 lived in just a few of the other territories, these were: the UK, France, Spain, Switzerland (home to CERN where the program that formed the conceptual basis of the World Wide Web was written), Italy, Germany, Belgium, the Netherlands, Denmark, Austria, Norway, Sweden, Finland, Taiwan, Republic of Korea, Democratic People's Republic of Korea, Japan and Australia.

> It is my belief that universal access to basic communication and information services is a fundamental human right
>
> (Pekka Tarjanne, undated)

5.2.30 Internet use 2002

Although when compared with Figure 5.32, Figure 5.33 shows the scope of the web spreading, it is spreading far from evenly. Compare how much more growth there has been in China than India, and perhaps compare that change with the relative sizes of the female service sector

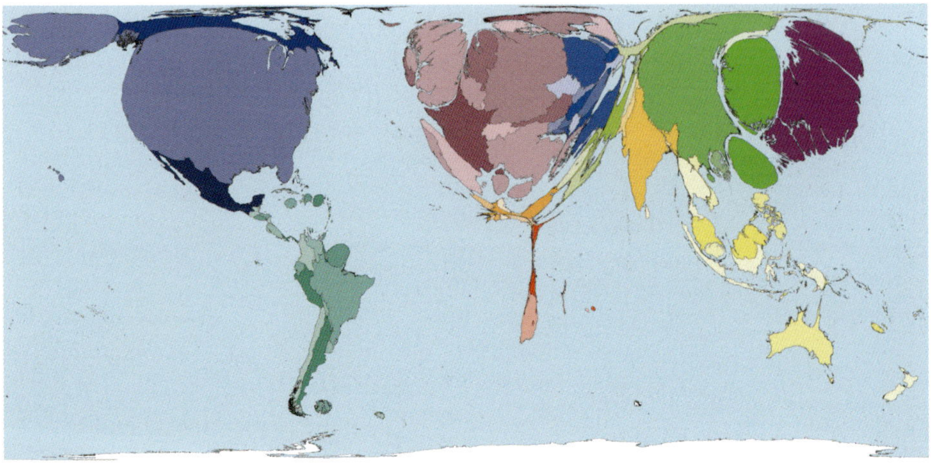

Figure 5.33 Territory size shows the proportion of worldwide Internet users who lived there in 2002 (Worldmapper Map 336)

in both those territories (Figure 5.18). Note too how little Africa has grown in relative size, and almost all the growth that there has been has been to the extremities of that continent. In contrast, many Caribbean islands are bulging with activity, suddenly no longer as remote as they once were. This is the map of who could be looking at the Worldmapper (and any other) website.

Text Box 5.31: Internet Users 2002

During the 12 years from 1990 to 2002, people using the Internet increased in number by 224 times. By 2002 there were 631 million Internet users worldwide.

Another change has been the distribution of these Internet users. In 1990 Internet users were mainly found in North America, Western Europe, Australia, Japan and Taiwan. By 2002 people living in Asia Pacific, Southern Asia, South America, China and Eastern Europe were notable Internet users. Some Internet users are also shown in Northern Africa, Southeastern Africa and the Middle East.

> The great mass of software, information and systems sold in the global marketplace excludes not only the cultures of poor countries – but also the very possibility that new technologies can be of use in the projects and lives of the poor.
>
> (Aníbal Ford, 2001)

5.2.31 Who's looking at us?

In Figure 5.34 is the map of who actually is looking at the Worldmapper website. It is dominated by affluent English-speaking nations and others where many have English as a second language. Interestingly the United States, for all the reports of myopia of its people,

Figure 5.34 Territory size shows the proportion of hits on the Worldmapper website that are made in that territory (Worldmapper Map 366)

appears slightly more interested than does Western Europe. Almost no-one in Africa or South Asia has seen the maps you have just been looking at, and just a handful of people have seen them in South America, East Asia and the Asia Pacific. How this map will change in the future we do not know, but having access to an infrastructure like the Web does not imply equal use of all it has to offer. Far more people have the luxury of time to browse and read in the richest parts of the world as compared with those living on most of this planet.

Text Box 5.32: Who's Looking at Us?

The Worldmapper project aims to communicate information that is collected about how we live together in the world, using maps. This map shows that the most people who have referred to the Worldmapper website live in the United States, the UK, Germany, Canada, Australia and the Republic of Korea.

Viewing is affected by Internet access and language spoken (at the time of data collection the website was only available in English). The map of Internet users in 1990 looked similar to this map – by 2002 the Internet users map had changed considerably. Perhaps a similar pattern will happen with Worldmapper visitors . . .

> it is of the greatest importance that the peoples of the earth learn to understand each other as individuals across distances and frontiers.
>
> (Pearl Buck, 1938)

5.3 The champagne glass of income distribution

Some of the huge disparities in the way we live should now be clear; the uneven nature of the distributions is epitomized by the 'champagne glass of income' shown in Figure 5.35. The height of the graph is split into five, each section representing one-fifth of the world population; the area of the 'champagne glass' shows how much of the incomes around the world go to the richest fifth (82.7 per cent) through to the poorest fifth (1.4 per cent). The stem of the glass is getting thinner. In 1960 the income of the wealthiest fifth was 30 times greater than that of the poorest fifth; now it is more than 80 times greater. The 'champagne glass' is looking less like a champagne glass. As some of those richest fifth are sure to agree, a champagne glass would normally have a taller section to contain the liquid. A glass drawn to represent wealth (this includes liquid and solid assets which may be passed between generations) would look more like a 'T' shaped stand than any sort of drinking vessel. Recently the World Institute for Development Economics Research (part of the United Nations University) reported that the richest 1 per cent of people in the world own 40 per cent of the world's wealth (Randerson, 2006).

These maps have provided just a glimpse of some aspects of the conditions of human life worldwide. This story has shown births and deaths, work and earnings, travel and spending, and the increase in ease of communications between people. These cartograms are scientifically (in the natural science sense) interesting as they result from the beauty and elegance of solving an algorithmic problem. These cartograms are also scientifically (in

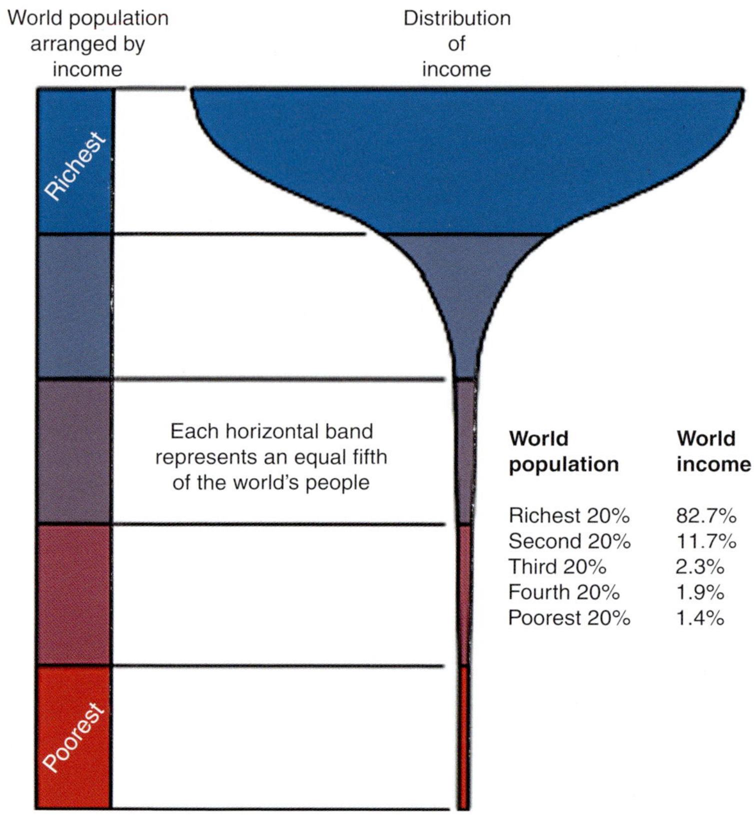

Figure 5.35 Sourced from Gordon (2004). This graph is redrawn based on the original

the social science sense) interesting because they show how we live now – they deliver a clear message about the current state of the world. Whilst each map shows static(istics) we can easily infer links between those of us who live in different places, and the connections between the topics that are mapped.

Visualization is powerful, and can oblige us to consider what is corrupt, immoral and profane about how life has come to be so ordered, so cheap and so unjust. And this is just a bland interpretation of data available to all that has been reported to United Nations agencies by governments. It is some of the best world data available at this point in time. In-depth journalist and special investigations are not needed in order to see a lot of what is most unfair and despotic in this world. Our interpretations are affected by our pre-existing and newly thought through ideas, standards and beliefs, in our case that people deserve equal chances, opportunities, and respect; and are of remarkably equal ability but rarely given anything like equal opportunities. This is not a particularly 'radical' view, given that it chimes with the Universal Declaration of Human Rights (1948), Article 1 of which reads: 'All human beings are born free and equal in dignity and rights. They are endowed with reason and conscience and should act towards one another in a spirit of brotherhood', or sisterhood.

Hopefully, the maps that these examples are drawn from provide us with a simple overview of how people living in parts of the world can be counted, compared, contrasted and probably connected. Often our ideas about the world are based primarily on more nebulous material that might include stereotypes, news reports and personal accounts. These maps add to that and our imaginations of the world because, rather than picking a few stories of interest, they attempt to find a space for everyone living in the world. We started this essay by quoting Rosa Luxemberg, who strove to improve the world, rather than excusing herself with beliefs about what is inevitable, or the idea that she was too small to make a difference (much like others who are now heroes and heroines of history books). Rosa argued that we should challenge the smoke-screens that can conceal what is happening. Towards this aim maps can be a useful tool as just a small part of the process of revealing what some have known for a long time: 'All that is solid melts into air, all that is holy is profaned, and man is at last compelled to face with sober senses his real conditions of life, and his relations with his kind' (Section I of the Communist Manifesto, 1848). In other words how you thought the world is ain't necessarily so. Hopefully, through this project, we have begun to show that there are thousands of different ways of looking at this planet and its people.

References

Communist Manifesto (1848) Section 1, *Bourgeois and Proletarians.* Available at: http://en.wikipedia.org/wiki/The_Communist_Manifesto

Dorling, D. and Thomas, B. (2004) *People and Places.* Bristol, Policy Press.

Fanon, F. (2001) *The Wretched of the Earth.* London, Penguin.

Gastner, M. T. and Newman M. E. J. (2004) Diffusion-based method for producing density equalizing maps. *Proceedings of the National Academy of Science USA* 101: 7499–7504.

Gordon, D. (2004) Eradicating Poverty in the 21st Century: When will Social Justice be done? Inaugural lecture given by Professor Dave Gordon on 18 October 2004. Available at: http://www.bristol.ac.uk/poverty/child%20poverty.html#inaugural

Kasser, T. (2002) *The High Price of Materialism.* Cambridge, MA, MIT Press.

Kidron, M. and Segal, R. (1981) *The State of the World Atlas.* London, Pluto Press.

Luxemberg, R. (1871–1919), quotation cited in Wallerstein, I. (2003) *The Decline of American Power.* London, The New Press, pp. 43–44.

Randerson, J. (2006) World's richest 1% own 40% of all wealth, UN report discovers. *The Guardian*, 6 December 2006, p. 24.

Smith, D. and Bræin, A. (2004) *The State of the World Atlas.* London, Earthscan.

Universal Declaration of Human Rights (1948) Sourced on 14 January 2007 from the UN website: http://www.un.org/Overview/rights.html

6

Re-visiting the Use of Surrogate Walks for Exploring Local Geographies Using Non-immersive Multimedia

William Cartwright

School of Mathematical and Geospatial Science, RMIT University

Surrogate travel (or sometimes referred to as movie maps) is the term given to the use of interactive multimedia products that allow users to 'travel' through environments. They are built with a variety of multimedia products, but they have in common the goal of providing tools that allow buildings, towns, cities and natural environments to be appreciated and understood without actually 'being there'.

This chapter describes surrogate travel and gives, by way of background, a brief description of two key surrogate travel cartographically related products – the *Aspen Movie Map* and the *Domesday* project. It then explores other forms of surrogate travel packages that have been built with non-immersive multimedia, discrete and distributed. Then it describes surrogate travel products developed by the author: the *GeoExploratorium*, the *Townsville GeoKnowledge* project and two Virtual Reality Modelling Language (VRML) products – the *Jewell Area prototype* and the *Melbourne Historical Buildings Demonstration product.* Finally, it provides a synopsis of the results from evaluations undertaken to ascertain the effectiveness of this approach for imparting a better understanding of local geographies.

6.1 Introduction

'Surrogate travel', or 'moviemaps;, is the ability to move around, allowing the user to laterally move through a recorded or created place (Mediamatic, 2004). This has been achieved by

Geographic Visualization Edited by Martin Dodge, Mary McDerby and Martin Turner

'building' installations, computer applications and multimedia products that provide users with the ability to 'move' through a pre-recorded or interactive 'space'. There are basically two approaches that can be taken – to build products where the user is taken on a guided tour (this requires no interactivity) or to allow the user to control how they move through or navigate through a space (which must be explicitly interactive) (Naimark, 1991). Collections of images, still or moving, individual or panoramic, are assembled and the user is invited to become an 'armchair traveller' (no interactivity) or to explore the space (via interactive collections).

The success of surrogate travel products depends upon two things: how the actual images are assembled and their ability to provide an adequate 'picture' of an area or space; and navigation, the key ingredient that enables users to properly move through spaces. Navigation can enhance the way in which a user moves through an 'information space'. Thoughtful design and graphic impact are essential elements of any presentation package, but if users do not know where they are in a package and how other elements relate, then their use of and interaction with the information provided will be at best limited and in the worst case non-existent. It is therefore essential not only that navigation of interaction packages be assessed, but also that information is provided so that users 'know' where they are in a (virtual) space and how that space would appear in reality.

The author's interest in exploring the potential of surrogate travel began when the potential for such products was seen in 1986, with the BBC *Domesday* product (Goddard and Armstrong, 1986; Openshaw and Mounsey, 1986). The *Domesday* double laservision videodisc provided an innovative multimedia 'picture' of Britain in the 1980s. It was jointly produced by the BBC (British Broadcasting Commission), Acorn Computers and Philips to commemorate the 900th anniversary of William the Conqueror's tally book. Part of the so-called Community videodisc was 'surrogate walks' through both urban and rural townships. Users of the system could 'walk' down streets, make turns wherever they wished and inspect the interiors of buildings, passageways and even rooms. Images photographed from the screen showing an example 'walk' are provided in Figure 6.1.

This was not the first time that surrogate travel had been used in multimedia products. The Aspen Movie Map, devised and undertaken by the MIT Machine Architecture Group in 1978 (Lippman, 1980; Negroponte, 1995), with Andrew Lippman as principal investigator

(a) (b)

Figure 6.1 *Domesday* videodisc (source: http://www.binarydinosaurs.co.uk/Museum/Acorn/domesday.htm)

Figure 6.2 Screen shot from the Aspen Movie map (source: http:/www.rebeccaallen.com/v1/work/list.php?isResearch=1)

(Wikipedia, 2006), used computer-controlled videodiscs to allow the user to 'drive' down corridors or streets of Aspen, Colorado, USA. Every street and turn was filmed in both directions. Users could enter buildings and view individual rooms. The interface is shown in Figure 6.2.

6.2 Queenscliff Video Atlas

Following this inspiration, the author developed the *Queenscliff Video Atlas* (Cartwright, 1987) for videodisc. The *GeoExploratorium* contained information about buildings of architectural significance in the historic township of Queenscliff, situated on the western headland of the entrance to Port Phillip Bay, Victoria, Australia, providing a multimedia historical and geographic information base.

A surrogate walk was included in the package. Maps, photographs, aerial photographs and videotape were used to capture the streetscape of Queenscliff, plus views from the jetty and of the Queenscliff coast. The contents of the videos include:

- Queenscliff street blocks; and
- views from the bay.

As opposed to still photography, block-by-block filming was thought necessary to give a more informative view of the living conditions of the Queenscliff residents. The filming sessions provided a collection of sequences containing individual street blocks in the Queenscliff Township. Initially, all video footage was captured on BVU videotape. Filming of every block in Queenscliff was completed, 23 blocks in all, using a video camera set up on the back of a flat-bed truck. The following specifications were adhered to at all times:

- filming was shot at an angle of 45° between the house fronts and the street;
- filming of the blocks was done in an anti-clockwise direction viewed from the front of the film vehicle; and
- the speed of the truck was kept at a constant 18 km/h.

A collection of still photographs of individual properties was also taken to capture:

- buildings of historical significance in the Queenscliff community; and
- a comprehensive photographic information resource.

Images from this product are provided in Figure 6.3 (note: these images are photographs from the screen and thus are slightly distorted and show some reflections).

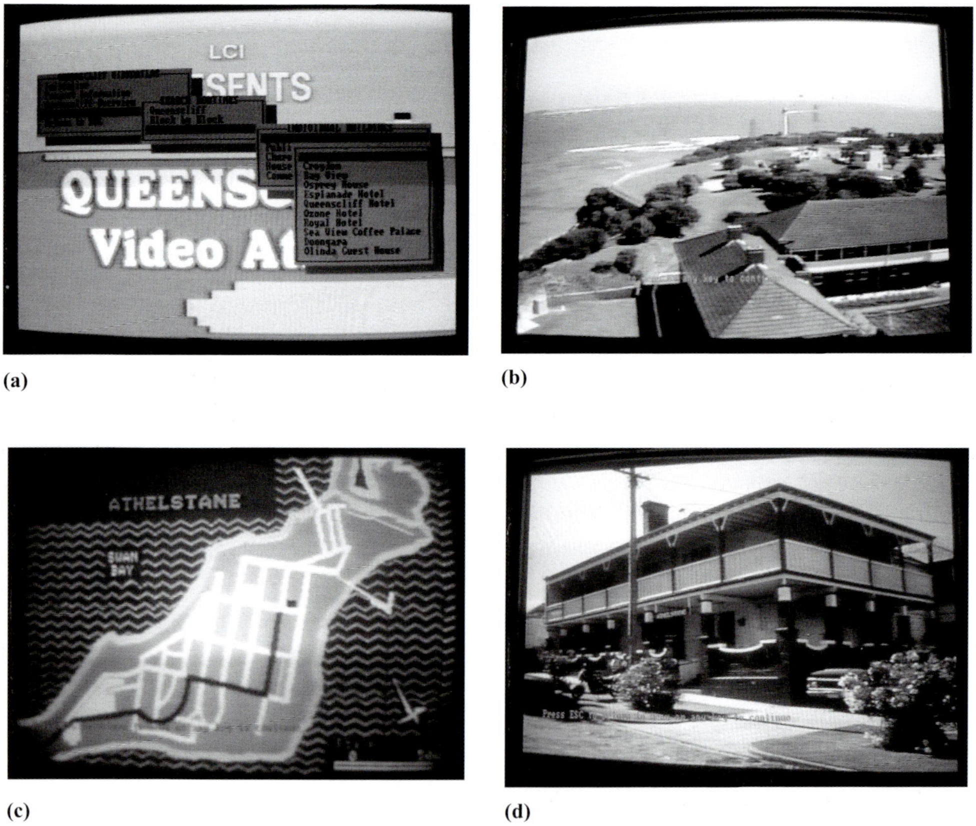

Figure 6.3 Queenscliff Video Atlas

Images were taken using 35 mm transparencies to show the frontages of each property. The images are all approximately the same scale, encompassing a view of the property as seen from the street. The total collection of building images was in excess of 100. Once the transparencies had been collated and selected for inclusion, they were captured on videotape, the same as the rest of the visual material. This involved the mounting of a slide projector beside the video camera in the filming studios. Each slide was captured in a 20 s videotape sequence, and later edited to six-frame segments on the master tape.

6.3 GeoExploratorium

Later, the *GeoExploratorium* was developed for Queenscliff (Cartwright, 1997, 1998) as a discrete/distributed interactive multimedia product that included surrogate walks – via still images and videos. It was developed as a combination discrete/distributed multimedia product. The discrete unit contained a coordinated collection of different types of information that can be readily accessed on demand. A package of mapping, photographic and video screen frames was supported by a textual base that offers information about the images displayed.

The discrete product was developed using Macromedia *Director* and then processed with Macromedia *Afterburner*, creating .dcr files that could be embedded into a World Wide Web page and then read with the Web browser (with Macromedia *Shockwave* and Apple *QuickTime* plug-ins). Due to the .dcr file being embedded in a Web page, the movement from discrete to distributed elements is transparent to the user. Figure 6.4 shows the initial user Web page for the prototype.

Surrogate walks were provided through the 'core' components of the product. This could be undertaken by 'browsing' the product via aerial photographs, maps or videos of each of the streets. Archival photographs include both terrestrial and aerial versions. The black and white photographs include historic streetscapes and individual buildings. The videos were the same as those used in the *Queenscliff Video Atlas*. Video was output from original videodisc at 30 frames per second (fps) and digital video made at 15 fps at a 240 × 180 pixel resolution, and then compressed using MPEG. It was found that producing movies at a larger resolution did not produce more usable products, just bigger ones that took up more storage space and were slower to run.

Colour aerial photographs and still photographs of properties were used in the prototype. The aerial photographs are at a scale of 1:10 000 and suit enlargement to depict individual street blocks within the township. These blocks were captured on video from the photographs to coincide with the moving video footage taken in Queenscliff and that from the maps above. A number of black and white vertical and oblique photographs were copied from originals at the Queenscliff Hydrographic Survey Office, and these show varying views of the Bellarine Peninsula, down to the entrance of Port Phillip Bay, including Queenscliff.

Accessing the colour aerial photograph collection moves the user to a two-part photographic overview of the township. The directional arrows at the bottom right of the screen can be used to toggle between the upper and lower images. Users can move to small-scale photographs, cadastral maps or back to the main menu by clicking on the appropriate hot spots. One of two large-scale photograph pages can be seen in Figure 6.5.

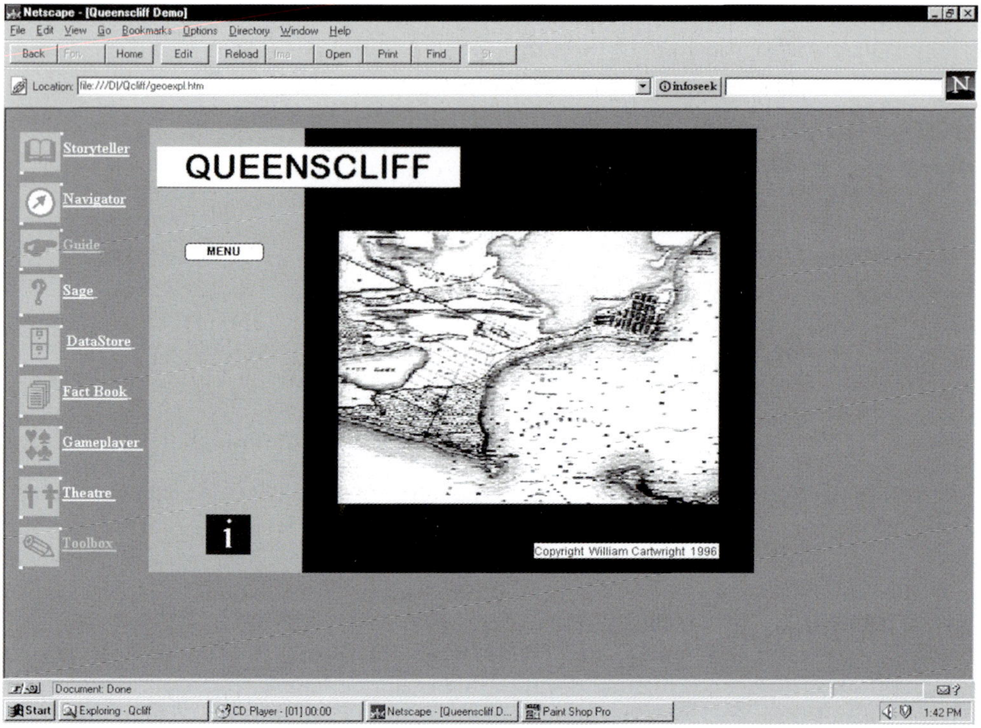

Figure 6.4 Prototype Web page. The Netscape browser (with *Shockwave* and *QuickTime* plug-ins) provides users with a transparent link between discrete and distributed elements

Figure 6.5 Small-scale aerial photograph of the northern part of Queenscliff

Figure 6.6 Large-scale aerial photograph of the northern part of Queenscliff

Once the user moves to the larger scale aerial photograph (Figure 6.6) the directional arrows can again be used to move from block to block. A move to the cadastral maps, invoked by clicking the 'Cadastral' icon, provides the user with a screen image of the cadastral section relating to the same area as that covered in the aerial photograph. The 'up' button allows for the smaller-scale photographs to be viewed, if an overview of the photograph coverage is required.

A site plan of each of the historical houses is also provided. The user has several options: to move from site plan to site plan using the directional arrows at the bottom of the plan; to go to the collection of building photographs; to view block-by-block movies (by clicking the appropriate dot on the map); or to return to the menu. Figure 6.7 illustrates the information available for individual building information. The 'clickable' icons, photographs, text and map 'hot spots' lead to more building plans, house photographs and videos of block-by-block streetscape footage.

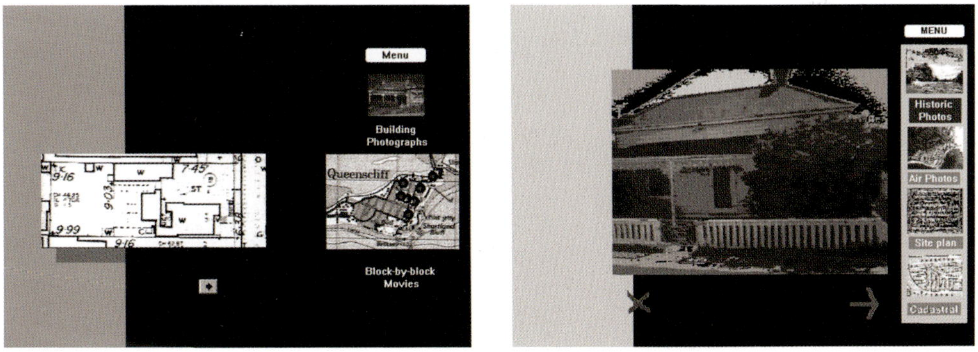

Figure 6.7 Individual house information

Once the user moves to the individual buildings section, the collection of approximately 50 photographs can be viewed. The directional indicators are used to move back and forth through the collection, and the icons to the right of the main photograph allow movement to other sections of the discrete unit, as well as moving back to the main menu.

As well as browsing the package this way, users could also use metaphors to further explore the package in a more focused, personalized manner. A number of conceptual metaphors were applied in a set which includes the Storyteller, the Navigator, the Guide, the Sage, the Data Store, the Fact Book, the Gameplayer, the Theatre and the Toolbox. These metaphors have been acknowledged by Laurini (2001) as suitable access metaphors for visualizing and accessing urban information. For surrogate walks, the use of the Storyteller, Navigator and Theatre metaphors are appropriate. Their implementation is elaborated upon in the following paragraphs.

Interactive storytelling can be used to enhance the information to which users gain access. Some users may wish to be 'told' a story while they view a map or graphical depictions on a screen. The story can be told using digital sound or by allowing the users to read text from the screen. Storytelling, whilst not expected to be used vigorously, needs to be made available for those users interested in finding out more about particular phenomena by being painted 'word pictures'. The Storyteller metaphor provides a number of 'stories' about Queenscliff and the region. As new stories, press releases and general news about Queenscliff becomes available, the Web page can be continually updated to present the most current information available. Owing to the nature of this metaphor, the actual stories will be in a constant state of flux, offering the user many different stories from the area. Once the user moves to the Storyteller Web page, links can be followed to areas of personal interest to the user.

For some 'power users' the use of navigation tools will be unwanted, but for novice or naive users the Navigator would allow them to move through the package in the most efficient manner. Final navigation strategies would be developed on the basis of user needs, modified according to actual usage patterns. The Navigator provides an overview of the houses in Queenscliff through the use of a video 'surrogate drive' around all of the streets in the township. The user chooses a block from a list on the navigator Web page. Figure 6.8

Figure 6.8 Block-by-block video

illustrates a video 'block-by-block' coverage of streets indicated on the accompanying map. Buttons and slide bars with functions that cater for fast-forward, fast rewind and pauses can control videos. The video package uses standard *QuickTime* controllers. This sample page shows a typical video page for one of the blocks of the township. The Web page contains a location map, showing this particular block as a green dot and the other blocks with movies as a red dot. Clicking the red dot associated with that a particular block can access other blocks.

The use of the Theatre metaphor involves functionality and thus allows the user to be engaged/pleased with the experience. This means that the user must understand the activity well enough to do something. Some users may prefer this type of 'discovery' activity to traditional human–computer interface methods. This metaphor enables everyday activities, life in general and items specific to certain problems to be depicted. The Theatre metaphor (Figure 6.9) allows users to explore the township. It provides an insight into the Township of Queenscliff and an overview of the Stage (the township and Port Phillip Bay), the Cast (buildings and the infrastructure of the township) and the Script (elements that impose upon and affect changes to both the Stage and how the Cast exist or function on that Stage).

Once the user moves to the Theatre, two sets of *QuickTime* movies are available: an overview of the township through two movies taken from the lighthouse in the centre of Fort Queenscliff; and two movies taken from Port Phillip Bay showing the lighthouses and the lifeboat shed and pier. Both videos show Queenscliff via the Theatre metaphor. Figure 6.10(a) is the overview of the township. The two *QuickTime* movies are looped, but users can stop playback using the standard controls. Figure 6.10(b) illustrates the views from Port Phillip Bay. These videos operate in the same manner as those of the Township.

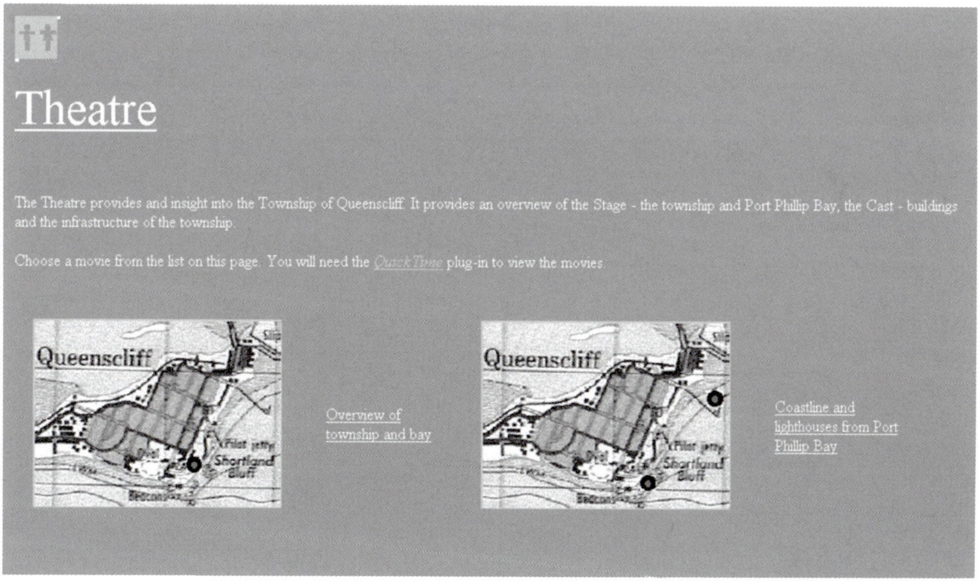

Figure 6.9 The Theatre Web page

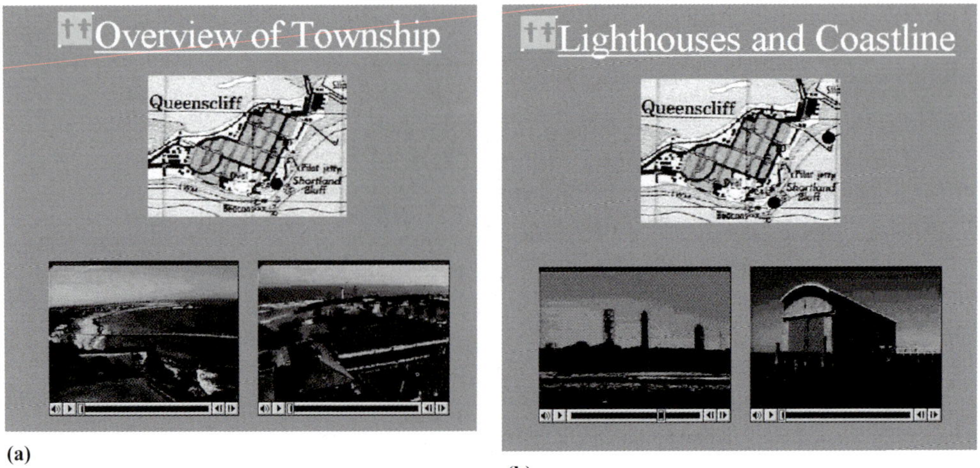

(a) (b)

Figure 6.10 The Theatre metaphor resources

6.4 Townsville GeoKnowledge Project

The prototype was designed to be delivered via a Web browser with minimal plug-ins – for *Flash* and *QuickTime* movies. As part of the package, users can undertake a 'surrogate walk' around the city. Once the user 'clicks' on one of the hot spots they are presented with a choice of images of the city, initially through a map interface (Figure 6.11) and then with a more detailed street intersection photograph collection (Figure 6.12). As well as individual images, users are also provided with a number of panoramas that were taken at key sites that overlooked the city. An example of a panorama is shown in Figure 6.13.

At each street intersection the user is provided with a choice to continue walking in the same direction, turn left or right or go back the way they came. The map to the right of the screen (Figure 6.14) 'keeps pace' with the user and indicates both the position and the direction of observation. If extra information, photographs or access to further information in the package is possible, information icons appear on the photograph. By clicking on the icon the user can access these supplementary information resources.

Whilst the user 'walks' along streets where additional information like panoramas, extra building photographs, etc. are available, icons indicate where these can be accessed. The access icons are shown in the photograph in Figure 6.15. Figure 6.16 shows how additional photography is provided.

6.5 Jewell Area prototype

As part of a research programme to develop and test web-delivered tools to support community collaborative decision-making, an interactive three-dimensional tool was developed using the Virtual Reality Modelling Language (VRML). It was designed to be delivered via the Web, and to be used at home or at Internet cafes located in the application area, and

Figure 6.11 Entry into the surrogate walk

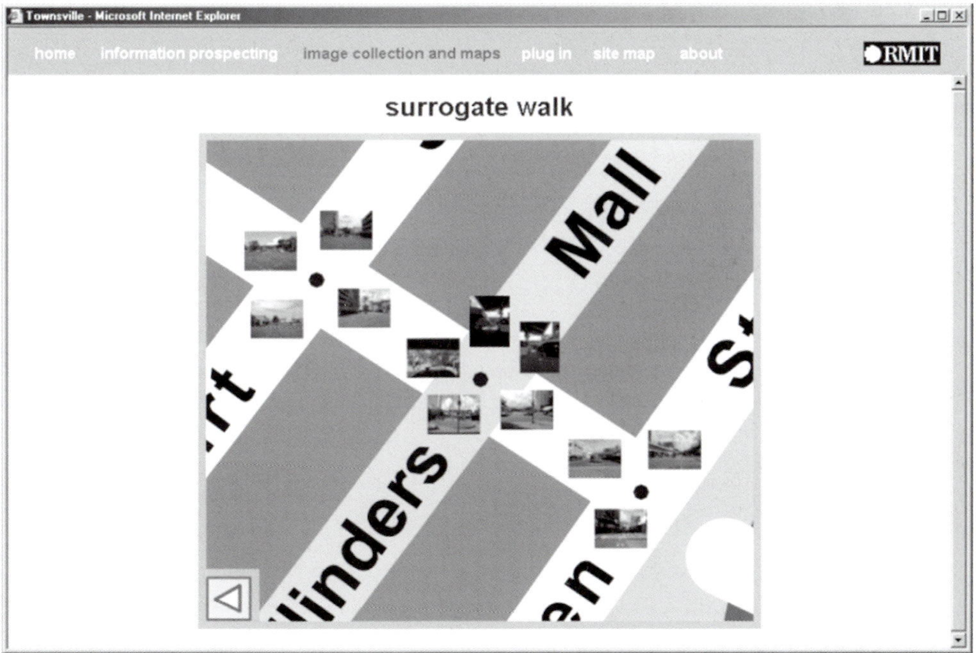

Figure 6.12 Image collection choice at street intersection

Figure 6.13 Images of the city. In this example a panorama image exists and the user is able to choose between an interactive panorama or to view the images that compose that panorama, as in this example

Figure 6.14 At each street intersection users can choose to move left or right, look backwards or move forward to the next street intersection and associated photograph set

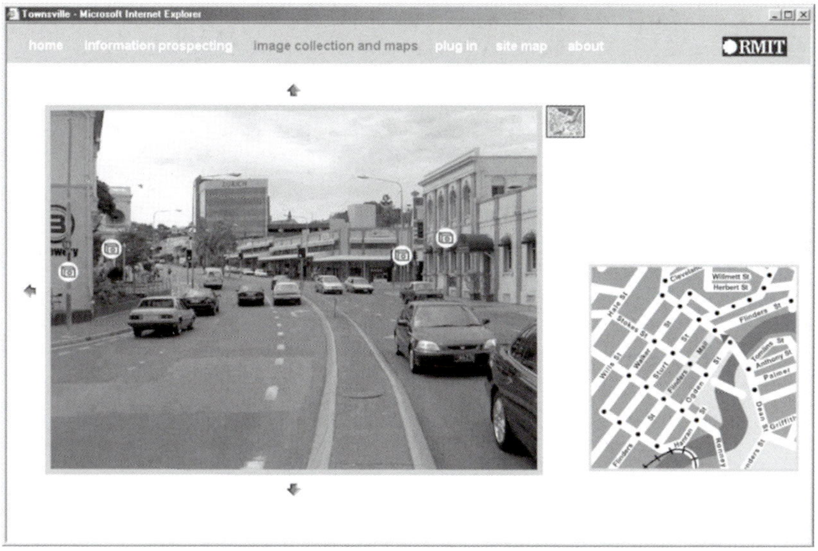

Figure 6.15 Supplementary information access

thus to be easily accessible by the general community. It could also be used at meetings to support collaborative decision-making deliberations (Cartwright *et al.*, 2004).

VRML was chosen as a development tool as it allowed open, extensible formats to be used and the 'built' worlds could be constructed in web browsers that included a VRML plug-in. VRML is an extensible, interpreted language and an industry-standard scene description language. It is used for three-dimensional scenes, or worlds, on the internet. To produce three-dimensional content two-dimensional components are defined/drawn and the viewpoint specified. Once this is defined the drawing package renders the three-dimensional

Figure 6.16 Additional photography available at a site

(a) (b)

Figure 6.17 VRML world

image onto the screen. VRML code defines objects as frameworks that are rendered. This makes file sizes very small. The appearance of rendered surfaces can also be modified using different textures. By using the computer's fast processing speeds, and specifying multiple, sequential viewpoints, 'walkthroughs' or 'flythroughs' can be produced.

All buildings in the study area were surveyed to ascertain position, use and building height. Also, each building façade was photographed for use in 'stitching' the images onto the sides of VRML primitive shapes. All buildings in the study area were subsequently inserted into the model.

From previous stages of this research (Cartwright *et al.*, 2004) a naive world was built that only contained basic building outlines and some landmark building detail. Also, at the end of each street 'end-of-the-world' images were added to ensure that the world did not 'end' at the edge of the model. Single images were captured at the end of each street and then 'pasted' at the edge of the VRML world to give the impression that the world continued beyond the extent of the model. The world and 'end of the world images' are shown in Figure 6.17.

6.6 Melbourne Historical Buildings Demonstration Product

An associated and complimentary VRML model is being developed for appreciating 'lost' city buildings of part of the Central Business District in Melbourne, Australia. This model will 'fuse' together existing buildings of significance with the 'missing' buildings – those removed hastily in the building boom of the 1950s, 1960s and early 1970s. The aim of producing the model was to develop a simple, accessible demonstration prototype that could be used to familiarize historians with the potential that three-dimensional simulations provide for better appreciation of what the city might have been if all significant buildings remained intact (Figure 6.18). It was also developed as a vehicle to support applications for funding to extend the prototype to cover the entire Central Activities District. As the prototype has developed techniques for sourcing imagery (current and historical), capturing and processing images of standing buildings, actually building the model using the VRML and delivering a usable web-delivered product, to extend the model to cover the entire Central Business District would hinge only on the access to adequate historical imagery; it

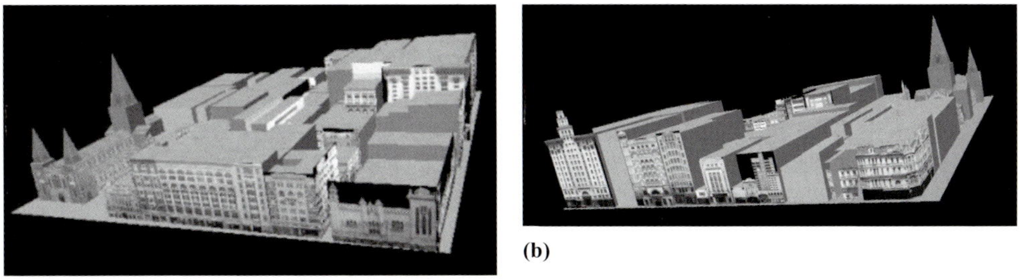

(b)

(a)

Figure 6.18 Composite VRML world of existing and lost heritage buildings in part of Melbourne (Australia)

would not depend on developing techniques for model development. Therefore, to develop a model for the entire Central Activities District would be a fairly straightforward operation (Cartwright, 2006a). It will be evaluated to ascertain its usefulness for historians, architects and the general public.

6.7 Testing the user's perception of space and place

As a multimedia installation of this type changes the way in which users access and use geographical information, a simple evaluation technique is inappropriate. As well as gauging the success of the use of the product, the ability of such a product to change the user's viewpoint of geographical reality, and thus a particular view of space, needs to be assessed. By using an interactive multimedia installation that encourages exploration in ways that individual consumers feel most comfortable, a better (if not different) perception of the world is provided.

This section provides information regarding evaluations that have been completed for:

- the *GeoExploratorium*; and
- the *Townsville GeoKnowledge* product.

6.7.1 GeoExploratorium

Questionnaires were sent to 17 reviewers. The reviewers were selected from the wider mapping sciences community and included a number of university professors who are involved with both digital mapping and multimedia production, a map curator and a school teacher who uses the Queenscliff area extensively for school projects with students. As with all questionnaires, problems with poor response rates usually eliminate the possibility for obtaining credible results with small samples. However, this review received 11 responses, a 65 per cent response rate.

A 'reviewer profile' was used to ascertain information about how the reviewers used mapping products in their profession as teacher/lecturers, researchers, map curators or as cartographers. They use a wide range of methods to produce maps: digital mapping packages, computer graphics packages, multimedia authoring software, web authoring, animation software and direct programming. Their use of geographical information resources is diverse and includes:

- printed paper maps;
- computer monitor displays from mapping, GIS and CD-ROM packages;
- Web-delivered map information;
- aerial photographs and remotely sensed imagery; and
- other artefacts like drawings, plans, animations, micro formats, photographs, video and film.

Other media they used in conjunction with 'main stream' products were graphics, diagrams, sound (including verbal instructions), television, tactile maps and compact disc-interactive.

These profiles included the types of products used regularly, their usual method of map production, the types of geographical information products they commonly use and the other methods they would use to support the communication of geographical information.

A questionnaire elicited comments about the metaphor set employed. Reviewers were asked to rate on a scale of 1 (strongly disagree with the statement) to 5 (strongly agree) their agreement or disagreement with comments about the metaphor set. The reviewers agreed/strongly agreed with the statement that other support or ancillary information helps users of geographical information better understand the information being depicted. They generally thought that maps alone are the best communication devices for geographical information. The metaphor set was supported as being a useful adjunct to the use of maps (mainly agreement or strong agreement with the statement). All reviewers agreed (most of them strongly) that the combined use of both discrete and distributed (web) products is a good way of providing map and ancillary information in the context of the *GeoExploratorium.*

In a second section of the questionnaire reviewers were asked to make general comments in five categories. The responses were sought to ascertain why users would choose this type of product over existing products and how it should be used. Comments were also requested about the metaphor set and how it might be improved. Lastly, the reviewers' thoughts regarding the actual demonstration product were requested to see how the product might be improved in terms of the general content and the way it operated. The answers to the first question illustrate the use of the product for better understanding the town. The question asked was: *In your opinion, what would motivate users to use something like the demonstration product to complement maps for the 'discovery' of geographical information?* The responses to this question are listed in Table 6.1.

6.7.2 Townsville GeoKnowledge product

Subjective methods are typically used to determine the level of media quality required in applications. At this stage of evaluation, an attempt was made to implement a more formal

Table 6.1 General comments – question 1

Reviewer	Comments
1	All of the non-referenced information can be linked to the map, enriches map information
2	A need for particular information – either their own information need or one generated artificially (e.g. through a game of some sort)
	Following from above, financial incentive to obtain information
	For some, curiosity. For some, that it is fun
3	This is an excellent product for the person interested in moving to a new city and wanting a general understanding of the new place
	Integration of graphical media allows a much richer 'feel' for an area without a site visit
4	The *GeoExploratorium* offers great possibilities for users to go through a geo-related database including all the media capable of describing 'land information'. The new 'tool' would help the user by correlating all the information linked to his territory
5	Educational project, travel to and around area, add value to
	Existing knowledge
6	Need for information (e.g. new visitor to town)
	Knowledge of existence of the demo (e.g. being told by library staff)
	Eagerness to experience new technology (e.g. by young people)
7	Financial gain – for business users
	Availability of tourist data – for travel planning
	Need for heritage/conservation/building data – for conservation studies
8	It is possible to draw panoramic maps, even large-scale ones, on the basis of topographic maps – I have worked with this technique even before the advent of the computer. The resulting drawings will not prepare users for the actual landscape, because they would realize insufficiently the result of cartographic abstraction; they would not realize what has been left out – so the main motivation for me would be to counteract the cartographic abstraction, and allow users to better anticipate what is in store for them in the landscape: actual shapes, vegetation, sounds, smells, etc.
9	If classroom situations were the norm, assigning students a task in which they were required to research a particular subject would be a motivating factor. Once using them, students would perhaps be encouraged to seek out other similar products
10	Public advertisement and display/demonstration of the product
11	The need to know more than where to go and how to get there. Any historical information or current access to information via local services is available through this product. The street directory only gives fundamentals

approach. The evaluation procedure demanded that it be 'built' on a sound educational theory and implemented through a proven *modus operandi.* The core educational theories of Bloom's learning behaviours, developed in his taxonomy of learning objectives (1956), were considered appropriate for developing tasks for formal evaluation of the prototype product. They were used to develop specific cartographic applications for testing. A summary of how Bloom's learning behaviours can be translated into cartographic applications and how these have been implemented in this cartographic product are provided in Table 6.2.

Table 6.2 Applying Bloom's taxonomy to cartographic products (note: the comments in columns 2, 4 and 6 are taken from Wakefield (1998) and are not the author's)

Bloom	Educational focus (from Wakefield, 1998)	Geographical Education focus	Educational materials (from Wakefield, 1998)	Cartographic materials	Measurable behaviours (from Wakefield, 1998)	Measurable geographical usage behaviours
Knowledge	Focuses upon the remembering and reciting of information	Learning geographical facts, for example learning country names, capital cities and then reciting this information	Events, people, newspapers, magazine articles, definitions, videos, dramas, textbooks, films, television programmes, recordings, media presentations	Maps, atlases, remotely sensed images, gazetteers	Define, describe, memorize, label, recognize, name, draw, state, identify, select, write, locate, recite	Define, describe, name, identify, recite, locate
Comprehension	Focuses upon relating and organizing previously learned information	Taking geographical information, formed as a 'mental map' and then re-organizing this information into a more coherent resource. Classification of information, for example taking the population sizes of countries and then re-ordering countries according to population size	Speech, story, drama, cartoon, diagram, graph, summary, outline, analogy, poster, bulletin board	Compilation of a thematic map, design of a cartogram	Summarize, restate, paraphrase, illustrate, match, explain, defend, relate, infer, compare, contrast, generalize	Compile, summarize, illustrate, map, generalize

Application	Using methods, concepts, principles and theories in new situations	Using geographical knowledge and producing a thematic map that illustrates classified information. Information is symbolized and a map product realized, for example producing an iso-demographic map that illustrates comparative population sizes by changing the actual size of countries depicted. Relative sizes of countries are determined by their population size, rather than their landmass	Diagram, sculpture, illustration, dramatization, forecast, problem, puzzle, organizations, classifications, rules, systems, routines	Thematic map. Mathematical element: application of map projections. Art element: map design. Science element: underlying distribution theory. Technology input: choice of production technology/ software and delivery technology/ communications systems	Apply, change, put together, construct, discover, produce, make, report, sketch, solve, show, collect, prepare	Design, project, classify, depict

(*Continued*)

Table 6.2 Applying Bloom's taxonomy to cartographic products (note: the comments in columns 2, 4 and 6 are taken from Wakefield (1998) and are not the author's) (*Continued*)

Bloom	Educational focus (from Wakefield, 1998)	Geographical Education focus	Educational materials (from Wakefield, 1998)	Cartographic materials	Measurable behaviours (from Wakefield, 1998)	Measurable geographical usage behaviours
Analysis	Critical thinking which focuses upon parts and their functionality in the whole	Using map information to compare region-to-region, city-to-city, for example using the population map described in the previous cell to compare the relative populations of various countries in a visual way	Survey, questionnaire, an argument, a model, displays, demonstrations, diagrams, systems, conclusions, report, graphed information	Thematic map – paper or delivered electronically. May contain a tool to assist usage, for example a map legend for the paper map or an interactive 'how to' tool in an interactive multimedia product	Examine, classify, categorize, research, contrast, compare, disassemble, differentiate, separate, investigate, subdivide	Compare, classify, measure, calculate, assemble, disassemble, map
Synthesis	Critical thinking which focuses upon putting parts together to form a new and original whole	Taking a number of cartographic artefacts and using them to assemble data about a certain geographical region, for example using an atlas, which contains maps, graphs, diagrams	Experiment, game, song, report, poem, prose, speculation, creation, art, invention, drama, rules	Maps, measuring tools, geographical visualization systems, Geographic information systems, cartographic information systems	Combine, hypothesize, construct, originate, create, design, formulate, role-play, develop	Buffer, combine, re-combine, differentiate, integrate, layer, order, filter, select, map

		and textual information and 'assembling' information about a geographical region. Or, alternatively, a GIS				
Evaluation	Critical thinking which focuses upon valuing and making judgements based upon information	Taking 'assembled' information about a geographical region (from many cartographic artefacts) and making a decision about the relative growth, economic position, environmental status, etc.	Recommendations, self-evaluations, group discussions, debate, court trial, standards, editorials, values	Considered and informed decision-making using an array of cartographic artefacts, for example developing 'what if' scenarios using geographic information systems	Compare, recommend, assess, value, appraise, solve, criticize, weigh, consider, debate	Compare, report, assess, value, appraise, consider, project, grade, demark, and speculate

A small expert user/producer group of eight candidates, 18–25 years old, was used to evaluate the product. The choice to use a smaller and more focussed group was made because it was easier to manage and because, for instance, Virizi (1992) has noted that 80 per cent of usability problems are uncovered with four or five test participants, so the use of a smaller group was deemed to be a sound one. They were first asked to answer the questions using a paper map of the selected area and then with a three-dimensional Web information source, the *Townsville GeoKnowledge* project interactive multimedia product (Cartwright *et al.*, 2003). The questions asked are provided in Table 6.3.

After the tasks had been performed, the candidates were asked to make general comments comparing the traditional map to the three-dimensional Web product. The purpose was to determine whether the medium changed their view of the geography of Townsville. Then general comments relating to product improvement were solicited. The diagram in Figure 6.19 shows how each of these tasks was linked.

Candidates first used the paper map, and then repeated the procedure using the surrogate walk product. At the end of each section, the candidates commented on how easy it was to perform these tasks with the artefact used – Easy, Fairly Easy, Moderately Difficult, Difficult, Hard, Very Hard or Impossible. Once they had done this with the paper map, the candidates went through the same procedure using the interactive multimedia package. One of the focus elements of this evaluation was to investigate if perceptions of geographical places are dependent upon the medium used.

There were definite changes in how the candidates viewed their allocated tasks when using the paper map and when completing the same tasks with the interactive multimedia product. Questions relating to the Knowledge section were rated, in the worst cases, as impossible/very difficult to answer with the paper map. This changed to fairly easy/easy with the multimedia product. For Comprehension hard/difficult moved to moderately difficult/easy. Application tasks were rated hard/moderately difficult when the paper map was used and moderately difficult/fairly easy with the multimedia product. Analysis-related questions were impossible/moderately difficult with the paper map and fairly easy/easy with the multimedia product. There was no change with the Synthesis tasks. Evaluation-specific questions were rated impossible/moderately difficult with the paper map. This perception changed to moderately difficult/easy with the multimedia product. In general terms, five of the 'umbrella' task areas were easier to perform, albeit two of the five tasks only changed slightly, with the multimedia product, which indicates the effectiveness of 'rich media' products.

Finally, the candidates were asked to comment on how their perception of what constituted the town had changed once they had used the interactive multimedia product. The ability to describe the town improved markedly when the interactive multimedia product was used. Initially, the candidates were either unable to adequately describe what constituted the town, or they just provided a general 'location of elements' statement. After they had used the interactive multimedia product, the candidates were able to comment on the structure of the town, its location by the sea, the fact that it had a harbour and that there were hills behind it. The general structure of the town – city centre with a few historic buildings and a suburban spread beyond – was also added to the description. The interactive multimedia product, with enhanced media attributes, enabled the users to get a better appreciation of the town. General comments are provided in Table 6.4.

The users commented that with the interactive multimedia product it was easier to imagine what the town looked like and to get a feeling for the topography. The functions of the

Table 6.3 Formal basis for the questionnaire – applying Bloom's Taxonomy to cartographic products

Application of Bloom's taxonomy to geographical education	Task	Question
Knowledge	General undirected map reading	Name the main streets in the central area of Townsville. Identify the harbours in the town, what are their names? Locate the railway station, the panoramic lookout point, the entrance to the harbour and the way to the airport
Comprehension	Directed map reading. Users are asked to locate specific items and then to summarize and generalize this information	Compile a list of points of importance to mariners using the port facilities. Summarize the elements that comprise the town. Generalize what constitutes the layout of the town
Application	Classification of information shown in the product and making general considerations related to the production of a 'second-generation' map product that would encapsulate the essential elements of the town	Considering the information shown on the map, how would you classify the basic elements of the town? If you were asked to produce a map that encapsulated the essential 'things' that comprise the town what would these elements be?
Analysis	Considering different topographical features. Measuring distances. Calculation of travelling times	Compare the central city area to the surrounding areas. How are they different topographically? Measure the distance from the panoramic lookout to the entrance of the main harbour. Calculate how long it would take you to walk this distance
Synthesis	Making informed judgements from the geographical information provided. Selecting information deemed to be important	Differentiate between the main elements of the town. What are they? If you were asked to act as a tourist guide to visitors to the town, which places would you take them to? Select two places that would provide the 'key' points to visit
Evaluation	Comparing the town to other towns. Evaluating the value of the town's attributes. Projecting possible scenarios	Comparing this town to where you live, what do you think are the main differences? What do yo think is the value of the harbour to the town. Speculate that if there was a flooding related to a tropical cyclone, which areas of the town would be most likely to be affected?

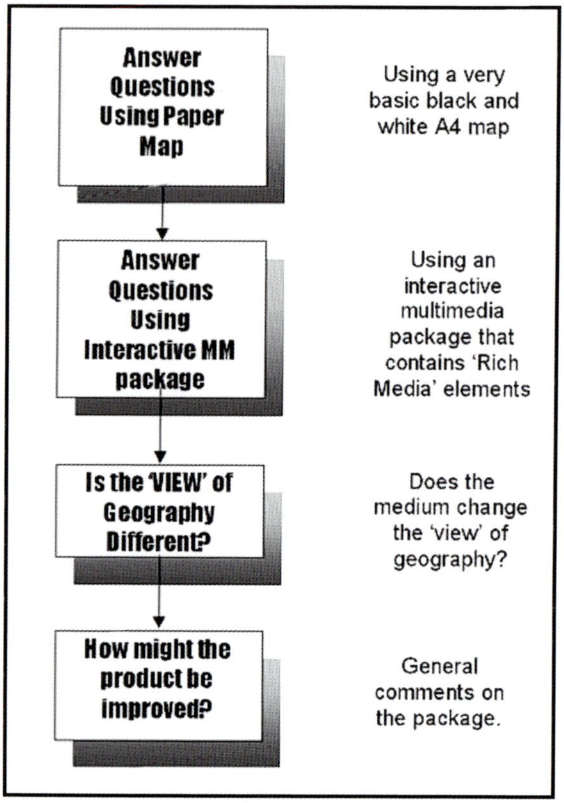

Figure 6.19 Evaluation components

various districts in the town were obvious and existing personal experiences could be enhanced. Generally, the candidates commented positively on the ability to gain an enhanced appreciation of the town with the interactive multimedia product.

6.7.3 Jewell Area prototype

The evaluation was based on the three primary way finding tasks specified by Darken and Sibert (1996). These are:

- naive search – where the navigator has no *a priori* knowledge of the whereabouts of the target, requiring an exhaustive search;
- primed search – where the navigator knows the location of the target, performing a non-exhaustive search; and
- exploration – where there is no target.

Table 6.4 Comparing the user's view

Candidate							
1	2	3	4	5	6	7	8
It is easier to imagine how the town (houses) look like. With the multimedia product it's easier to get a feeling for the topography	With the interactive product you can see the topography You can see what the town looks like	The multimedia product gave more information about the size, the settlement, the functions of the town and its districts	With the paper copy you can't imagine how the town looks as in the 'real world'. With the photographs in the interactive map you really have an overview of the town and can connect experiences	With the multimedia product the city looks more friendly/ sunny with less industry	No real change in perception	—	Could figure-out how the town looks due to panoramas, photographs

The naive search, as Darken and Sibert note, is rare in the real world, but common with first-time users of a virtual space. Naive search will therefore rely on certain wayfinding aids to support their movement through the virtual world. Darken and Sibert (1996, p. 4) proposed that the basic principles of organizing an environment to support wayfinding were to:

1. divide the large-scale world into small, distinct small parts, preserving a sense of 'place';
2. organize the small parts under a simple organizational principle; and
3. provide frequent directional cues.

This stage of the evaluation process will have as its goal the determination of what users need to know or understand about SPACE (general information about the 'area' being studied) and PLACE (dictated/determined by location and purpose-specific elements that are unique to the particular user and their usage requirements). What needed to be resolved were:

- users' concepts of space;
- their concept of place;
- how they navigate through space; and
- how they navigate through their personal place.

Twenty-nine candidates participated in the evaluation. The age range was 18–25 and all had competent to efficient map use skills. The test candidates were split into two groups: group 1, identified as having *a priori* knowledge of the area. Twelve candidates were identified as belonging to this group. Group 2 had relatively no knowledge of the area. Secenteen candidates belonged to this group. Then each of these groups was further split into two sub-groups: groups 1a and 1b, and groups 2a and 2b.

The candidates first completed a profile proforma to glean information about their proficiency in map and map-related tool use and also their perceived knowledge of the area.

The session operated thus:

1. Groups 1a and 2a were taken on a 'guided tour' of the area.
2. During this time Groups 1b and 2b undertook the evaluation/feedback of the VRML model.

GROUP			
1. Local knowledge		2. No local knowledge	
1a Tour 1	**1b Tour 2**	**2a Tour**	**2b Tour 2**

Figure 6.20 Groups

Then this process was reversed:

1. Groups 1b and 2b were taken on a 'guided tour' of the area.
2. During this time Groups 1a and 2a undertook the evaluation/feedback of the VRML model.

Candidates were asked to conduct two searches of the three-dimensional model of the study area, one as a general 'exploration' of the area and the other a task-related search. In the task-related search they were required to find specific buildings that are typical in the study area. These searches were (1) naive (the candidates had no knowledge about where they were located) and (2) primed (they knew the area after a walking tour prior to the evaluation). They were asked to view two 'virtual tours' of the study area, one as a general 'exploration' of the area and the other a task-related search. In the task-related search users were asked to identify the different building types that are typical in the study area. They were also asked to note the buildings that they thought were the key, or landmark, buildings.

Results

Candidates first considered whether the amount of detail provided was sufficient for them to understand the general geography of the area. Did it provide sufficient information for them to be able to make informed comments about potential developments in the area?

1. *The amount of detail is sufficient.* All groups found that there was sufficient detail to understand the area.

2. *There are adequate landmarks to assist in orienting oneself.* Landmarks in the area, as previously noted are mainly prominent buildings. All found that they had adequate landmarks to assist, except Group 2a. This group had no prior knowledge of the area, and they had been on a tour of the area prior to undertaking the evaluation. Here, the users identified that, even though they had been on a tour, they thought that extra information was required.

3. *Having all buildings in full detail is necessary.* All groups except 1a thought that there was sufficient detail.

4. *I could understand the area with less detail in this three-dimensional model, which would provide me with an adequate mental representation of the area.* Group 2b thought that more detail was required. This group had no prior knowledge of the area and they had not been on a tour prior to the evaluation. The other part of the 'no knowledge' group (2a) thought that the level of detail was sufficient, but they did not indicate full support of the amount of detail. Therefore, when users have little knowledge of the area, they will respond better to the model if a tour is conducted prior to actually using the tool.

5. *Less information/detail would still allow me to build a mental image of the area.* Again, those who had no knowledge of the area could not accept a model with less detail, and the members of this group who did not undertake the tour indicated that they could not work with less detail. This again supported the concepts that having knowledge of the area allows for a simpler model to be provided, and that a pre-use tour assists in better exploiting the model.

6. *Having all elements in full detail makes the image too complex (i.e. it has a negative effect, rather than improving the model).* All candidates generally disagreed with this statement. The level of detail did not make the tool more complex.

7. *I need the addition of street signs for me to orientate myself.* The inclusion of the street signs was supported, but less so by groups 1b and 2a. (Later comments about the inclusion of the street signs included: "Street signs also provide human scale in terms of height".)

8. *'End-of-the world' images make the three-dimensional image look more real.* These images were added to the model so that it did not appear to 'end' at the edge of the VRML world. All candidates supported the inclusion of end-of-the-world images.

9. *Adding light poles and wires makes the three-dimensional image look more like an inner Melbourne shopping strip.* This inner urban area has the usual trappings of overhead wires, poles and banners. The test prototype provides the option of having these items 'on' or 'off'. All candidates thought that the addition of these items was necessary.

10. *Changed environmental conditions make the three-dimensional world more appropriate for better visualizing local conditions.* The prototype model allows for the environmental conditions to be changed – sunny to overcast, day to night. These items can be chosen by selecting the appropriate radio buttons in the interface. However, the test candidates did not think that the addition of this function enhanced the tool use.

11. *The area consists mainly of small shops.* All candidates thought that this was the general concept of the area, both before and after the tour. Group 2a thought that the area consisted of more than 'just shops'. This perception of the area was considered in the next question.

12. *The area consists of shops and some significant buildings.* All candidates agreed with this statement. Therefore, it is thought that all of these elements must be provided in the model.

Lastly, candidates were asked to identify what they thought were the landmark buildings in the area. That is, if only some buildings could be shown in full detail and others in outline mode only, which buildings must remain in full detail to allow you to properly navigate through the area. As well, these 'landmark' buildings, plus all other buildings as outlines, would enable candidates to build a mental image of the area. They were asked to consider that the model must provide them with sufficient information for them to be able to make informed comments about potential developments in the area.

Figure 6.21 Detail of the Temperance and General building, situated at the north-east corner of the study area

Landmark recall was best for those groups who undertook the tour, and worst for the group with little local knowledge and no tour. Candidates indicated that all landmark buildings must remain in the model.

6.7.4 Heritage model

Initial impressions of the model indicate that it works effectively using the combination of Microsoft's Internet Explorer browser and the The BitNet Management VRML browser plug-in. Whilst odd initially, the black and white model allows the city buildings to be adequately visualized. Buildings still standing are easily recognized and their 'rebuilt' neighbours provide information that was hitherto unavailable in a composite model. As can be seen in Figure 6.21, some of the building images are still to be sourced. As they are found, they will be processed and subsequently added to the model. However, the final verdict about whether the model 'works' needs to be received after the product is evaluated. This will ask users two questions – stage 1 of the evaluation process will ask 'is it usable?' and stage 2 will ask 'Is it useful?'

6.8 Further development work

Work continues on this project and a three-dimensional map shop VRML interface (Cartwright, 2006b), providing links to the surrogate travel media. In the completed prototype the information provided to the user was designed to contain the same information as that in the *Townsville GeoKnowledge* product, but delivered via a VRML-built interface.

(a)

(b)

Figure 6.22 (a) Access to information via a virtual *map shop*; (b) map drawers provide links to additional Web-delivered information

Users are able to 'walk' around the shop and discover information in a serendipitous way. Information is provided via an interactive central pillar, in and on top of map drawers and as interactive panoramas on the walls. The interface is shown in Figure 6.22(a), with detail provided in Figure 6.22(b).

6.9 Conclusion

This chapter has outlined the underpinning concepts of research into the provision of a 'different' interactive multimedia package for the exploration of geographical space. It has provided a general overview of surrogate travel and provided examples of inspirational projects – the *Domesday* project and the *Aspen Movie map*. It then described the *Queenscliff Video Atlas*, the *GeoExploratorium* and the *Townsville GeoKnowledge* product and provided examples of the surrogate walk elements of each package, followed by an outline of the results from an evaluation of the *Townsville GeoKnowledge* product based upon Bloom's *Taxonomy of Educational Objectives* (1956).

The use of surrogate travel products can provide a means for allowing users to experience how a city operates by 'walking' through an interpreted or represented world. If properly designed and presented, users can travel through a virtual space in a serendipitous manner, and explore images of a city in a natural manner. They can turn, move and generally navigate through the presented image set in a fairly natural manner. The use of surrogate travel multimedia packages provides a powerful means for appreciating environments through virtual exploration.

Acknowledgements

The concept and design of the *Townsville GeoKnowledge* project was by William Cartwright. Data collection, HTML programming, *Dreamweaver*, *Flash* and *Swish* were by Susanne Sittig and Andrea Lidl (exchange students from the University of Applied Sciences, Munich,

Germany). Extra HTML programming was provided by Jess Ngo. *Photoshop*, *Dreamweaver* and *PixAround* work was by Susanne Liebschen.

An extension to this project is being developed with contributions for image capture and panorama generation by staff and students from the Department of Geography and Regional Research, the University Of Vienna, Austria. A group from the University of Vienna travelled to Australia in 2004 to work on the project as part of an overseas exercise that forms part of their academic programme.

References

Bloom, B. S. (1956) *Taxonomy of Educational Objectives: The Classification of Educational Goals. Handbook 1: The Cognitive Domain*. New York, David McKay.

Cartwright, W. E. (1987) Paper maps to temporal map images: user acceptance and user access, *Proceedings Information Futures: Tomorrow TODAY Conference, Victorian Association for Library Automation*, Melbourne, pp. 68–87.

Cartwright, W. E. (1997) The application of a new metaphor set to depict geographic information and associations, *Proceedings of the 18th International Cartographic Conference*. Stockholm, International Cartographic Association, June, pp. 654–662.

Cartwright, W. E. (1998) The development and evaluation of a web-based 'geoexploratorium' for the exploration and discovery of geographical information, *Proceedings of Mapping Sciences '98*. Fremantle, MSIA, pp. 219–228.

Cartwright, W. E. (2006a) Using 3D models for visualizing 'the city as it might be', *Proceedings ISPRS TC II Symposium*. Vienna, ISPRS TC II, July 2005.

Cartwright, W. E. (2006b) Exploring the use of a virtual map shop as an interface for accessing geographical information. In *Geographic Hypermedia*, Stefanakis, E., Peterson, M. P., Armenakis, C. and Delis, V. (eds). Lecture Notes in Geoinformation and Cartography Series. Berlin, Springer, pp. 73–95.

Cartwright, W. E., Williams, B. and Pettit, C. (2003) Geographical visualization using rich media and enhanced GIS, *Proceedings of MODSIM Conference*, Townsville, July.

Cartwright, W. E., Pettit, C., Nelson, A. and Berry, M. (2004) Building community collaborative decision-making tools based on the concepts of naive geography, *GIScience 2004*. Washington, DC, Association of American Geographers.

Darken, R. P. and Sibert, J. L. (1996) Wayfinding strategies and behaviours in large virtual worlds, *CHI96 Electronic Proceedings*, www.acm.org/sigchi/chi96/proceedings/papers/Darken/Rpd_txt.htm (accessed 1 November 2000 and 3 December 2007).

Goddard, J. B. and Armstrong, P. (1986) The 1986 Domesday Project. *Transactions of the Institute of British Geographers*, 11(3): 290–295.

Laurini, R. (2001) *Information Systems for Urban Planning: A Hypermedia Co-operative Approach*. London, Taylor & Francis.

Lippman, A. (1980) Movie-maps: an application of the optical videodisc to computer graphics, *Proceedings of the 7th Annual Conference on Computer Graphics and Interactive Techniques*, Seattle, WA, pp. 32–42; http://delivery.acm.org/10.1145/810000/807465/p32-lippman.pdf?key1=807465&key2=7516214411&coll=portal&dl=ACM&CFID=68710763&CFTOKEN=19206607

Mediamatic, www.mediamatic.nl/magazine/8_2/Huhtamo-Armchair-E2.html (accessed June 2004).

Naimark, M. (1991) Elements of realspace imaging: a proposed taxonomy. *SPIE/SPSE Electronic Imaging Proceedings*, Vol. 1457, San Jose, CA.

Negroponte, N. (1995) Affordable computing, *Wired* July: 192.

Openshaw, S. and Mounsey, H. (1986) Geographic information systems and the BBC's Domesday interactive videodisk, *Proceedings Auto Carto*, London, Vol. 2, pp. 539–546.

Virizi, R. A. (1992) Refining the test phase of usability evaluation: how many subjects is enough? *Human Factors* 34(4): 457–468.

Wakefield, D. V. (1998) Address to the Governor's Teaching Fellows, Athens Georgia, 19 November 1998; `www.lgc.edu/academic/educatn/Blooms/criticalthinking.htm#Applying% 20Bloom's%20Taxonomy` (accessed 26 May 2003).

Wikipedia (2006) Aspen Movie Map; http://en.wikipedia.org/wiki/Aspen_Movie_Map, accessed 4 April 2006.

7

Visualization with High-resolution Aerial Photography in Planning-related Property Research

Scott Orford

School of City and Regional Planning, Cardiff University

7.1 Introduction

The visualization of the built environment in geography, planning and architecture has generally been concerned with visualizing built form. Traditionally this has been related to CAD/CAM (computer aided design/computer aided manufacturing) systems and GIS, but recently there has been significant growth in the use of scientific visualization such as multimedia, animation, virtual reality and web-based technologies (Orford, 2005). However, an older technique for visualizing the built environment has been through photography and, in particular, aerial photography. Although well established, their use has been marginalized in an era of computer graphics and computer simulations, despite the fact that they can offer important insights into a city's function and form, particularly over time. This is changing with the advent of high-resolution aerial photographs of many of the world's key urban areas becoming easier to access from both commercial and non-commercial organizations.

Aerial photographs are among the most important, widely available and commonly used kinds of remotely sensed images. Remote sensing is the identification or survey of objects from a distance using instruments aboard aircraft and spacecraft to collect information. The photographic camera remains the oldest and most common of such devices. One of the

Geographic Visualization Edited by Martin Dodge, Mary McDerby and Martin Turner

principal uses of aerial photographs has been to construct and update topographic maps and produce spatial measurements such as ground distances and terrain elevations. In town and country planning (such as in the UK), they have also been used in a variety of natural resource and appraisal surveys, typically providing a contextual backdrop to development plans or for monitoring land-use change over time. Table 7.1 provides a summary of some of the traditional uses of aerial photography in planning, based on early work by Stone (1964) and supplemented by more recent examples. The fact that Stone's work still provides a good summary of the uses of aerial photography in planning after 40 years indicates the lack of development in this area. For instance, Chamberlain (1992) provides a summary of the principal uses of aerial photography by Hertfordshire County Council, UK in the 1990s and all these relate directly to the different uses in Table 7.1.

Part of the reason for the lack of development in the uses of aerial photography in planning is a reflection of the cost of obtaining good-quality, up-to-date images. This situation is changing rapidly, however, with the advent of new and improved sources of digital and up-to-date aerial photographs from companies such as UK Perspectives, Getmapping and Cities Revealed in the UK, Pictometry International Corp. in the United States, and more recently from internet portals such as Google Earth. These are providing new opportunities for the use of aerial photography in planning teaching and research and also in planning practice as well as allowing the general public access to high-quality images. This chapter begins with a brief introduction to the history of aerial photography and an overview of the technical and conceptual literature. It gives a brief summary of some of the new sources of aerial photography, such as Google Earth and Microsoft Virtual Earth. The chapter then discusses some of the uses, traditional and emerging, of aerial photography in planning-related property research. First, the use of aerial photography in the construction and maintenance of property databases from basic building enumeration to property attribute extraction is examined. Second, the potential use of aerial photography in computer-assisted mass appraisal (CAMA) is discussed with reference to the British Valuation Office Agency and local taxation. Third, the use of aerial photography in visualizing residential densities and its use in increasing residential capacity is explored. Lastly, emerging debates about the use of high-resolution aerial photographs in extracting information on people and property in order to inform planning is discussed with reference to the effects of increasing surveillance on society.

7.1.1 A brief history of aerial photography and planning

The history of aerial photography is closely related to the history of manned flight and twentieth-century conflict (Paine and Kiser, 2003). The first known aerial photographs were taken in 1858 by photographer Gaspard-Félix Tournachon from a balloon outside Paris. The first aerial photograph that is still in existence was taken over Boston in 1860 by James Wallace Black, also from a balloon. Aerial photographs were also taken using kites from 1882 onwards with George Lawrence, an early pioneer of this technique, using it to photograph San Francisco shortly after the 1906 earthquake. The first aerial photographs taken from an aeroplane were taken near Le Mans, France in 1908, and this quickly became the dominant method of aerial photography, being substantially developed during the First World War (Lo, 1976). The Second World War saw rapid development in the technology of aerial photography and after the war many countries started a programme of collating aerial

Table 7.1 A guide to the interpretation and analysis of aerial photography in planning (Adapted from Stone, 1964 courtesy of Blackwell Publishing)

	Examples of uses			
Main applications	Identification	Characterization	Measurement	Planning
Transportation	Elements of transportation present	Characteristics of routes; types of vehicles	Congestion	Transportation routes and traffic flows
Coastal configuration	Depths alongshore; distribution of beaches	Beach material; shoreline	Shape, configuration of shoreline	Coastal erosion; flood management
Drainage	Artificial drainage; underground drainage	Flood risk areas	General direction of flow	Flood management
Surficial configuration	Drainage basins; relative relief; areas of erosion/deposition	First- and second-order drainage divides	Slopes	
Surficial geology and underlying structures	Areas of rock outcrop; fluvial formations; glacial landforms; extractions (quarries/mines, etc.)	Major rock types		Mineral sites monitoring
Natural vegetation	Vegetation types	Areas based on vegetation	Estimate density; chart land-use change; hedgerow loss; tree loss	Protection of habitats
Rural land-use – agriculture	Areas in agricultural production; outline field and farm ownership; local farming practices	Subsistence and commercial areas	Crop species; cultivation patterns; chart land-use change	Farm subsidy accountability; protection of habitats
Rural land-use – non-agricultural	Rural settlement patterns	Recreational land-uses; military uses; semi-industrial offensive/defensive activity	Land-use change	Protection of habitats

(*Continued*)

Table 7.1 A guide to the interpretation and analysis of aerial photography in planning (Adapted from Stone, 1964 courtesy of Blackwell Publishing) (*Continued*)

	Examples of uses			
Main applications	Identification	Characterization	Measurement	Planning
Military installation and effects				
Urban features	Outline built-up areas; urban encroachment	Characterize neighbourhoods; land-uses (e.g. residential, commercial, industrial)	Street patterns; residential densities; dwelling characteristics; urban growth/land-use change	Protection of green belts; regeneration/urban renewal
Industry	Heavy/light industry; transportation routes; storage facilities; heat and power units			
Archaeology	Undiscovered sites; existing sites	Sites; historical periods	Shape, distribution; density	Manage and conserve existing sites
Forestry management	Clearance patterns; crown extent	Broadleaf/needle trees	Shapes of stands; densities and heights of stands	Harvesting/ replanting
Disaster management	Area at risk of fires, flooding and storms	Low, medium, high risk areas	Response times	Terrorism; earthquakes; general emergency evacuation procedures

photographs to help support massive post-war reconstruction programmes. The beginning of the space age in the late 1950s saw the development of aerial photography shifting from aeroplanes to satellite platforms with commercial satellite imagery available from the 1970s onwards.

With the wholesale emergence of GIS in the 1980s, commercial mapping companies in the UK and the United States invested considerable resources into developing digital aerial photograph databases that could be used to extract map detail (Jones, 2003). By 1995, the first commercial imagery databases in the UK were being compiled and published by The GeoInformation Group under the product label 'Cities Revealed', initially released for central London but then for other urban areas as well. Britain's national mapping agency, the Ordnance Survey (OS), also began a systematic programme of aerial photographic surveys in 2002 as part of the development of their digital framework product OS Mastermap. In addition, many local authorities in Britain, the United States and elsewhere have commissioned aerial photographic surveys on a regular basis (Cassettari, 2004), although access to these resources is often difficult and controlled (Chamberlain, 1992). Finally, in 2005 aerial photography experienced a worldwide resurgence in interest with the advent of Google Earth and other similar internet portals providing high-resolution images free of charge.

7.1.2 Aerial photography: a technical background

In order to understand and appreciate some of the debates on the uses of aerial photography for visualization in planning and property research, it is first necessary to have a basic comprehension of some its technical aspects. This section will only outline some of the basic concepts and the reader is recommended to the work by Lo (1976), Konecny (2003) and Paine and Kiser (2003). Aerial photographs can be categorized into two main types: vertical and oblique. In a 'true' vertical photograph the axis of the camera is orthogonal to the ground at the time of exposure. In an 'unintentionally tilted' vertical photograph the axis is no more than 3° from the vertical. Most vertical aerial photographs fall into this category. In an oblique photograph the camera's axis is tilted between 30 and 60° from the vertical. If the horizon is visible, the photograph is a high oblique; if it is not visible then it is a low oblique.

When taking vertical aerial photographs, flight paths are arranged in parallel strips called swaths that have an overlap of 20–30 per cent. Along the flight path there is usually a forward overlap of 60 per cent between photographs. The overlapping swaths permit the photographs to be joined together and also allow stereomodels (three-dimensional models) to be viewed using a stereoscope (Lo, 1976). The aircraft flies a straight-line course at constant speed and at a pre-determined height – the height determining the eventual scale of the photograph. As a rule flights are only made in clear, cloudless skies when sun's altitude is higher than 30° but lower than 60°, preferably during the middle of the day when the quality of sunlight is better. Winter is preferred for topographical surveys as there is less foliage (called 'leaf off' aerial photography), although this can cause problems in areas prone to prolonged snow-cover as this can hide important features as well as winter often being associated with poor weather conditions and extensive cloud cover.

It is common for vertical aerial photographs to be geo-referenced to a conventional grid-referencing system (either manually using control points of automatically by GPS) and transformed into an orthophotograph or orthophoto. An orthophoto is a geo-referenced aerial photograph that has been geometrically corrected using a process called orthorectification, such that the scale of the aerial photograph is uniform and the photograph is planimetrically correct. This means correcting the photograph for inaccuracies caused by distortion and displacement (Paine and Kiser, 2003). Distortion is any shift in the position of a feature on a photograph that alters the perspective characteristics of the feature. Common causes of distortion include lens distortion and aircraft movement. Displacement is any shift in the position of a feature on a photograph that does not alter the perspective characteristics of the photograph. Common causes of displacement include camera tilt and the effect of topographic relief (Lo, 1976). The latter can affect the scale across the photograph since features on higher elevations will appear bigger on photographs than the same features on lower elevations. This displacement needs to be corrected to ensure that the scale of the photograph is uniform, although this correction can produce visible radial displacements of buildings, trees, bridges and other structures on the orthophoto, particular of large-scale urban areas. This can cause mis-match problems if the orthophoto is used in a GIS and overlaid with other feature data, such as digital building outlines.

Orthophotos have typically been used for photogrammetry and photo-interpretation. Photogrammetry is the art and science of obtaining reliable quantitative measurements from photographs such as distances, angles, areas, volumes, elevations, slopes and the sizes and shapes of objects (Konecny, 2003). Photo-interpretation is the determination of the nature of objects on a photograph and the judgement of their significance. It requires an elementary knowledge of photogrammetry since, for example, the size of an object is frequently an important consideration in its identification. Similarly, photogrammetry is always closely related to photo-interpretation since there is always a need to know what one is measuring (Lo, 1976). Nonetheless, work on aerial photographs is no substitute for field work when interpreting features on the photograph. The end result of both photogrammetry and photo-interpretation is frequently a thematic map (Paine and Kiser, 2003).

7.1.3 Google Earth and other internet portals

Google Earth was launched in June 2005 and sources hundreds of thousands of individual satellite images and aerial photographs from over 100 sources. Two principal sources of satellite images have been from the Landsat programme and Digital Globe's Quick-Bird programme (Nourbakhsh, 2006). The photographs have been mosaiced together and wrapped around a three-dimensional virtual globe that can be downloaded from the internet (http://earth.google.com). The technology was developed by Keyhole, a digital mapping company that was purchased by Google in October 2004. To many commentators (e.g. Goodchild, Chapter 2) Google Earth represents the first incarnation of 'Digital Earth' conceptualized in a speech by Al Gore (Gore, 1998). Virtual tours can be made with users zooming in from space towards a desired destination. Users can zoom in either by scrolling with the mouse or by typing in a place name or postcode into a search box and the view can be tilted and rotated as required. At the time of writing Google Earth also provides additional data layers including three-dimensional buildings in 38 US cities, and road networks in the US, Canada and western Europe which can be overlain on top of the images. Users can add their own data to Google Earth using KML (Keyhole Markup Language) an XML-based language that specifies a set of features (placemarks, images, polygons, three-dimensional models, textual descriptions, etc.) for display in Google Earth.

The resolution and currency of the photographic images varies across the Google Earth's surface with key urban areas in Western countries often having a resolution as high as 15–30 cm with rural areas and less developed countries typically having a resolution of 15 m or less (Nourbakhsh, 2006). The average refresh cycle of the images is around 18 months, although this is shorter in some key urban areas. Indeed, because different images are mosaiced together from a variety of sources, a single city or region may have consecutive images taken from different months which can be noticeable if adjacent images show different weather conditions. Not all cities are depicted in high resolution and Google's database contains more imagery of the United States than elsewhere (Trimbath, 2006). This US bias is evident in place name searches and the default setting to US customary units of measurement rather than standard international units. There are also some problems with the accuracy of photogrammetric measurements obtained using Google Earth's tools (Goodchild, Chapter 2).

In July 2005 MSN launched MSN Virtual Earth (also known as Microsoft Local Live), which is similar to Google Earth but does not yet cover the whole planet (Biever, 2005). MSN Virtual Earth also includes oblique photographs, called the 'birds-eye' view, of several US and UK cities. Figure 7.1 is a comparison of Google Earth and MSN Virtual Earth for the same area – a rural landscape a few miles southwest of Bristol, UK. The resolution for this area in MSN Virtual Earth is very high, allowing a clear view of a farm and out-buildings. The corresponding photograph from Google Earth shows a much lower resolution image with the farm barely visible in the mosaic of fields. The difference in the resolution of the images is a good example of the inconsistency in aerial photography provision by internet portals, with different portals sourcing images for the same location from different companies. This raises reliability and data quality issues when using images from internet portals for research and decision-making in planning and also poses potential problems surrounding data standards and ownership.

The added value of the oblique photographs is demonstrated in Figure 7.2, which shows a property in Bristol orthogonally and obliquely from the four cardinal compass directions. Photo-interpretation is often easier using an oblique photograph because the profile view

Figure 7.1 A comparison of differences in resolution of images between MSN Virtual Earth (left) and Google Earth (right): a farm 5 miles to the southwest of Bristol, UK, October 2006 (source: author screenshot from http://earth.google.com and http://maps.live.com)

Figure 7.2 Vertical and oblique aerial photography available from MSN Virtual Earth, October 2006 (source: author screenshot from http://maps.live.com/)

may make some objects more recognizable. In this instance, the number of stories of the building is revealed and this can be important in some planning situations such as when calculating residential energy efficiency rates for individual dwellings. Also, oblique photos may make some objects visible that are obscured from a vertical perspective, such as structures hidden by a protruding roofline. Low-level oblique aerial photographs can also show the relative scale and sizes of buildings and the intricacies of urban areas and the landscape 'that the maps cannot convey, in which trees are more important features than building' (St Joseph, 1977, p. 180). However, because scale varies across the image, oblique aerial photographs cannot be used in photogrammetry.

7.2 Applications of aerial photography in planning-related property research

7.2.1 Property attributes data extraction and collection

Detailed data on the housing stock is fundamental to many local planning authority functions. However, attribute data for individual properties is difficult to collect and maintain with many authorities relying upon secondary sources. These can include: sample surveys, such as housing needs and house condition surveys; information from other sources, such as local taxation registers and land registry records; information from property professionals such as valuation officers, property surveyors and local real estate agents; or information from comprehensive national surveys such as population censuses. There are several problems common with data from these sources: the data are collected infrequently and can often be out of date; the data are based upon samples of houses rather than the total housing stock; the data may apply only to particular sectors of the housing stock (e.g. the public sector); the data are not available for individual properties but as counts for an area; the data do not include the information required by the local authority; collecting the data is often very expensive and time-consuming, even for a sample of the housing stock. In addition, even in a stable housing market properties are constantly changing. People die or move away and the property is passed on. Old properties are demolished or abandoned and new ones are built. The use of properties can change, from residential to commercial and vice versa, or they are sub-divided or amalgamated into new dwellings. All these factors mean local authorities and quite often central government departments can have poor quality information on individual properties. For instance, although the British Valuation Office Agency (VOA) holds the most extensive set of property attribute data in the UK, the vast majority of these data are held on paper records (VOA, 2002a) making them both difficult and costly to access and maintain.

Government departments are therefore looking towards new techniques for collecting property information and high-resolution aerial photography is increasingly being promoted and used as an important source of such data. In conjunction with high-resolution GIS framework data and other ancillary data sources, aerial photography can enhance existing dwelling data and provide new information (Orford and Radcliffe, 2007). Methods are being developed that can extract building information automatically from aerial photography, such as building footprints, roof types and photogrammetric measurements (Elaksher, Bethel and Mikhail, 2003). Aerial photographs can also show how the property fits into the landscape and relates to its neighbours. Crucially, these data can be collected relatively cheaply and unobtrusively as Diane Leggo, director of local taxation at the VOA, has recently commented: 'aerial photography and photogrammetry matched with geographical positioning systems [allow] significant data gathering without inspections' (cited in the *Daily Express* newspaper, 29 May 2006, p. 9).

This data gathering 'without inspections' means that aerial photography is a very effective method of gathering property information in rural and sparsely populated areas. Here, it can be often too expensive to undertaken conventional property surveys due to the distances involved in travelling between properties. Instead, high-resolution aerial photography can provide the necessary information for a fraction of the cost. Harper (1997) discusses how aerial photographs can provide insights into the nature of rural areas by revealing subtle features which are not apparent from ground level. This can include properties hidden from view that have been built without planning permission or building permits but can be identified on aerial photographs. Even in more densely populated areas, aerial photography can show features of a property that may not be apparent from a visual inspection from the road, for instance unreported improvements such as a new garage or swimming pool. Aerial photographs can also allow the visual analysis of the rear elevation of a property rather than simply the front elevation, which is characteristic for physical surveys. As will be discussed later, these uses can have important relevance in respect to local taxation issues and also issues relating to privacy.

Aerial photographs are especially useful in places where very little property data, or indeed any socio-economic data, exists such as in the developing world (Corker and Nieminen, 2001). Here, they are often used to map buildings, chart development and provide a basic enumeration of population. A topical example in this area is the recent work on charting the global growth of gated communities, a phenomenon that is occurring in many countries. Aerial photographs are good at showing the dimension and borders of gated communities

(Roitman, 2005) and analysing street patterns within them (Grant and Mittelsteadt, 2004). Aerial photographs have been used to identify gated beach resorts in Lebanon and to chart their establishment (Glasze and Meyer, 2000). In Durbin, South Africa, Durlington (2006) has used aerial photography to map changes in the landscape of gated communities from the 1970s onwards. He has shown that, although these communities are advertising themselves as eco-friendly, they essentially wreak havoc on the environment during their construction.

7.2.2 Computer-assisted mass appraisal

Computer-assisted mass appraisal (CAMA) is a world-recognized approach to mass valuation of property for tax purposes. CAMA is an umbrella term for computer software and methodologies that are used to estimate the values of thousands of properties within a particular area that are stored within a database. The mass valuations are based upon the values of a sub-sample of properties that have been inspected by a professional valuer. A statistical model (usually based upon multiple regression) which specifies the value of a property as a function of its attributes (such as dwelling type, floor area, lot size etc.) is estimated for this sub-sample. This model is then used to estimate the value of all the remaining properties in the database, based upon their attributes. This method allows government agencies, planners, real estate agents, property developers and other property experts to gain a clear understanding of how the housing stock is valued across a particular area and, if kept up-to-date, how the dynamics of the housing market are varying through time. It is used in many countries as a method of maintaining and up-dating local taxation records.

Ensuring that there is 100 per cent coverage requires both a comprehensive map and database of tax-payers and their properties. Owing to the lack of accurate cadastral maps in many countries (Bird and Slack, 2004), aerial photographs in conjunction with GIS are often used to provide digital base maps and, as discussed above, the property attributes. For example, in Northern Ireland, a new local taxation based upon property values is set to be introduced in April 2007. This is the called the council tax, which has been in operation in the rest of the UK since the early 1990s. Over 700 000 homes across Northern Ireland were valued using CAMA, the first time this method has been used in the UK. Data came from several sources, including photogrammetric measurements from aerial photographs which were also used to update domestic property attribute data at the same time. The photogrammetric measurements were particularly useful in cases where other sources of attribute information were missing. It also allowed attribute data from other data sources to be checked and verified. The British government is currently investigating if the council tax in England could be re-valued using similar methods (VOA, 2002b).

7.2.3 Visualizing, measuring and re-developing residential densities

High-resolution aerial photography is a good tool to visualize urban sprawl and also to investigate and examine issues relating to housing densities (e.g. Burchfield*et al.*, 2006). In particular it can be an effective tool for allaying public fears of increasing residential densities, an issue of increasing importance in Western countries such as the United States and the UK, where planners and designers are promoting 'smart growth' and are advocating a return to

Figure 7.3 Examples of aerial photography in the 'Visualizing Density' catalogue. Densities shown: 9.2 units per acre (top left), 8.4 units per acre (top right), 7.3 units per acre (bottom left) and 5.3 units per acre (bottom right), October 2006 (source: author screenshot from www.lincolninst.edu/subcenters/VD)

high-density developments and compact site plans (Hayden and MacLean, 2000). One of the persistent obstacles to compact development is the public's aversion to density. Misplaced concerns over density often prevent the construction of urban infill projects or the revision of zoning regulations that would allow for compact growth. Campoli, Humstone and MacLean (2003) argue that, while there is published work addressing the topic of density measures that often state the desirability and benefits of density, they do not specifically address the problem of measured vs perceived density. Low-altitude oblique aerial photographs, however, are especially useful for the visualization of the character of urban areas and have the potential to contrast images of differing residential densities (Figure 7.3).

In the *Visualizing Density* catalogue complied by Campoli and Maclean (2005), a collection of over 300 aerial photographs illustrate more than 80 neighborhoods in locations across the United States. The images are arranged by density level, measured as dwelling units per acre. The catalogue is arranged on a continuum from low density (less than one unit per acre) to high density (134 units per acre). It features a broad array of housing in many different configurations, demonstrating that living closer together can take many forms. By illustrating the physical form of density, Campoli and Maclean argue that the density problem can be viewed as a design problem, shifting the publics' and planners' concerns away from density numbers and toward appropriate design approaches. The purpose is to increase the readers' familiarity with density numbers as they relate to neighborhood form, and to enable viewers to visualize different design approaches to achieve density.

Figure 7.4 An example of aerial photography used for identifying suburban in-fill development, Brinsley, UK, October 2006 (source: author screenshot from http://maps.live.com)

In addition to illustrating high residential densities, aerial photographs can be used to identify land in existing residential areas that could be developed at higher densities. This is of particular interest to UK local authorities where there has been pressure to reduce urban sprawl and increase development on brownfield sites. Residential infill development is nothing new in British towns and cities, where there is a history of replacing large properties with smaller units, usually flats, at higher densities. However, recent renewed calls for higher-density building by developers, planners and policy makers may increase this type of development, especially in light of the Planning and Compulsory Purchase Act (2004) which stipulates that gardens over 30 m metres (99 ft) long can be purchased for (appropriate) residential development. Figure 7.4 is an example of how aerial photography (together with other data such as accessibility to amenities and transport nodes) has been used by planners and developers to identify properties in suburban areas suitable for infill development. In this example from a suburban neighbourhood in Brinsley, UK, the property inside the box was acquired by a developer, together with half of the neighbouring garden. The property and the garden were then redeveloped at a higher density, with the potential to develop the backs of the remaining gardens along the street (personal communication, property owner, March 2006).

7.3 Aerial photography, property and surveillance

The development of aerial photography is strongly associated with issues of surveillance. Its technical development in the First and Second World Wars and the development of satellite remote sensing from the 1950s onwards were strongly linked to the collection of military intelligence. The advent of cheaper and accessible high-resolution aerial photographs in the past decade has only served to increase the concern over surveillance and privacy, especially in the context of the growth and integration of digital data on people and places. The emergence of Google Earth and other internet portals has caused concern in some quarters that images of sensitive military, government and industrial sites are freely available to anyone in the world (Nourbakhsh, 2006). What is more interesting is the reaction by the public and the media to the availability of these images and the apparent uses of them by government departments and the commercial sector. In particular, there is a concern that 'people might object to a picture of their house being freely available' (Biever, 2005, p. 29) and, though Google has stated in defence that 'the same information is available to anyone who drives by or flies over a piece of property' (cited in Biever, 2005, p. 29), the advent of new technologies may makes issues of privacy become more salient.

In the UK at least, the news that local authorities and the Labour Government may be using high-resolution aerial photographs for identifying areas that could accommodate an increase in residential densities and in-fill development has caused controversy. A headline in a popular national tabloid newspaper, 'Labour use a *spy in the sky* to target your back garden' (*Daily Express*, 29 May 2006, p. 9; added emphasis), is typical of the reactions. Similar headlines appeared in other national newspapers, some of which had, a few weeks earlier, been offering 'FREE giant aerial photograph of your home!' (*Daily Mail*, 4 May 2006, p. 1). Predictably, opposition political parties to the Government also shared a similar negative stance to the proposals: '[Labour] is creating a database of every garden in the country to help cram more development into suburban communities, regardless of local opinion', asserted Caroline Spelmen MP, Conservative Secretary of State for Local Government (cited in the *Daily Express*, 29 May 2006, p. 9).

The announcement that aerial photographs could be used as source of data for calculating local taxation in the future also provoked a negative reaction from politicians and social commentators: 'Northern Ireland is now being used as a testing ground, from the trial of the *Big Brother* computer database to the levying of a new house price tax, to the use of *spies in the sky* to peer down at home extensions and gardens ... using aerial photography to *invade people's privacy* and lay the ground for a new stealth tax on home improvements' (Caroline Spelmen MP, 2006; cited in *Belfast Today*, 2 May 2006, p. 4; added emphasis). Shami Chakrabarti, the director of the UK pressure group Liberty, had similar sentiments: 'It's ludicrous. It's not only a waste of resources it's a shockingly disproportionate *interference with people's privacy*' (cited in *The Times*, 1 January 2006, p. 1; added emphasis). Similar issues relating to invasions of privacy have been recorded in the United States, and again these have been linked to the use of aerial photography in property taxation and reassessment and, in particular, unrecorded property improvements (Showley, 2006).

It seems, then, at least in some quarters, that the use of high-resolution aerial photography in property planning and research is not seen as a good thing. Despite the fact that government departments routinely collect data on house sales, home improvements (through the development control/building permit planning processes) and housing developments, the perception that high-resolution images may be used to enhance these data has caused a disproportionate amount of controversy. This is partly because aerial photographs appear to present an accurate representation of the world and this makes them seem more intrusive than other types of (abstract) data on people and property. In this way, internet portals such as Google Earth may be regarded as a 'mirror world' (Gelernter, 1991), where computer screens become a mirror of the real one and users gain a whole picture of the world through systematic 'zooming in and poking around' (p. 15). Hence the debate is not necessarily one directly concerned with increasing tax or increasing density but instead is deeply rooted in issues of privacy and surveillance (e.g. Curry, 1997; Lyon, 2001, 2003).

Surveillance is any collection and processing of personal data for the purposes of influencing and managing people and society (Lyon, 2001). Surveillance is intrinsic to modern society. It is linked to the growth in computing power, the weakening of face-to-face social relations (such as communications and transactions) and the need to minimize risk and anticipate future situations to allow society and the economy to function efficiently. One result of surveillance is 'simulation' in which individuals (and places) are modelled and profiled using the fragments of information that have been collected by various public and private sector agencies. It is these digital simulations that public and private sector agencies use to represent, organise and manage the real world (Lyon, 2003). In this respect, the concept of the 'mirror world' is somewhat misleading as what appears on the computer screen tends to be simulation of reality rather than a reflection. A similar argument can be levelled at aerial photographs since, as these images have typically been transformed and manipulated, they can suffer from various effects of displacement and distortion and are often stitched together from a variety of sources. They can be difficult to interpret, even by an expert, whilst the images on Google Earth and other internet portals vary in their age and their resolution. Thus even high-resolution aerial photographs do not show a true picture of reality but just another digital representation. The result is that the impact of surveillance on society tends not to be as clear and deterministic as 'mirror world' would have us believe, but instead is much more uncertain and unpredictable (Lyon, 2001).

7.3.1 Surveillance, social sorting and the housing market

One uncertain and unpredictable outcome of surveillance, and particularly simulation, is the way in which it sorts individuals and how this sorting can manifest itself within the city. Lyon (2001, p. 1) argues that surveillance 'sorts people into categories, assigning worth and risk, in ways that have real effects on their life chances'. One way that this sorting and categorization can occur is through the mechanisms of the housing market in which a socially sorted population emerges within spatially and structurally differentiated housing sub-markets (Orford, 1999). Traditionally, buyers and renters of property were restricted in their choice by high costs of the search process – that is the gathering of information on properties on the market for sale or rent. Information on potential properties was filtered by a tacit knowledge of the market, by real estate agents and the local media and, together with the constraint of having to physically visit a property in order to assess its suitability, reduced the number of properties in the choice set. To a degree this reduction in choice had the effect of sorting people into areas which in turn helped create housing sub-markets. However, the recent growth in publicly available information on people and property could be changing these traditional housing market dynamics.

Key to this is the evolution of internet-based neighbourhood information systems (IBNIS) that provide information and construct images of different neighbourhoods within a city (Burrows, Ellison and Woods, 2005). These images are important as they can influence the lives of local residents and the attitudes and behaviour of others. In particular, it can affect decisions of whether to move to or from a particular area. Whereas images and perceptions of neighbourhoods were traditionally constructed from local sources – local residents, newspapers, real estate agents – new technology means that the way in which they are now constructed, disseminated and consumed has undergone a revolution. Neighbourhood images on IBNIS are a simulation of reality rather than a true reflection. The information used

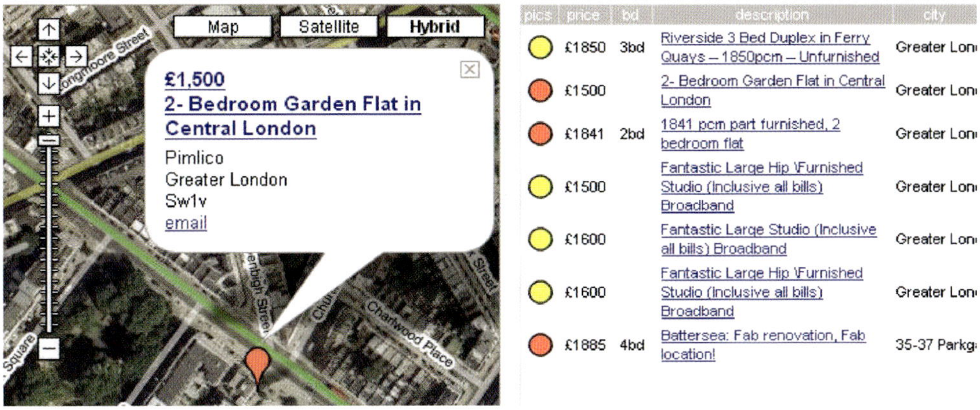

Figure 7.5 An example of an aerial photograph linked to a property listings website, October 2006 (source: author screenshot from www.housingmaps.com)

to construct these images tends to include data on property prices, geodemographic profiles, crime rates, local taxation rates, accessibility measures to local amenities and services and the educational standard of local schools. These statistical 'images' of neighbourhoods are becoming enhanced and contextualized by high-resolution aerial photographic images, allowing people to 'see' what the data are describing. IBNIS are often linked to real estate websites which in turn provide information on individual properties for sale or rent in the neighbourhood (Figure 7.5).

With IBNIS and real estate websites, the search costs are lowered and more properties can be considered. The information on potential properties is now filtered through search engines and internet portals and buyers and renters can make decisions as to whether to view a property after analysing on-line property details and IBNIS data including external images of the property from high-resolution aerial photographs. These photographic images can help indicate the 'status' of a street or neighbourhood by revealing makes of car parked in driveways, the presence of swimming pools in back gardens, the density of development, the amount of greenery/vegetation in the street or neighbourhood, the general quality of the street environment, the indication of children living in the immediate neighbourhood (e.g. by play equipment in the garden, which may be deter some people, such as elderly buyers), the quality of neighbouring houses and the state of neighbouring gardens, etc. These characteristics can be very influential for a potential buyer/renter and cannot be gained solely from statistical data provided by IBNIS or a real estate agent. The result in this lowering of search costs and the availability of information on property and neighbourhood may mean that members of the public will be motivated 'to sort themselves out' (Burrows, Ellison and Woods, 2005, p. 36) within the housing market to a greater degree than has happened before.

7.4 Conclusion

The chapter has provided a broad overview of some of the visualization applications of aerial photography in planning with reference to property-related research. Although it is an old tool, aerial photography is experiencing a resurgence of interest in planning and beyond, principally due to an increasing availability of high-resolution images. This is partly due to falling costs and increasing competition from commercial companies using aircraft as platforms for collecting aerial photographs and also from the growth in high-resolution images from satellite platforms. The most important development, however, has been the release of Google Earth in 2005, which provides free public access to worldwide images, some at very high resolutions. The immediate impact of Google Earth on the internet community and beyond has shown that many people and organizations are interested in using aerial photography either in their own work, or as a source of information to help in making personal decisions. In terms of planning and property research, Google Earth has certain flaws that can limit its usefulness. The photographs come from different sources and are of different ages and hence Google Earth does not provide a snap-shot of an area at a particular point in time. It varies in its resolution and this is especially significant in rural and sparsely populated regions in which the resolution is often too low for meaningful visual analysis. There are also problems with regard to using Google Earth in constructing photogrammetric measurements.

This is not to say the Google Earth has not had an impact on the planning community. It is arguable that it has encouraged planners to reconsider the utility of aerial photography in their own area of work, with commercial companies providing the images (Jones, 2003). Google Earth is also an important source of information in parts of the world where access to cartographic and socio-economic information is limited. In terms of property research, there would appear to be a growth in the uses of aerial photographs, particularly in terms of property attribute collection, property valuation and local taxation. Whilst aerial photographs are currently enhancing existing data sets and data collection methods, the growth in computing power and automatic photogrammetric techniques points the way to a more comprehensive and systematic approach to constructing property databases. These developments have not gone unnoticed and there is a general negative feeling, in the media at least, against local and national governments using high-resolution aerial photographs in collecting information on property. Indeed, their use has been couched in terms of spying and surveillance, whereas other more comprehensive and exhausting social surveys, such as the Census of Population, tend not to be. The personal use of aerial photographs has not been seen in a bad light and in a lot of cases has been received positively. This is despite the fact that the impact of internet portals such as Google Earth on how people view the world and make decisions based on this view is poorly understood. As a GIS commentator in the US stated: '[w]e are starting to rely on Google and Microsoft for our view of the world' (cited in Biever, 2005, p. 29). When integrated with other data on the internet, aerial photographs have the potential to present images of places that are digital simulations rather than true reflections and this can have important implications when people, planners and policy makers use these images to help make decisions.

References

Biever, C. (2005) Will Google Earth help save the planet? *New Scientist* 187: 29.

Bird, R. M. and Slack, E. (eds) (2004) *International Handbook of Land and Property Taxation.* Cheltenham, Edward Elgar.

Burchfield, M., Overman, H. G., Puga, D. and Turner, M. A. (2006) Causes of sprawl: a portrait from space. *Quarterly Journal of Economics* 121(2): 587–633.

Burrows, R., Ellison, N. and Woods, B. (2005) *Neighbourhoods on the Net: The Nature and Impact of Internet-Based Neighbourhood Information Systems.* Bristol, The Policy Press.

Campoli, J. and Maclean, A. (2005) *Visualizing Density.* Cambridge, MA: Lincoln Institute of Land and Policy. Available atL www.lincolninst.edu/subcenters/VD/

Campoli, J., Humstone, E. and Maclean, A. (2003) *Above and Beyond: Visualizing Change in Small Towns and Rural Areas.* Chicago, IL, Planners Press.

Cassettari, S. (2004) Photo mapping of Great Britain: a growing opportunity? *The Cartographic Journal* 41: 95–100.

Chamberlain, K. (1992) Aerial surveys for Hertfordshire County Council. *Photogrammetric Record* 14(80): 201–205.

Corker, I. and Nieminen, J. (2001) Improving municipal cash flow – systematic land information management. *International Conference on Spatial Information for Sustainable Development,* Nairobi, Kenya, 2–5 October. Available at: www.fig.net/pub/proceedings/nairobi/corker-nieminen-TS6-1.pdf

Curry, M. R. (1997) The digital individual and the private realm. *Annals of the Association of American Geographers* 87(4): 681–699.

Durlington, M. (2006) Race, space and place in suburban Durban: an ethnographic assessment of gated community environments and residents. *GeoJournal* 66: 147–160.

Elaksher, A. F., Bethel, J. S. and Mikhail, E. M. (2003) Roof boundary extraction using multiple images. *Photogrammetric Record* 18: 27–40.

Gelernter, D. (1991) *Mirror Worlds, or the Day Software puts the Universe in a Shoebox.* Oxford, Oxford University Press.

Glasze, G. and Meyer, G. (2000) Workshop gated communities – global expansion of a new kind of settlement. *DAVO-Nachrichten* 11: 17–20.

Gore, A. 1998. *The Digital Earth: Understanding our Planet in the 21st Century.* California Science Center, Los Angeles, California, on 31 January 1998. Available at: www.digitalearth.gov/VP19980131.html

Grant, J. and Mittelsteadt, L. (2004) Types of gated communities. *Environment and Planning B: Planning and Design* 31(6): 913–930.

Harper, D. (1997) Visualizing structure: reading surfaces of social life. *Qualitative Sociology* 20(1): 57–77.

Hayden, D. and MacLean, A. (2000) *Aerial Photography on the Web: A New Tool for Community Debates in Land Use.* Cambridge, MA: Lincoln Institute of Land and Policy.

Jones, A. (2003) Aerial photography for mapping the UK. *The Cartographic Journal* 40(2): 135–140.

Konecny, G. (2003) *Geoinformation: Remote Sensing, Photogrammetry and Geographic Information Systems.* London: Taylor and Francis.

Lyon, D. (2001) *Surveillance Society: Monitoring Everyday Life.* Buckingham, Open University Press.

Lyon, D. (ed.) (2003) *Surveillance as Social Sorting: Privacy, Risk and Digital Discrimination.* London, Routledge.

Lo, C.-P. (1976) *Geographic Applications of Aerial Photography.* New York, Crane, Russak and Co.

Nourbakhsh, I. (2006) Mapping disaster zones. *Nature* 439: 787–788.

Orford, S. (1999) *Valuing the Built Environment: GIS and House Price Analysis.* Aldershot, Ashgate.

Orford, S. (2005) Cartography and visualization. In *Questioning Geography,* Castree, N., Rogers, A. and Sherman, D. (eds). Oxford, Blackwell, pp. 189–205.

Orford, S. and Radcliffe, J. (2007) Modelling UK residential dwelling types using OS Mastermap data: a comparison to the 2001 census. *Computers, Environment and Urban Systems* 31: 206–227.

Paine, D. and Kiser, J. D. (2003) *Aerial Photography and Image Interpretation*, 2nd edn. Hoboken, NJ: Wiley.

Roitman, S. (2005) Who segregates whom? The analysis of a gated community in Mendoza, Argentina. *Housing Studies* 20(2): 303–321.

Showley, R. M. (2006) Prying eyes? Computerized aerial photos could ease county assessor's job, but privacy issues loom. *The San Diego Union-Tribune* 8 October. Available at: www.signonsandiego.com/uniontrib/20061008/news_mz1h08aerial.html

St Joseph, J. K. S. (1977) *The Uses of Air Photography*. London: J. Baker.

Stone, K. H. (1964) A guide to the interpretation and analysis of aerial photos. *Annals of the Association of American Geographers* 54(3): 318–328.

Trimbath, K. (2006) Google mapping software gives engineers the earth. *Civil Engineering* 76(1): 35.

VOA (2002a) *Data Quantity/Quality Audit – Property Attributes Data*, CTR(E) IA 11120. Available at: www.voa.gov.uk/publications/CouncilTaxIas/021211-ctre-ia.htm

VOA (2002b) Council Tax news – Council Tax reveal 2007 – VOA considers the biggest computer assisted revaluation in the world. Available at: www.voa.gov.uk/publications/CouncilTaxIas/Documents/ctre-reval-news.htm

8

Towards High-resolution Self-organizing Maps of Geographic Features

André Skupin and Aude Esperbé

Department of Geography, San Diego State University

This chapter introduces the use of high-resolution self-organizing maps (SOM) to represent a large number of geographic features on the basis of their attributes. Until now, the SOM method has been applied to geographic data for both clustering and visualization purposes. However, the granularity of the resulting attribute space representations has been far below the resolution at which geographic space is typically represented. We propose to construct SOMs consisting of several hundred thousand neurons, trained with attributes of an equally large number of geographic features, and finally visualized in standard GIS software. This is demonstrated for a data set consisting of climate attributes attached to 200 000+ US census block groups. Further, overlays of point, line and area features onto such a high-resolution SOM are shown.

8.1 Introduction

This volume demonstrates the range of approaches currently pursued in the field of geographic visualization. Geographic visualization has clearly captured the public's imagination. Evolutionary changes in the creation, distribution and interaction with cartographic depictions have powerfully converged in early realizations of the *digital earth* concept (see Chapter 2). Further convergence of various technologies and methodologies is likely, including trends towards high-resolution imagery (see Chapter 7 by Orford) and locations captured

Geographic Visualization Edited by Martin Dodge, Mary McDerby and Martin Turner

using GPS or cell phones (see Chapters 15 and 16). One example of such convergence is the extension of space–time paths to n-dimensional attribute space (Skupin, 2007).

Geographic visualization does not merely represent the reemergence of cartography, to which GIS was once thought to have delivered a deadly blow. Instead, the confluence of various academic and market forces towards formation of such new disciplinary categories as information visualization and visual data mining and the success of products and services like MapQuest and Google Earth suggest the emergence of a *cartographic imperative*, a *need to map*, aimed at making sense of voluminous, multi-facetted data. This imperative delivers a powerful impulse to create meaning-bearing visualizations, even of non-georeferenced data and of the non-spatial elements of geographic data. In order to achieve this, one is, however, required to shed notions of cartography as being essentially about attaching symbols to geometry in order to communicate geographic reality. On a much more fundamental level, such approaches as spatialization remind us that cartography is all about transformation (Tobler, 1979) and that the impact of visualization often derives from novel combinations of transformative processes.

As reflected in this volume, geographic visualization at its core grew out of the cartographic tradition of representing geographic objects in a representational space derived through projection of locations from a curved two-dimensional space (i.e. *latitude* and *longitude*) into a planar map space (i.e. x and y). New techniques of representation have emerged, such as parallel coordinate plots (PCP) and self-organizing maps (Kohonen, 2001), and are now being linked to form increasingly powerful means for discovering interesting patterns and relationships in large, multidimensional geographic databases. However, the geographic map tends to be the element binding it all together, the one with which all other representations interact and are bound to, and to which users will ultimately refer. One major reason for this is the well-deserved recognition given to the possible effects of spatial autocorrelation – or the First Law of Geography (Tobler, 1970) – in any such geographic investigation. In fact, the effects of spatial autocorrelation, such as the appearance of spatial clusters, are the very subject of many investigations. However, apart from such effects, one might argue that an additional impetus for referring back to the geographic map is the sheer richness provided by it. Ultimately, this richness derives not simply from a choice of symbols for point, line, area and text objects – because that would apply to many non-geospace representations as well – but from the finely grained geometric base to which such symbols become attached. The inherent geometric detail or resolution provided by a geographic map tends to remain unmatched by alternative representations.

Out of these considerations the theme of this chapter then emerges, to advance the convergence of intense computation with the cartographic tradition, to spatialize an n-dimensional geographic attribute space with a resolution detailed enough to mimic geographic maps, and to apply a range of transformations towards eventual visualization. This is demonstrated through a visualization derived from climate attributes associated with more then 200 000 US census block groups.

8.2 Self-organizing maps

First introduced a quarter-century ago, the self-organizing map has become a popular method for visual modelling of complex, n-dimensional data. A number of excellent overviews of the method, edited volumes, as well as a comprehensive monograph on the subject, exist (Deboeck and Kohonen, 1998; Oja and Kaski, 1999; Kohonen, 1982, 1990, 2001), so in this chapter the method is introduced in only the briefest terms. A SOM is an artificial neural network in which neurons are arranged as a low-dimensional, typically two-dimensional, lattice such that each neuron has either four or six neighbours (i.e. square or hexagonal neighbourhood). Each neuron is associated with an n-dimensional vector of weights. Input data presented to the neuron lattice are of the same dimensionality. For example, 30 population attributes associated with 200 000 census enumeration units would correspond to a training data set consisting of 200 000 thirty-dimensional vectors.

During training, one input vector at a time is presented to all the neurons and the most similar neuron is determined, typically based on a Euclidean similarity coefficient. The n weights of that so-called best-matching unit (BMU) then get adjusted towards an even better match. More important – and essential for the self-organizing nature of a SOM – is that weights of neighbouring neurons around the BMU are likewise adjusted, up to a certain neighbourhood size and with a diminishing magnitude best described as distance decay. Over the course of many such training runs, the low-dimensional lattice of neuron vectors begins to replicate major topological structures existing in the n-dimensional input space.

A trained two-dimensional SOM can itself be visualized in various forms, including the display of weights for a particular variable as colour shading across the neuron lattice. This is also known as component plane display and an example is included later in this chapter. One could also opt for a display based on multi-dimensional computation, such as clustering of neuron vectors using hierarchical or k-means clustering (Skupin, 2004). A very popular choice has been to visualize n-dimensional differences among neighbouring neurons using the so-called U-matrix method (Ultsch, 1993). Finally, the original input vectors or other vectors, if they contain the same variables and underwent identical preprocessing, could also be visualized. This involves finding for each the BMU from among the trained neurons and placing point symbols and text labels at that respective BMU's location.

Further computational and visual transformations may be desired, but existing SOM software is in fact severely limited in that respect. The vast majority of examples of SOM – at least when used for visualization purposes – make a choice among a limited number of available SOM software solutions. Extremely popular has been the SOM software created by the Neural Networks Research Centre at the Helsinki University of Technology. One important reason for this popularity is that the software is freely available, including access to the source code. *SOM_PAK* (Kohonen *et al.*, 1996a) is a collection of programs written in C, which can be compiled for different platforms, although Windows executables are also available. It implements the standard SOM training algorithm and was used for all examples presented in this chapter. Its visualization functionality is, however, rudimentary. This was a major reason for our implementation of GIS-based storage and visualization of a trained SOM. From the same source as *SOM_PAK* comes the equally free *SOM Toolbox for Matlab* (although it requires Matlab to already be installed), which includes various visualization options. However, compared with graphic design or GIS software, it is much harder to allow a user's imagination drive the control and transformation of these visualizations. That is why the majority of visual examples of *SOM Toolbox* applications found in the literature have a fairly uniform appearance. That is also the case for most commercial SOM software, like *Viscovery SOMine* (www.eudaptics.de).

8.3 High-resolution SOM

This section first presents the main arguments for wanting to build SOMs consisting of a very large number of neurons. Some examples for large SOMs already exist and are discussed in the context of finally introducing our proposed high-resolution SOM, most notably using GIS-based storage and visualization of neuron geometry.

8.3.1 Rationale

The rationale for creating high-resolution SOMs derives from the desire to: (a) represent macro and micro-structures existing in n-dimensional data; (b) use a trained SOM as a base map; and (c) leverage GIS technology.

When strictly used for clustering, a SOM will typically consist of only up to a few dozen neurons, especially if the attribute space portion occupied by an individual neuron is interpreted as a single cluster. Thus, a three-by-three neuron SOM trained with demographic attributes for 51 geographic objects – 50 states of the USA and the District of Columbia – would come to represent the attribute space in nine clusters, which in the case of attributes of geographic features can be readily visualized in *geographic space* (Figure 8.1). Such a SOM can also inform colour design and provide a convenient, logical legend layout (see bottom right of Figure 8.1). However, such an extremely low-resolution SOM is barely useful for visualization of input features in *attribute space*. In fact, SOM software tends to have problems with multiple input features being mapped onto the same neuron, due to overplotting of symbols and labels. For example, in the *SOM Toolbox* only one feature vector can be

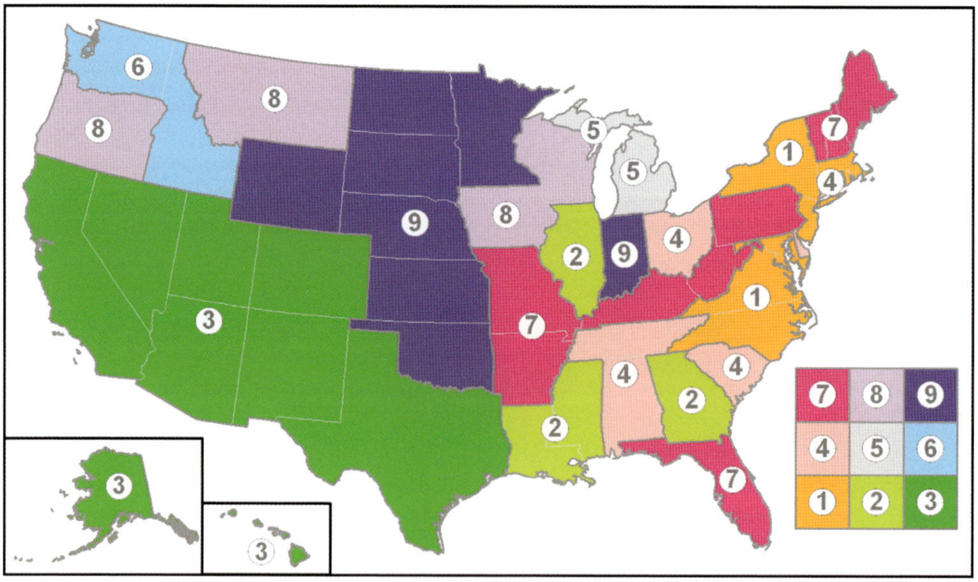

Figure 8.1 Use of a low-resolution SOM for clustering and geographic visualization. A three-by-three neuron SOM is trained with demographic data for the United States

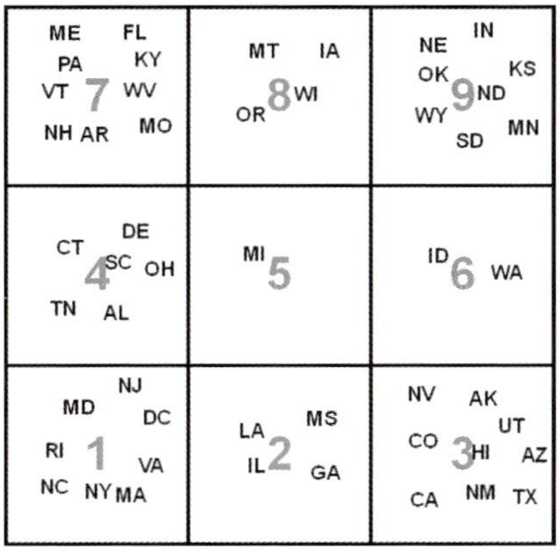

Figure 8.2 Mapping of 51 features onto low-resolution SOM (nine neurons), including disambiguated geometry through random placement inside winning neuron polygons

labelled at each neuron location and any further input vectors at that location remain basically invisible. One solution is to map features randomly near the respective best-matching neuron (Skupin, 2002), as seen in Figure 8.2. Fifty-one geographic objects are here mapped onto the same nine-neuron SOM. However, this is a purely graphic solution and, similar to the mixed-pixel problem known in raster GIS, the model provides no means to actually distinguish n-dimensional differences among vectors assigned to the same neuron. In more general terms, one can say that a low-resolution SOM only allows visualizing global or macro-structures existing in n-dimensional data, while finer structures remain hidden.

Akin to resolution effects in raster GIS, the approach proposed here is to provide a larger number of map units (i.e. neurons). For example, a 20-by-20 neuron SOM provides 400 different map units onto which the 51 geographic objects can be mapped (Figure 8.3). It is important to note that such a SOM at this point stops functioning as a clustering method,

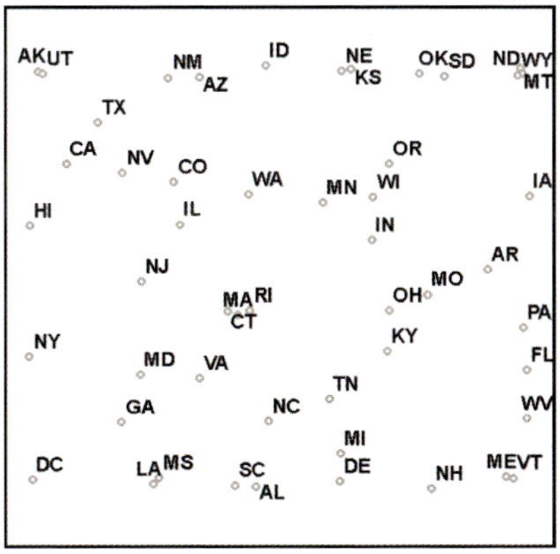

Figure 8.3 Mapping of 51 features onto higher-resolution SOM (400 neurons), with much reduced need for disambiguation of geometry

because a 400-cluster solution for 51 objects is not very useful. Instead, the SOM allows detailed two-dimensional layout of the geographic objects. Those states still assigned to a single neuron (e.g. Louisiana (LA) and Mississippi (MS)) are too similar to be distinguishable even at this level of granularity.

The detail provided in high-resolution SOMs makes it possible to have them play the role of a base map onto which various other data could be mapped. This is particularly true due to the fact that a SOM does not directly represent the input vectors as such, in contrast to such methods as multidimensional scaling (MDS) or spring models. Instead it creates a low-dimensional output *model* of the n-dimensional input space. That model can be applied to other data, as long as they have the same dimensionality. Once those data are mapped onto the SOM, other features can be attached. For example, if a SOM is constructed from multi-temporal demographic attributes of geographic objects, one could link individual temporal vertices to form trajectories and then visualize previously unrelated attributes onto those trajectories (Skupin and Hagelman, 2005). From clustering to labelling of neuron regions, a number of transformations have been proposed that all depend on a view of high-resolution SOMs as base maps (Skupin, 2004).

Most SOM software solutions provide limited support for effectively storing and transforming large neuron lattices and derived data, such as trajectories and surfaces. One alternative is to leverage the ability of GIS to deal with large, low-dimensional geometric data sets. Within GIS one can first choose among the various geometric data models, have access to various database solutions and perform a wide array of transformations, from interpolation to overlay operations. Finally, in the hands of a cartographer, GIS can produce attractive visualizations with a large degree of automation (for example for the complex task of feature labelling), while still performing data-driven visualization. Use of GIS can thus make high-resolution SOMs a much less daunting proposition on many levels.

8.3.2 Examples of high-resolution SOMs

Most SOM implementations are based on lattices of no more than a few hundred neurons, and typically much less than that. A few examples for large SOMs exist though. Most of these were in fact created by the research group around the method's inventor, Teuvo Kohonen. In the mid-1990s they mapped more than 130 000 newsgroup postings onto a SOM that eventually consisted of 49 152 neurons, though in a two-stage process that began with a much smaller SOM of 768 units, from which the larger SOM was interpolated and further training was then applied (Kohonen *et al.*, 1996b). By far the largest SOM known was created from the text of almost 7 million patent applications (Kohonen, 2001). Training was a three-step process, during which progressively finer SOMs were created, beginning with a 435-neuron SOM and eventually leading to a model consisting of 1 002 240 map units. Training took 6 weeks on a six-processor computer system.

Training speed is not merely a function of the number of neurons, but also of the model's dimensionality. Text documents tend to be represented with much longer vectors than other data. The demographic data visualized in Figures 8.1–8.3 includes 32 attributes, while Skupin's visualization of AAG conference abstracts represented each abstract as a 741-dimensional vector (Skupin, 2002). At that time, training of a 4800-neuron SOM with the conference abstracts took 3 hours. Training of a much higher-resolution, yet very

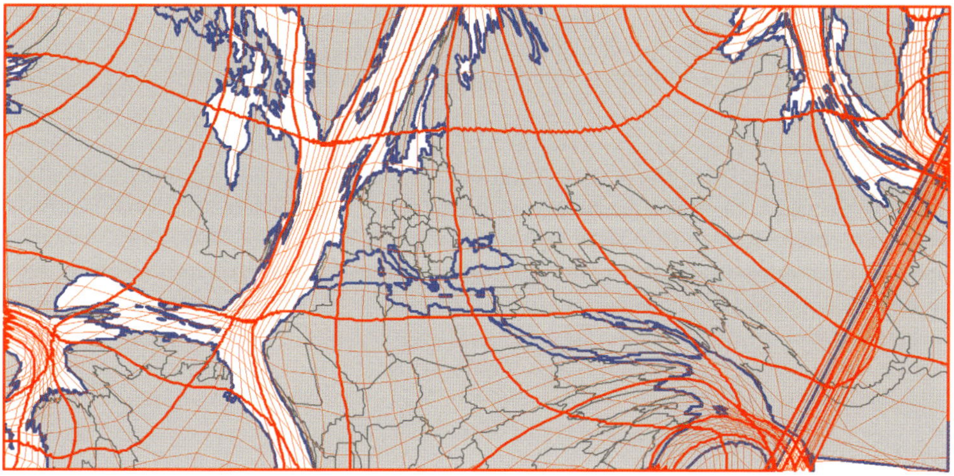

Figure 8.4 Multiple layers projected onto a high-resolution SOM (125 000 neurons) trained with geographic coordinates (from Skupin, 2003)

low-dimensional, SOM can be quite fast. An extreme example is probably the projection of geographic coordinates (without consideration of any other attributes) into a SOM space consisting of 125 000 neurons (Skupin, 2003), which took 48 hours on an 800 MHz Pentium III PC and resulted on an odd new form of map projection (Figure 8.4). Note that there are many more factors influencing the speed of SOM training, including the number of training cycles and the specific SOM algorithm used (e.g. the later stages of training for the patent SOM used a variation known as the *Batch Map*).

8.3.3 Proposition

The core idea advocated in this chapter is to use the SOM method to project geographic objects into a finely grained display space in order to provide a different, yet equally rich and holistic, perspective on geographic phenomena than that provided in traditional map space. While the latter is based on location given in geographic coordinates, the former will be constructed from the objects' attributes.

When dealing with non-georeferenced data, such as text documents, a high-resolution spatialization can become the centre of a visualization system because it is often the first and only such visual depiction and has the potential for becoming the central access mechanism for complex data and, with widespread acceptance, even developing iconic and reference status for a large user group, for example in the visualization of scientific knowledge domains (Shiffrin and Börner, 2004). This is different for georeferenced data, where the geographic map naturally maintains a central role, owing to the already discussed spatial autocorrelation effects. However, a detailed visualization derived from just the non-spatial attributes can provide an alternative perspective on geographic phenomena. This point is driven home by another aspect of our proposal, which is to juxtapose geographic and attribute space depictions while deliberately applying uniform designs and thus allowing the data to 'speak' about commonalities and differences between the two visualizations.

GIS can play a central role in implementing high-resolution SOMs. While SOM training functionality is generally not provided in GIS (with the exception of some functions available in *IDRISI* software), the low-dimensional neuron lattice of a trained SOM can readily be represented using GIS data models. There is also plenty of flexibility when it comes to using multiple data models, like vectors and rasters. For example, the vector data model may be used to represent the locations of vectors mapped on the SOM or the trajectories of features over time. Component planes – each representing a single variable retrieved for all neurons – could be represented using a polygon structure, but for very large SOMs it becomes more efficient to represent component planes as rasters interpolated from neuron centroids. From the interpolation of these component landscapes to the dissolving of boundaries during cluster visualization, GIS provides a large toolset readily usable for spatialization. Another advantageous aspect is the traditional integration of spatial and non-spatial attributes in GIS databases.

8.4 High-resolution SOM for Climate Attributes

The high-resolution SOM demonstrated here is related to one presented by Skupin (2007), in which demographic attributes for all 200 000+ census block groups were spatialized in a SOM. Space–time paths captured in different cities and based on different modes of transport were then projected onto that SOM by tracing the sequence of transected block groups. One future direction of that effort is to combine demographic attributes with other human and physical attributes towards a rich, attribute-based model of geographic space. Ultimately, we would like to create a single SOM combining all of these attributes, which requires integrating attributes originating in very different domains and stored in different formats and attaching them to identical geographic features. The latter could be of uniform shape and size, like raster cells, or one could use varied features, like polygons of different shape and size. To that end, the implementation described in this chapter demonstrates how a set of physical attributes – specifically climate attributes – are summarized for census block groups, which are then spatialized. This allows experimentation with a number of interesting aspects, including performing complex overlay procedures for transferring attributes from raster grids to several hundred thousand polygon features.

Owing to the relatively smooth variation of climate attributes across space, using only the climate data allows for more detailed observation of differences between the geographic and attribute space visualization. With dominant spatial autocorrelation effects, neighbouring regions in geographic space will tend to have similar climates and they should thus remain in close proximity in the spatialization. Where that is not the case, one is either dealing with pronounced geographic structures, characterized by rapid change of attribute values across space, or with distortion caused by the dimensionality reduction technique.

8.4.1 Climate source data and preprocessing

The data chosen for this study consisted of 11 climate attribute attached to point locations in the contiguous states of the United States (48 states plus the District of

Columbia). Data were obtained from the Web site of the National Climatic Data Center (www.ncdc.noaa.gov/oa/ncdc.html). The point attributes used included annual averages of:

(1) the numbers of days classified as cloudy, clear, or sunny;

(2) humidity;

(3) precipitation;

(4) snowfall;

(5) average, minimum, and maximum temperature;

(6) average and maximum wind speeds.

Note that the attribute data obtained consisted of only just over 200 points distributed across the contiguous United States. Future studies will include a much larger point data set, thus providing a better match between the granularity of these data and that of the block groups and the high-resolution self-organizing map.

The methodology then called for interpolation of all attributes to continuous raster grids, followed by a zonal average computed for each block group. This created unexpected challenges to the creation of appropriate source data before SOM training could even occur. Note that block groups represent a detailed tessellation of geographic space into areas of varied shape and size. Each block group is an aggregation of several census blocks into units containing around 1500 people, with a range of around 600–3000 people. Thus, block groups in rural areas can be quite large, while urban block groups literally consist of only a few city blocks. Given this potentially very small size, the interpolation method and method-specific settings have to be carefully chosen. To that end, all attributes underwent a rigorous process of cross-validation, where one point observation at a time is removed, interpolation is performed and predicted and known values are compared. When this is done for all points, a summary measure based on a root mean square error (RMS) can be computed. This was performed for several dozen combinations of interpolators and settings. In all cases, some variation of kriging produced the best result, though with different models (e.g. spherical, Gaussian, etc.), reflecting different patterns of spatial variation for the various attributes.

The most difficult lessons learned in the preprocessing of SOM training data related to the limits of current commercial off-the-shelf (COTS) GIS software in performing spatial analysis on very large data sets. Throughout this process, *ArcGIS 9.1* was used, including the *Geostatistical Analyst* extension. Given the potentially very small size of block groups, attributes were at first interpolated at a pixel resolution of 1 km^2. Then, a zonal average was attempted to be computed for each of the 200 000+ block groups. However, even at a pixel size of 1×1 km, standard zonal operators fail for a large number of urban block groups, because they do not contain a single pixel centroid. This was circumnavigated by converting pixel centroids to points, resulting in very large point files. There are numerous ways to then implemented point-to-polygon transfer of zonal attributes, most of which work reasonably well for small subsets, like for a single city or county, consisting of up to a few thousand block groups. However, even for those subsets, overlay operations take a significant amount of time. For larger data sets, execution would theoretically take several days, but overlays quickly run into limits determined by available RAM. Resolution of interpolated raster grids was eventually reduced to 10 × 10 km, which made overlay operations feasible. Failure of block groups to contain any pixel centroids was solved by assigning attributes of the nearest point features.

Every block group thus had a set of 11 climate attributes associated with it. These were then normalized to a 0–1 range and the order of block groups was randomized. This accounts for the fact that the standard SOM training algorithm takes one input vector at a time, presents it to the lattice of SOM neurons, finds the most similar neuron and then updates weights of that neuron and its neighbourhood. The size of that neighbourhood is larger early during training and then begins to shrink and the magnitude of changes made to neuron weights likewise decreases over time. Input vectors presented early thus have more influence on the training process than later vectors.

8.4.2 SOM training and transformation

An ASCII text file containing normalized climate attributes for block groups, with random order of block groups, was presented as input to SOM training. The number of block groups was the principal factor in determining the number of neurons. Ideally, we would like to obtain a unique two-dimensional point location for each of the 200 000 block groups. Therefore there should be at least as many neurons as block groups. Despite the expected density effects, with denser attribute space regions being represented with more neurons, there will still be a fair number of neurons capturing multiple input vectors. This is mostly due to the fact that some neurons will have to represent empty portions of the input space, although in highly contracted form. In addition, there is the problem of edge neurons, where the expanded/contracted representation of input space tends to be less reliable and many input vectors tend to get captured. Given these concerns and computational resources, it was decided to train a SOM consisting of 250 000 neurons (500 × 500). This will not translate into a square SOM, but into a rectangular two-dimensional lattice, due to the use of a hexagonal neighbourhood, with rows dropping into the 'gaps' below (see also Figure 8.5).

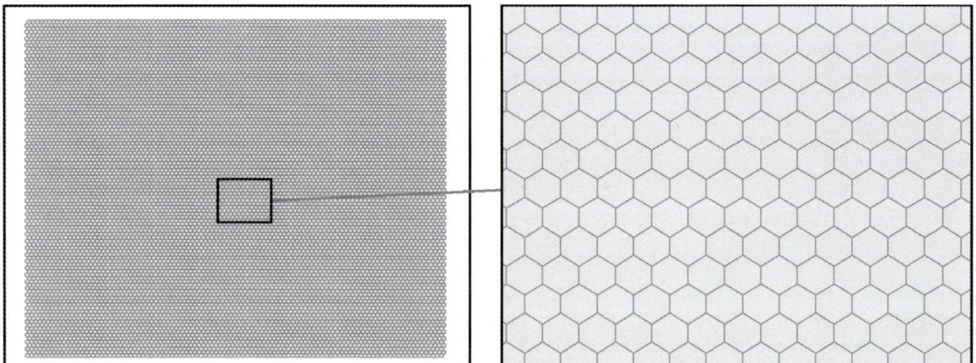

Figure 8.5 Neuron geometry in a high-resolution SOM with hexagonal neighbourhood (10 000 neurons)

Using *SOM_PAK*, initial weights of neurons were set randomly and training then proceeded in two stages. During the first stage, 20 000 training runs were performed, with an initial neighbourhood size of 250. This stage serves to represent major, 'global' structures among the input data. The second stage then aims at shaping the representation of regional and local structures. It consisted of 100 000 training runs, with a starting neighbourhood size of 25 neurons. Training took 47 minutes for the first stage and 313 minutes for the second stage (wall clock time), on a single-CPU 2.3 GHz Xeon processor system.

In retrospect, given the number of neurons and the number of input vectors, a larger number of runs might have been preferable, as seen in some of the visualizations below. However, one important lesson to be conveyed in this chapter is that, in order to visualize n-dimensional data, one will often proceed in an iterative manner, especially when it comes to detecting anomalies in the data and for setting training parameters. Experience shows that visualization is in fact not only uniquely suited to generating knowledge about the mapped domain as such, but it is also a powerful tool for testing and refining visualization methods and data. For example, note below the discussion on the effects of the wind speed variable on the trained SOM.

SOM_PAK was also used to 'map' input vectors onto the trained SOM. Every block group climate vector was compared with all neuron vectors to find the most similar neuron. *SOM_PAK* thus produced two output files. One is the trained SOM, also known as the codebook file and consisting of a list of all neurons and their final weights for all variables. The other output contains information about the best-matching neuron found for each input vector.

While *SOM_PAK* has only rudimentary visualization capability in the form of PostScript output, its codebook format has become a standard read by a number of SOM software solutions, including the *SOM Toolbox for Matlab* and *Viscovery SOMine*, where one can then perform such operations as display of individual component planes, U-matrix, and sometimes limited clustering. However, there tends to be virtually no user control over design specifics, like colour schemes and other symbol choices, and the display of features mapped onto the SOM is extremely limited and virtually useless when dealing with large numbers of features. This lack of control over the visualization is arguably the main reason for the widespread uniformity and lack of visual appeal associated with most SOM-based visualizations described in the literature, with output from the *SOM Toolbox* being particularly prevalent. On the other hand, transformation of SOM output into a form agreeable with standard GIS allows tight control over visual appearance and has the added advantage – compared with most graphic design software – of still being data-driven and thus quickly adaptable to SOMs of different size and type. We decided to convert the codebook file of neuron vectors into the *ESRI Shapefile* format, with neurons represented as hexagonal polygons and neuron weights for all attributes placed in the associated dbase file. *SOM_PAK*'s initial output of the best-matching neuron for each input vector was transformed into a *Shapefile* containing a unique point location for each vector, based on random placement inside the respective best matching neuron.

8.4.3 Visualizing the SOM

The use of climate data in this experiment is specifically aimed at observing and understanding issues arising with a high-resolution SOM. This must precede future studies, in

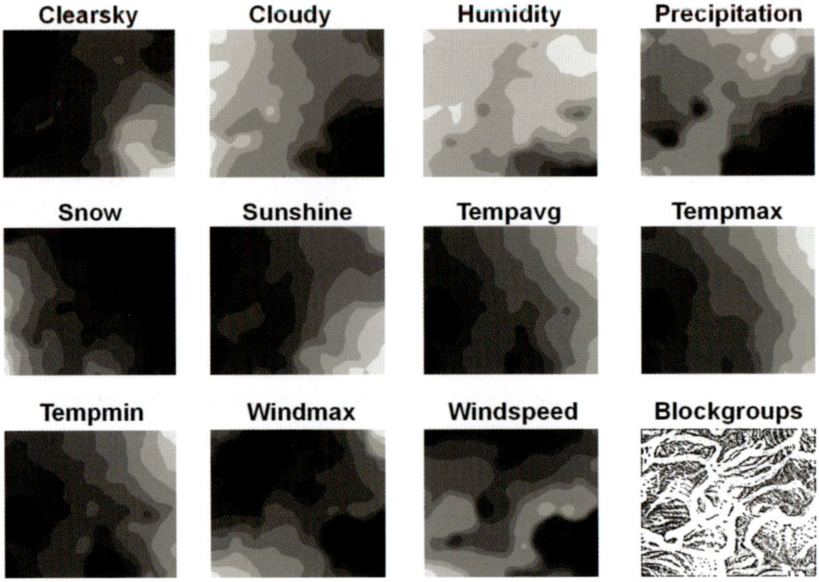

Figure 8.6 Component planes of a high-resolution SOM of 250 000 neurons constructed from climate data for 200 000+ census block groups. Block groups mapped onto SOM (bottom right)

which such SOM can then become the basis for discovery of patterns and relationships that are simultaneously 'valid, novel, potentially useful, and ultimately understandable' (Fayyad *et al.*, 1996). The *Shapefiles* generated from *SOM_PAK* code book files can be visualized in an uncomplicated manner in standard GIS software, as shown in Figure 8.6. Each of the 11 component planes is simply shown through greyscale shading of all 250 000 neurons (lighter shading indicates higher weight for an attribute). Side-by-side display allows relationships between variables to be observed. For example, in the upper-right corner of the SOM one observes high values for humidity, precipitation and temperature, but low amounts of snowfall, and a medium number of days with sunshine. Equally straightforward as the display of component planes is the point symbol display of all 200 000+ block groups (lower-right corner of Figure 8.6).

That point display is an example of the suggestive power of visualization and the danger inherent in it. Notice how the apparent arrangement of densely occupied and empty portions in the spatialization suggests the existence of linear separation regions in attribute space, which separate 'clusters' from each other. While one might find aesthetic value and even beauty in this display of 'data mangroves', they in fact turn out to be largely artefacts caused by data preprocessing. Some in-depth explanation of this issue is in order, less for what it may teach us about SOM, but more as an example of how visualization has the power to not only deceive us, but also enable us to uncover and explain those deceptions.

Compare the block group display with each of the component planes and try to find some correspondence, which might indicate that the corresponding variable possibly had an overriding influence on SOM training! It would appear that patterns in block group point locations are most closely related to patterns in the average wind speed variable. In that component plane one observes sudden changes in wind speed, with some wind speed classes forming narrow bands, even though all classes have equal width for that attribute. Those narrow bands correspond to the dominant linear separation features in the point display. None of this would be problematic, *if* the wind speed variable, as observed in nature, indeed shows such a staggered structure as opposed to the pattern mostly being an artefact of some data transformation process. Unfortunately, the latter turned out to be largely the case here. The correspondence between block group locations and wind speed prompted us to look more closely into that variable. It turned out that average wind speed in this data set had a very small absolute range of values when compared with other variables. That should not cause any problems, since all variables were scaled to a 0–1 range. However, in the process of interpolation all 11 attributes had become represented as integer raster grids in order to limit data volumes for rasters spanning all of the contiguous United States at a 1 km^2 resolution. For attributes with small absolute range the combination of integer storage and 0–1 scaling meant the introduction of wide gaps along that attribute dimension, when compared with attributes with larger absolute ranges. Effectively, the SOM represents these gaps existing in the source data correctly (see lower right corner of Figure 8.6), although they were introduced through preprocessing rather than being representative of an actual climatic phenomenon.

8.4.4 Juxtaposition of visualizations in geographic space and attribute space

One argument raised earlier in this chapter was that high-resolution spatialization of geographic features based on their attributes may provide a useful alternative perspective on geographic phenomena. For instance, one would be able to juxtapose geographic and attribute space visualizations in a more equitable manner. Synchronization of symbology takes semiotic choices out of the equation and lets the two-dimensional layout speak for itself. However, such synchronization may first require further transformation of the spatialization geometry. This is where the rich set of out-of-the-box tools available in GIS software becomes especially handy. For example, when the goal is to create an alternative map of the lower 48 states, one can start with point locations for census block groups (Figure 8.6), create a two-dimensional Voronoi region for each block group, and then dissolve boundaries between Voronoi regions of block groups within the same state. With identical symbology attached to state polygons, the two maps can now juxtaposed (Figure 8.7).

Given the regular, predictable geographic patterns of climate derived from a coarse set of geographic samples, one would expect the SOM map (bottom of Figure 8.7) to mirror many of the structures found in the geographic map (top of Figure 8.7). Owing to the nature of the SOM update rule – with similar training vectors being attracted to and updating nearby neuron regions – major topological relationships will tend to get replicated in the low-dimensional neuron lattice. One would thus expect that states are represented as contiguous polygons in SOM space. In many cases, this does in fact come true in the SOM map, as in the case of such states as Florida (FL), Louisiana (LA) or California (CA). As another consequence of the preservation of topology, one would expect that states sharing a boundary in geographic space will also be neighbours in the SOM space. For example, notice how the clock-wise order of neighbours of the state of Indiana (IN) is replicated in the SOM (KY-IL-MI-OH). Topology preservation is sometimes even able to bridge the earlier mentioned chasms caused by the preprocessing of data. For example, the large polygon depicting

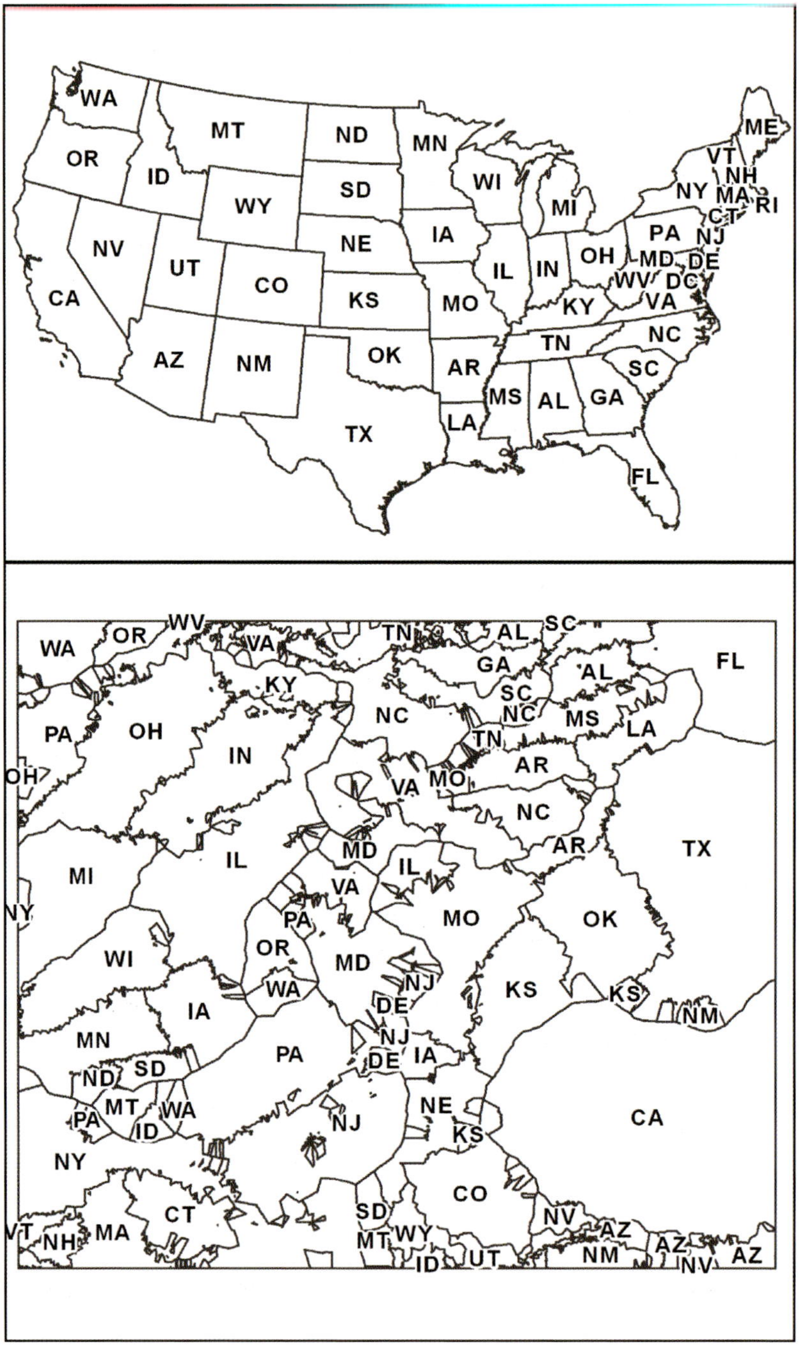

Figure 8.7 US states in geographic map (top) and spatialized based on a high-resolution SOM of 250 000 neurons trained with climate data for 200 000+ census block groups (bottom)

most of the state of Ohio (OH) combines block groups whose mapping had included distinct gaps (see region corresponding to Ohio in the block group visualization in Figure 8.6).

There are also a number of examples where the expected preservation of topological relationships does not occur, and illustration of this is also made easier by a GIS-based representation (Figure 8.8). Some states are represented by a number of disjoint polygons. For example, Pennsylvania (PA) becomes represented by four main polygons in the spatialization of states (left side of Figure 8.8). Notice, however, the distinct geographic organization of these four parts as representing mainly the western and eastern portions of the state (parts numbered 1 and 2). All four portions of Pennsylvania are themselves positioned next to neighbouring states (parts numbered 1–4 positioned next to Ohio, New Jersey, New York and Maryland, respectively).

There are also some decidedly odd neighbours in the SOM-based map of states. For example, notice the polygons representing Kansas and Nebraska appearing as neighbours of California (Figure 8.7). Zooming in on a part of that SOM region shows that there is a large gap between block groups in northern California/southern Oregon and those in southeastern Nebraska/northeastern Kansas (right side of Figure 8.8). Clearly, creating and using a high-resolution SOM is a difficult proposition and much remains to be learned about strategies for training such a SOM and for how to identify artefacts introduced by the computational process.

This is greatly helped by specific knowledge of the computational method used for dimensionality reduction. For example, the SOM method is known to preserve the density of input vectors. During the course of SOM training, the occurrence of multiple input vectors from one region of attribute space will cause that region to be represented by a large number of neurons, i.e. the region will be represented in expanded form. The opposite is true also, so that thinly 'populated' attribute space regions become represented by few neurons, i.e. attribute space appears compacted. In the case of census block groups this has pronounced effects. Owing to the role played by total population numbers in the delineation of census block groups (aiming at a total population of around 1500, as mentioned earlier), geographic areas exhibiting high population density will be represented by more block groups and therefore more neurons, compared with regions with lower population density. In our experiment, where state polygons are constructed from block group polygons, the SOM thus acts as a type of area cartogram! That is why high-population states like California (CA) and high-density states like Connecticut (CT) are represented as relatively large polygons, while low-density states like North Dakota (ND) or Idaho (ID) remain small (Figure 8.7). The reliability of this density preservation effect is, however, tempered by edge effects. Edge neurons capture relatively large portions of attribute space and the size of state polygons near the edges and especially corners (like Arizona (AZ) in the lower right corner of Figure 8.7) should thus be treated with caution.

8.4.5 Mapping n-dimensional clusters onto the climate SOM

Readers will at this point appreciate the difficulty of judging whether two-dimensional patterns and relationships visually observed in the SOM actually correspond to n-dimensional structures. For instance, our climate experiment led to multiple examples of the SOM generating shared borders between states that are not neighbours in traditional geographic space. It would be nice to more directly operate on n-dimensional data, while being able to project

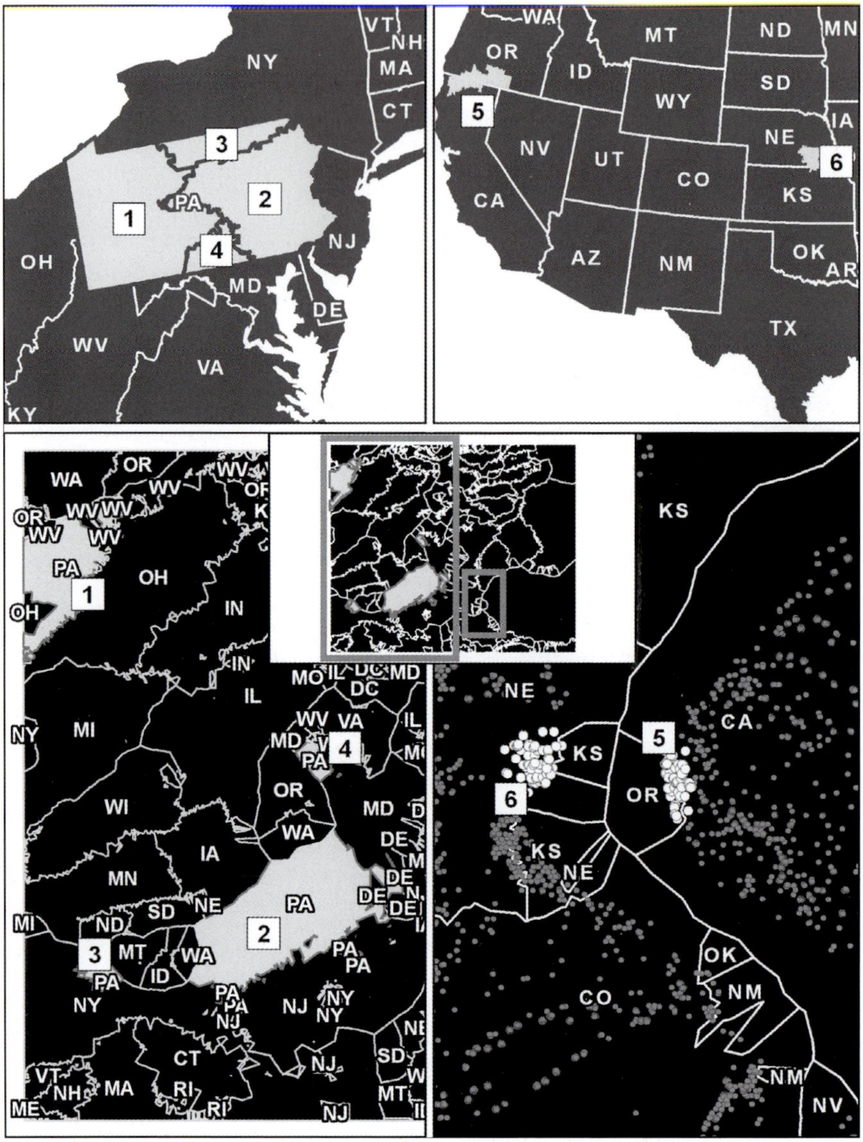

Figure 8.8 Investigating regions of interest in geographic map and spatialization. Break-up of Pennsylvania into several regions (left) and different geographic regions appearing in relative proximity in the spatialization (right)

such operations onto the geographic map and spatialization. Among such operations are the various clustering methods commonly applied in multivariate analysis.

In order to demonstrate this, we computed k-means clustering on all 200 000+ block groups, based on the very same input vectors used in SOM training. Given the typically compact nature of k-means solutions and the SOM's tendency towards preservation of neighbourhood relationships, one would expect that clusters computed with the k-means method will tend to form contiguous shapes after being projected onto the SOM. To test this, block groups in both the geographic map and the SOM-based map were joined into larger polygons based on shared cluster membership and overlaid on top of states (Figure 8.9). This direct comparison of the mapping of identical cluster solutions onto two different 'base maps' provides support for the need to perform further training on the high-resolution SOM. Clusters appear almost completely contiguous when mapped onto the geographic map (top of Figure 8.9), while some clusters appear broken into several parts in the SOM-based map (bottom of Figure 8.9). That is surprising, since the SOM and the k-means clusters ($k = 25$) were computed from exactly the same source vectors.

One would further expect that climate clusters and state boundaries should show virtually no correlation (except for such cases where a state boundary coincides with a physical feature affecting climate, like the ridge line of a major mountain range). Indeed, that is the case in the geographic map. It is true in many parts of the SOM as well. For example, notice how the progression of climate clusters as parallel bands in the southeastern United States (sequence of clusters 2–6–18–14–1–16) gets represented in both geographic and SOM-based visualization, virtually independent of state boundaries. However, there are cases in the SOM-based map where cluster and state boundaries coincide and these correspond to the same oddly neighbouring states mentioned earlier. For example, the 'northern' and 'western' boundaries of California in the SOM completely coincide with cluster boundaries. The large gap among two-dimensional block group locations along the same boundaries therefore seems to indeed be justified (see also bottom-right portion of Figure 8.8). The earlier mentioned break-up of Pennsylvania (left side of Figure 8.8) is supported by the association of the large western portion of Pennsylvania (no. 1 in Figure 8.8) with Ohio in cluster 15 and of the eastern portion (no. 2 in Figure 8.8) with New Jersey in cluster 17.

8.4.6 Mapping extreme weather events onto the climate SOM

All visual transformations illustrated so far were based on the very same input vectors used to train the SOM. However, it is also possible to map other data onto the SOM that were not part of the training process, based on two different approaches. One consists of finding best-matching neurons for non-training vectors, as long as those vectors consist of the same dimensions as the training vectors and identical preprocessing has been applied (e.g. scaling to 0–1 range based on the same minimum/maximum values). In our case, one could map climate data for geographic areas outside of the United States onto the SOM to identify global similarities. For example, areas along the Mediterranean coast would likely end up inside of cluster 24 'in' Southern California (see Figure 8.9).

Another option is to use a currently mapped geographic feature as a socket through which another geography feature can be mapped onto the SOM, based on shared geographic location. Skupin and Hagelman (2005) demonstrated this by first training a SOM with multi-temporal census data and then using single-time vectors as temporal vertices that define the trajectory of a geography feature. Another proposed trajectory mapping technique takes space–time paths, such as those captured by GPS, and projects them onto a spatialization based on the sequence of geographic features traversed (Skupin, 2007). We demonstrate this here for hurricanes that made landfall in the continental United States during the 2005 hurricane season (Figure 8.10). This involves determining the sequence of block groups

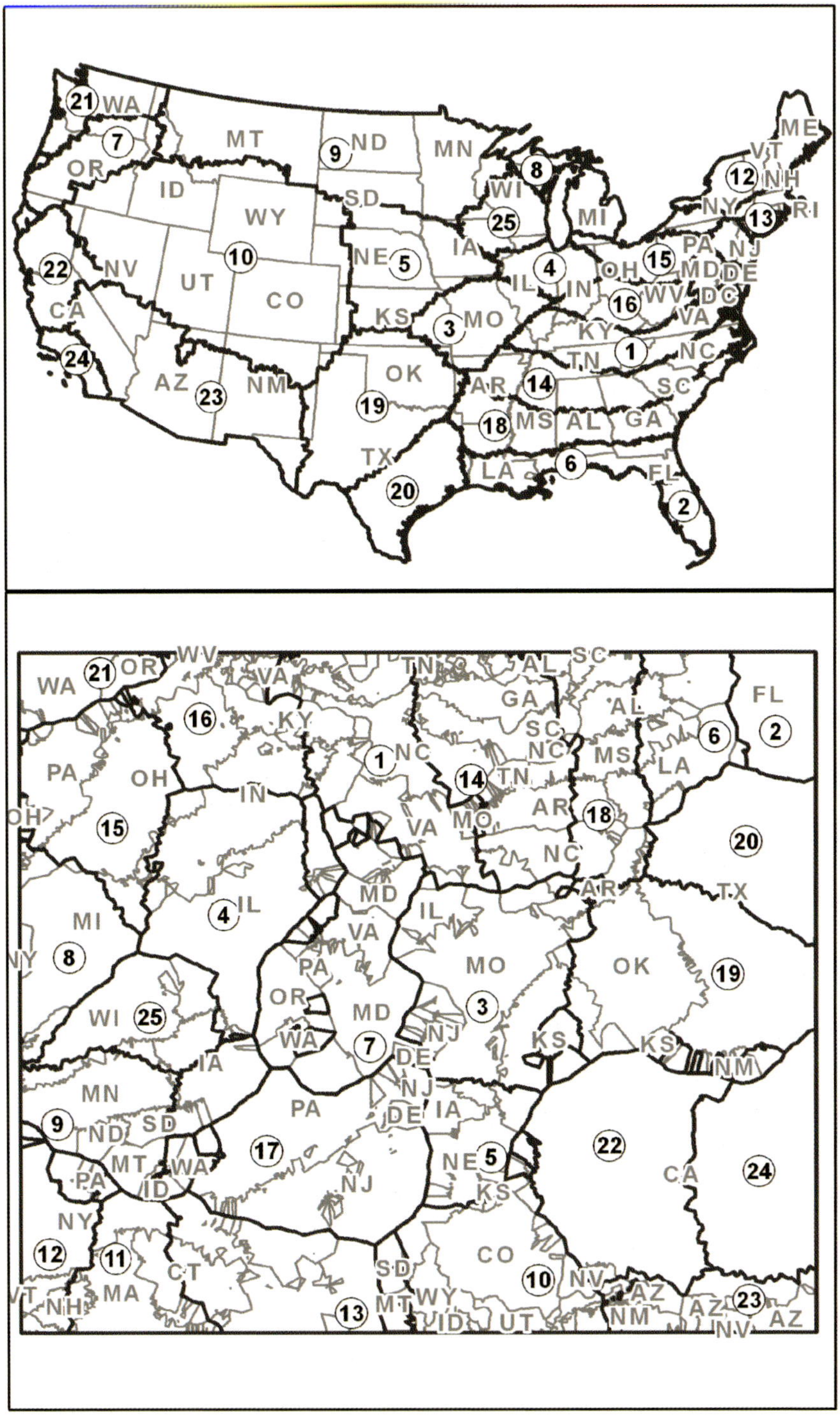

Figure 8.9 *k*-Means clustering of climate attributes for 200 000+ census block groups visualized in geographic map and spatialization ($k = 25$)

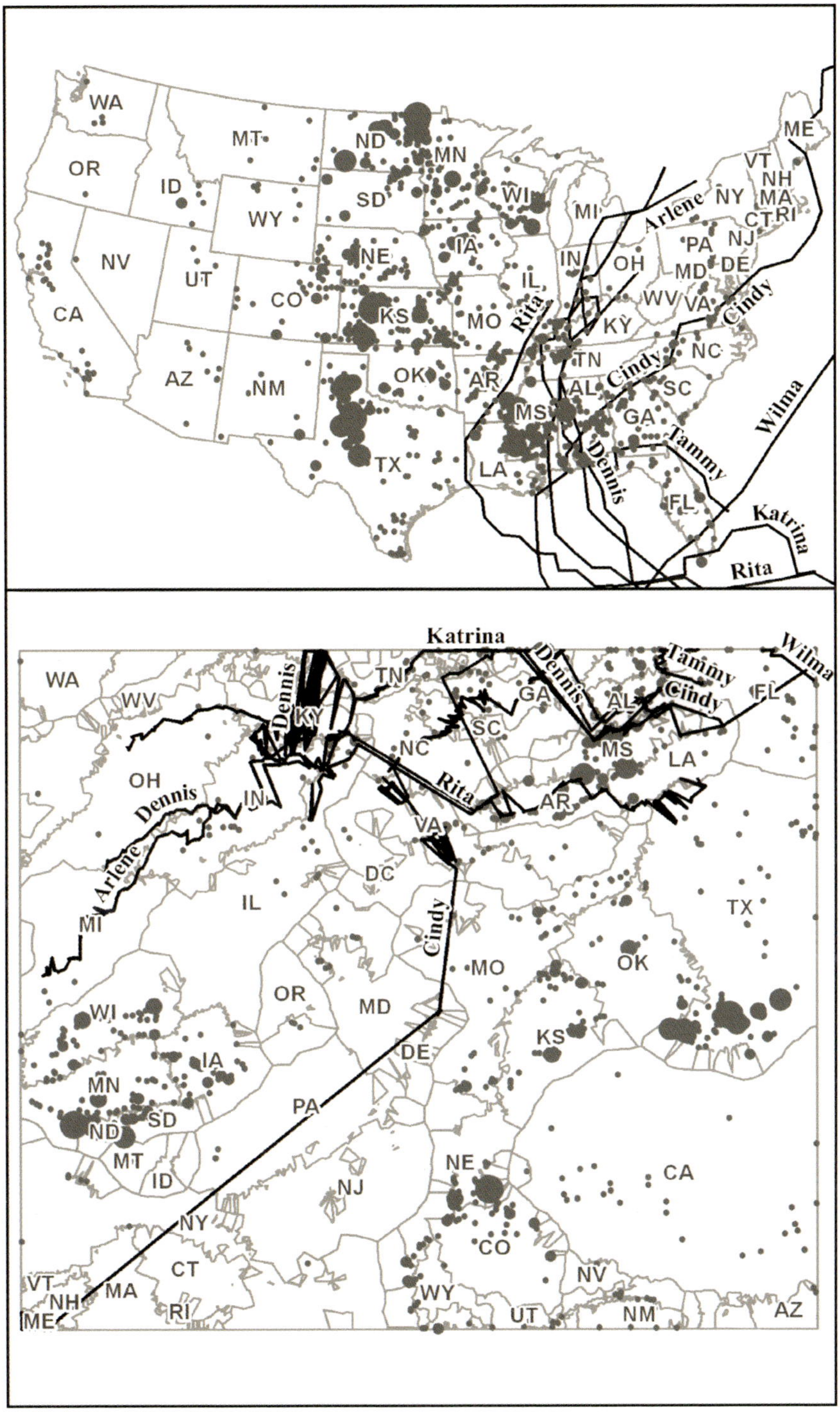

Figure 8.10 Tornado touchdowns and hurricane paths observed during 2005 mapped onto geographic map and climate-driven spatialization

traversed by each hurricane. While conceptually a straightforward overlay operation of hurricanes (represented as uni-directional routes) with block groups, it turned out to be extremely challenging in commercial GIS software, due to the very large number of block group polygons (200 000+). Mapping of point features onto the SOM is much easier, as then the block group sockets can be accessed via a simple point-in-polygon overlay. In this manner we mapped another data set of extreme weather events onto the SOM, the locations at which tornadoes touched down in 2005 (Figure 8.10).

While this mapping of hurricanes and tornadoes illustrates the technical principles driving this type of n-dimensional overlay, it does not generate any particular insight, due to the regular spatial pattern of climate in the continental United States. The technique becomes much more interesting when complex n-dimensional patterns are encountered across geographic space. Demographic data, such as those captured by the US Census Bureau, contain many examples of such complex patterns. Skupin (2007) describes a SOM of equal resolution as the climate SOM described here, but generated from demographic attributes for 200 000+ block groups. Actual geographic movement captured by GPS is then mapped onto the SOM. For example, the author's commute from his previous residence in the Mid-City neighbourhood of New Orleans to the University of New Orleans located near Lake Pontchartrain is mapped onto the SOM generated from nationwide data (Figure 8.11). Among the interesting patterns observed in this visualization are the relative compactness of neighbourhoods in attribute space (e.g. *Mid-City*, *French Quarter*, *New Marigny*), and that movement between neighbourhoods is accomplished either via large hyperjumps

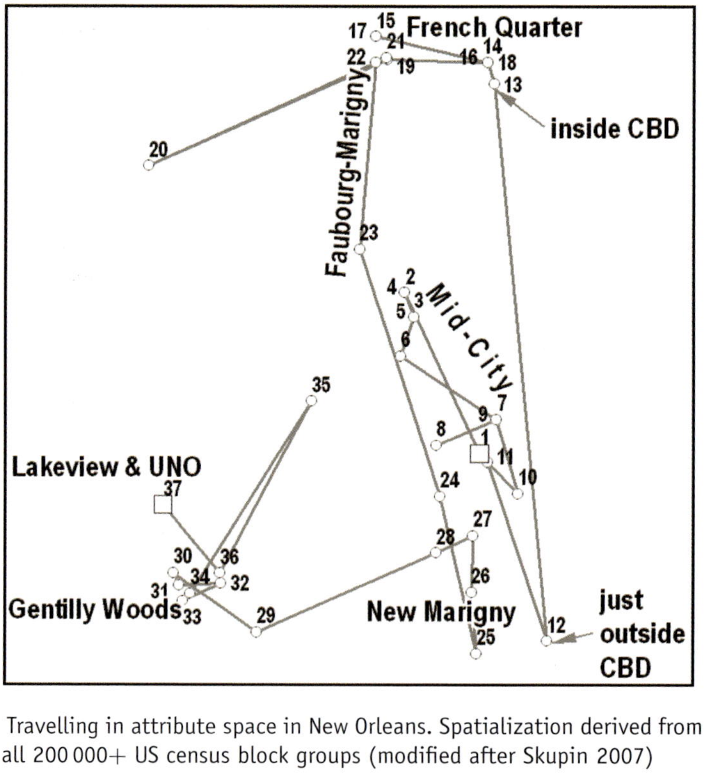

Figure 8.11 Travelling in attribute space in New Orleans. Spatialization derived from demographic attributes for all 200 000+ US census block groups (modified after Skupin 2007)

(e.g. movement between block groups 12 and 13) or via bridge block groups that are transitional in terms of both geographic and attribute space location (e.g. block group 23).

8.5 Summary and outlook

This chapter proposes the creation of high-resolution spatializations from the attributes of geographic features. Its main aim is to generate an alternative to the standard approach in geovisualization, where the geographic map tends to be the only stable element, while other visualizations are characterized by fluid geometry, topology and visual appearance. It is argued that the self-organizing map could be one method able to generate more stable base maps onto which various types of other data could be mapped through a series of geometric and attribute-based transformations. This approach is decidedly different from the current use of such tools as scatter plots, parallel coordinate plots and even of the SOM method itself.

We demonstrated this approach by first training a SOM consisting of 250 000 neurons with climate data generated for more than 200 000 geographic features, performing various transformations, and finally juxtaposing SOM-base maps with geographic maps. Numerous challenges were encountered, beginning with the difficulty of performing geographic overlays involving several hundred thousand objects in standard GIS software. Training using *SOM_PAK* was unproblematic, but a number of artefacts in the visualization – in particular the break-up of *k*-means clusters after projection onto the SOM – suggest that more training cycles should be applied, beyond the 120 000 cycles used in our experiment (20 000 and 100 000 in the two training stages, respectively).

Future work will include a more formal investigation of the distortions incurred by the training of a SOM consisting of several hundred thousand neurons with an equally large number of training vectors. On one hand, there is a need to develop recommendations for how to use standard SOM tools (e.g. *SOM_PAK*) in the context of high-resolution SOM. This must occur in recognition of the fact that geometric distortion as such is unavoidable when high-dimensional data are represented by low-dimensional geometry and that the SOM method is in fact able to bridge large dimensional gaps between source data and display space through density-driven effects of expansion and compression. On the other hand, overall distortion characteristics can be addressed by recent variations of the SOM method. For example, edge effect distortions are caused by topological heterogeneity in the neuron structure, with edge neurons having fewer neighbours than neurons further inside of the SOM. This source of distortions could be diminished by arranging neurons on a closed surface. A number of spherical SOM approaches have been proposed (Sangole and Knopf, 2002; Wu and Takatsuka, 2005), but have not yet been used much in practical SOM applications and have not involved large numbers of neurons. Another interesting question in the context of high-resolution SOMs is how the total number of neurons is to be determined. The standard SOM algorithm used in our climate experiment as well as spherical and other variants take the number of neurons as an input parameter, which therefore allows user control of SOM granularity. A very different, alternative approach involves the deletion or addition of neurons in response to certain threshold functions. This leads to so-called *growing SOMs* (Fritzke, 1999). Apart from these approaches addressing problems observed in both low- and high-resolution SOMs, there may be issues arising principally with SOMs of high and extremely high resolution that are as yet undefined. Systematic studies should be undertaken to explore this, which may in particular call for well-controlled synthetic data sets.

The use of climate data in a high-resolution SOM and the use of existing geographic objects as place holders of climate attributes was informed by the desire to extend the notion of *attribute space travel* (Skupin, 2007). One of the overarching goals of that research direction is to provide a methodological framework for dealing with the experience of geographic space in a computational manner. Socio-economic characteristics of a geographic place have an effect on one's experience with and interaction in that place, but the physical attributes, such as temperature and humidity, obviously play an important role as well. The experimental work presented in this chapter is meant as a first step towards an integrated visual modelling of physical and social attributes, in this case by attaching climate attributes to demographic enumeration units. Future work will include the actual combination of very different types of attributes, including climate, demographic, land use/land cover, and many others, and thus create rich spatializations of geographic objects. Such work may include irregular enumeration units – therefore incurring the cartogram effects described in this chapter – as well as regular tessellations of geographic space.

Acknowledgements

We gratefully acknowledge the assistance of Charles Schmidt in the pre-processing of climate data and of Martin Lacayo in creating geometric and attribute base data for the climate SOM.

References

Deboeck, G. and Kohonen, T. (eds) (1998) *Visual Explorations in Finance with Self-Organizing Maps.* Berlin, Springer.

Fayyad, U. M., Piatetsky-Shapiro, G., Smythe, P. and Uthurusamy, R. (eds) (1996) *Advances in Knowledge Discovery and Data Mining.* Menlo Park, CA, AAAI/MIT Press.

Fritzke, B. (1999) Growing self-organizing networks – history, status quo, and perspectives. In *Kohonen Maps*, Oja, E. and Kaski, S. (eds). Amsterdam, Elsevier.

Kohonen, T. (1982) Self-organized formation of topologically correct feature maps. *Biological Cybernetics* 43: 59–69.

Kohonen, T. (1990) The self-organizing map. *Proceedings of the IEEE* 78: 1464–1480.

Kohonen, T. (2001) *Self-Organizing Maps.* Berlin, Springer.

Kohonen, T., Hynninen, J., Kangas, J. and Laaksonen, J. (1996a) *SOM_PAK: The Self-Organizing Map Program Package.* Espoo, Helsinki University of Technology, Laboratory of Computer and Information Science.

Kohonen, T., Kaski, S., Lagus, K. and Honkela, T. (1996b) Very large two-level SOM for the browsing of newsgroups. *1996 International Conference on Artificial Neural Networks.* Springer, Berlin.

Oja, E. and Kaski, S. (eds) (1999) *Kohonen Maps.* Amsterdam, Elsevier.

Sangole, A. and Knopf, G. K. (2002) Representing high-dimensional data sets as closed surfaces. *Information Visualization* 1: 111–119.

Shiffrin, R. M. and Börner, K. (2004) Mapping knowledge domains. *Proceedings of the National Academy of Sciences* 101: 5183–5185.

Skupin, A. (2002) A cartographic approach to visualizing conference abstracts. *IEEE Computer Graphics and Applications* 22: 50–58.

Skupin, A. (2003) A novel map projection using an artificial neural network. *21st International Cartographic Conference*, Durban.

Skupin, A. (2004) The world of geography: visualizing a knowledge domain with cartographic means. *Proceedings of the National Academy of Sciences* 101: 5274–5278.

Skupin, A. (2007) Where do you want to go today [in attribute space]? In *Societies and Cities in the Age of Instant Access*, Miller, H. J. (ed.). Berlin, Springer, pp. 133–149.

Skupin, A. and Hagelman, R. (2005) Visualizing demographic trajectories with self-organizing maps. *GeoInformatica* 9: 159–179.

Tobler, W. R. (1970) A computer movie simulating urban growth in the Detroit region. *Economic Geography* 46: 234–240.

Tobler, W. R. (1979) A transformational view of cartography. *The American Cartographer* 6, 101–106.

Ultsch, A. (1993) Self-organizing neural networks for visualization and classification. In *Information and Classification: Concepts, Methods, and Applications*, Opitz, O., Lausen, B. and Klar, R. (eds). Berlin, Springer.

Wu, Y. and Takatsuka, M. (2005) Geodesic self-organizing map. In *Conference on Visualization and Data Analysis 2005/Proceedings of SPIE*, Vol. 5669, Erbacher, R. F., Roberts, J. C., Grohn, M. T. and Börner, K. (eds), San Jose, CA, 17–18 January 2005.

9
The Visual City

Andy Hudson-Smith

Centre for Advanced Spatial Analysis, University College London

Nothing in the city is experienced by itself for a city's perspicacity is the sum of its surroundings. To paraphrase Lynch (1960), at every instant, there is more than we can see and hear. This is the reality of the physical city, and thus in order to replicate the visual experience of the city within digital space, the space itself must convey to the user a sense of place. This is what we term the 'Visual City', a visually recognizable city built out of the digital equivalent of bricks and mortar, polygons, textures and most importantly data.

Recently there has been a revolution in the production and distribution of digital artefacts which represent the visual city. Digital city software that was once in the domain of high-powered personal computers, research labs and professional software is now in the domain of the public-at-large through both the web and low-end home computing. These developments have gone hand in hand with the re-emergence of geography and geographic location as a way of tagging information to non-proprietary web-based software such as Google Maps, Google Earth, Microsoft's Virtual Earth, ESRI's ArcExplorer and NASA's World Wind, amongst others. The move towards 'digital Earths' for the distribution of geographic information has, without doubt, opened up a widespread demand for the visualization of our environment where the emphasis is now on the third dimension. While the third dimension is central to the development of the digital or visual city, this is not the only way the city can be visualized, for a number of emerging tools and 'mashups' are enabling visual data to be tagged geographically using a cornucopia of multimedia systems. We explore these social, textual, geographical and visual technologies throughout this chapter.

9.1 The development of digital space

Digital space takes many forms. However in terms of the visual city, we are concerned with the creation of space that allows us to generate a visual understanding of our built

Geographic Visualization Edited by Martin Dodge, Mary McDerby and Martin Turner

environment. Knowledge of space is hard-wired into us, insubstantial and invisible; space is yet somehow there and here, penetrating all around us. Space for most of us hovers between ordinary, physical existence and something given. Thus it alternates in our minds between the analytical and the absolutely given (Benedikt, 1996). Our interpretation of space and the resulting sense of location and place that is engendered influence our perception of space in both real and digital terms.

Bell (2006) identifies three different kinds of space: visual, informational and perceptual. Visual space is unsurprisingly all that we can see. It is the array of objects that surround us creating, when viewed collectively, our environment. Each of the objects in any such space has a multitude of different attributes, from variations in light and colour to reflectivity. These objects create a reality which is a fully immersive environment in cartesian space, space that can be interrupted and explored in three dimensions. If these objects are broken down to singular levels, then each can be viewed as being made up of a combination of primitives. Primitives in turn are a collection of graphic tokens such as points, lines and polygons, forming a two- or three-dimensional arrangement, and it pays us to think of visual space populated by these tokens (Mitchell, 1994). If these points, lines and polygons can be recreated in digital space, along with their attributes, then digital space can mimic sufficient aspects of reality in terms of the urban dimensions necessary to create what we have called the visual city.

Informational space can be seen as an overlay to visual space and it is in this space that we communicate and receive information. From urban signage to oral communication, information is communicated in visual space. In terms of the visual city, information should not be viewed as a separate space but an additional attribute or in more prosaic terms a new layer. Digital information takes the form of an embedding of data within digital space. This combination of informational and visual space can be seen as forming the basis for Google Earth and other digital globes. With the addition of user-friendly communication to convey such informational space, an overlap occurs with the third form of space, that of social or perceptual space. Social space defines the user's identity and role in relation to other users in the social environment. In digital space, the social dimension is increasingly important and this is seen in the rise of social networks such as MySpace, Facebook and Twitter, to name but a few. Of interest is the fact that these social spaces allow either the creation of visual space in terms of multi-user, three-dimensional, environments such as the virtual world Second Life or more direct mashups which combine geo-located photographs of general users as displayed and accessed through Flickr within Google Maps. These applications are the key to the Visual City and we will come back to them in more detail later.

9.2 Creating place and space

In our research group, Centre for Advanced Spatial Analysis (CASA), we have built a three-dimensional model of Greater London which we consider represents a Visual City. The production of the model has only been made possible due to the development of three-dimensional GIS and related relevant tools in our case ESRI's ArcScene, three-dimensional Studio Max and various online visualization packages, most notably Google Earth. A reoccurring theme in the development of such three-dimensional city models is the way in which emerging technologies are enabling us to query, manipulate and construct our environment remotely. The Visual City can now effectively be streamed and developed over the internet, opening up a range of possibilities, not only for visualization, but also for displaying attributes of the population in the form of socio-economic geographic data, agent-based models of how cities function or even as actual users engaging with the software. Indeed it is fair to say that we are at a tipping point in city-based information systems in the way they are both used and created.

The goal of our Virtual London project is to develop a truly virtual city which can be occupied, queried and manipulated by citizens within a collaborative environment. This development route has entailed a combination of data capture, model development and optimization. The acquisition of suitable digital data is central to the development of Visual Cities and their use in the emerging online three-dimensional GIS systems. In terms of pure visualization the production of photorealistic models of the built environment is key to the creation of visual space, yet it is a time-consuming, manual process and one that up until recently was in the domain of professional photogrammetry. The standard approach to producing a photogrammetric reconstruction of the city has been through the use of calibrated images and matching control points. Figure 9.1 illustrates the development of one of the key buildings along the north bank of the Thames which is modelled using a combination of oblique photography from helicopter capture and ground-based imagery. The model took approximately two days to produce.

In today's Google-led world, which is based on releasing free software with high levels of functionality combined with low levels of required expertise, it is now possible to considerably reduce the time taken to produce such models. Google SketchUp is a unique program that is available in both professional and freeware versions. The differences between the free and professional versions are negligible and are only significant in terms of importing and exporting data. This now means that the public at large are able to photomodel through SketchUp and produce their own sections of the city. Google SketchUp is linked directly

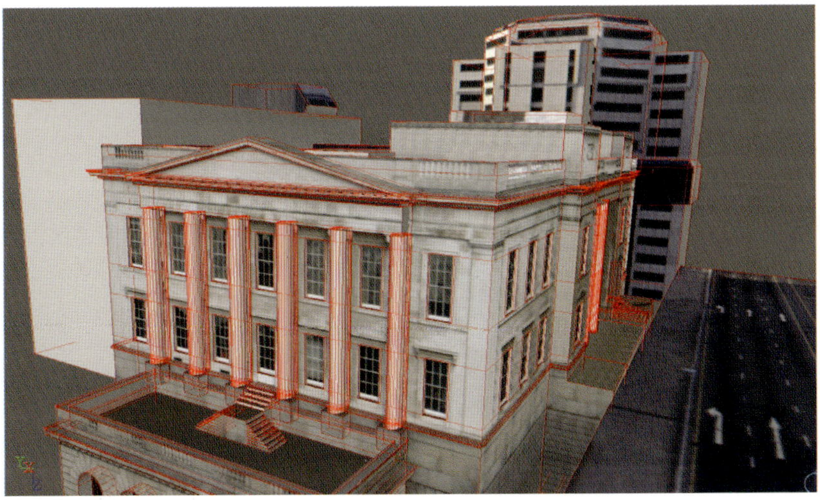

Figure 9.1 Photogrammetric modelling

with Google Earth, which we examine further a little later, and it is linked for a good reason – for users to develop free content.

Creating a Visual City, one that reflects the actual built form, is a huge task and therefore time-consuming. So far, the only groups that have been able to get close to representing the city visually are Games companies such as Sony and more recently Microsoft. Games such as 'The Getaway 3' on the Playstation 3 represent the cutting edge in city visualization. The Getaway originally appeared on the PlayStation 2 as a three-dimensional rendition of London covering approximately 10 square miles (16 square kilometres). The team behind the model produced a wire frame model based on a photographic survey of London and then projected the resulting textures onto the geometry.

The results of such developments are impressive but the costs are typically in order of tens of millions of dollars to produce, while the models are also only of use for gaming. They cannot be easily ported into contexts either where geographical analysis is required or where the public at large can interact with them, largely due to the nature of their construction. Therefore to reduce cost, Google released their SketchUp software so the users-at-large could produce the city themselves, block by block, building by building. Of note is the latest version of SketchUp which, at the time of writing, allows users to import and calibrate their own photography, directly modelling over the imagery. Although not as accurate as the traditional photogrammetry, it does allow rapid modelling and the widespread adoption of photorealistic content. Figure 9.2 illustrates a streetscape modelled in less than a day using SketchUp. The image is shown without textures to illustrate how architectural detail can be added to the model quickly and easily.

The release of Google SketchUp has in turn led to the development of the Google three-dimensional Warehouse, an online repository which is directly linked with the SketchUp program. Currently there are 252 user submitted buildings in London, ranging from landmarks to people's houses. Buildings are uploaded to the warehouse automatically from the programme, creating a quick and easy way to populate the city. This of course leads to duplication and worries about quality, ruling it out for real-world applications such as architectural impact analysis or planning applications work, but it does provide a quick and

Figure 9.2 Rapid modelling using Google SketchUp

effective visual insight into the city. The best of the models are selected by Google for viewing in the Community Layer of Google Earth, thus completing the cycle of the public-at-large creating the Visual City.

One of the additional techniques we have used extensively in our group at CASA to communicate a visual sense of the city is panoramic imagery. The use of panoramas is not a new phenomenon; indeed the first panorama was patented in 1787 (Wyeld and Andrew, 2006). Panoramic visualization is not three-dimensional *per se*, in that it consists of a series of photographs or computer-rendered views stitched together to create a seamless image. Rigg (2007) defines a panorama as an unusually wide picture that shows at least as much width-ways as the eye is capable of seeing. As such, it provides greater left-to-right views than we can actually see (i.e. it shows content behind the viewer as well as in front). It was not until 1994 and the introduction of Quick Time Virtual Reality (QTVR) for the Apple Macintosh that panoramic production became available on home computers. Software was released that allowed a series of photographs to be seamlessly stitched to form a single complete $360 \times 180^\circ$ view, and we illustrate an example panorama in Figure 9.3.

Although panoramas are essentially two-dimensional, they can be inserted into a three-dimensional scene to provide an instant sense of location and place. The field of view of a panorama equates to the coverage of a sphere. As such, by draping a panoramic view onto

Figure 9.3 Swiss Re $360 \times 180^\circ$ panorama

Figure 9.4 Panoramic images in Google Earth

a sphere and then moving the viewing field to the spheres centre, or nodal point, the view straightens out the lines in the image providing an exact replica of the human eye's line of sight from the location. The ability to drape onto a sphere allows the panorama in to be depicted in x–y–z three-dimensional space, and it can be embedded in other models of the Visual City. Figure 9.4 illustrates a panoramic sphere embedded within Google Earth.

The images are placed on the reverse face of the sphere, allowing the user to look inside; while wrapping around a user when they enter the nodal point of the view. The panoramas or 'urban spheres' as we call them, are open source files linking to imagery outside of Google Earth on sites such as Flickr, which quickly enables us to create a sense of location and place. Google Earth has been fundamental to the development of the Visual City and it is this to that we now turn.

9.3 Visual cities and the visual Earth

The World Wide Web has provided a revolution in the way we obtain, distribute and react to information and we now take for granted the ability to search, edit and publish information regardless of location. The first commercially available browser, Netscape, based on the earlier Mosaic, was released in 1994 and much has happened in terms of the way we now distribute, manipulate and visualize data since that time. It is arguable that we also take for granted to ability to zoom into any location on the globe and view various levels of informational and visual data in a three-dimensional environment. Yet it is barely 24 months, at the time of writing, since the original Keyhole Earth Browser was re-branded and launched as Google Earth.

Google Earth is the current buzz word in terms of geographic information and is covered in detail by Michael Goodchild in Chapter 2. The importance of Google Earth to the Visual

City is three-fold. First is the ability to view the city in two dimensions via high-resolution aerial imagery. Levels of detail vary according to location with 'Googleplex' (the Google Campus) providing the highest current resolution at 2.54 cm per pixel. The use of high-resolution digital imagery allows the user to gain a visual overview of a city from the air. This is a new emergence and as such has led to the rise of sites that track locations of sightseeing within Google Earth. Although the highest resolutions are limited to urban areas, Google Earth sightseeing is a global phenomenon. To hone this discussion back to the city, the move to the third dimension has probably the largest impact in terms of visualization of geographic information than any other. Although predominantly US and Japan based due to copyright issues on data, which we will return to later in terms of our own model, Google's three-dimensional cities are fundamental to the idea of the Visual City. They represent a significant development in the visualization of city environments, not only in terms of our ability to view building outlines and polygons but also due to their location in true geographical space. Thus geographical location provides the third issue of importance in Google Earth, which involves the ability to add data and visualize information using the three-dimensional Visual City as a backdrop or canvas to other data sources. The ability to visualize and overlay information opens up a number of applications for the Visual City, applications which were once in the domain of the professional user, it is these to which we now turn.

9.3.1 Applications in the visual city

With the rise of computing power has come an increase in publicly accessible GIS information and with it the ability to visualize in three dimensions, leading to a massive demand for city models. In terms of Virtual London, the fully functional complete model has been developed in different ways for different audiences. This is of some importance as each audience requires a different level of interaction and interface. While the visual use of the city is almost universally similar between different users, what changes is the level of data mining possible, the delivery method and the interface. Broadly it is possible to identify two main categories of use, firstly we have fully professional usage which includes the use of the model by architects, developers, planners and other professionals who are anxious to use its full data query and visualization capabilities. For example, an architect might place a building within the model and use this to assess a variety of issues from its basic visualization to the impact it might have on traffic and surrounding land use.

In terms of our London model, the fully professional application has been our main focus. The three-dimensional model has been rolled out to all 33 London boroughs, providing London with its first city-wide three-dimensional GIS system. This raises a number of issues in terms of software, hardware and expertise required to manage and view the model. As such we have rolled out alongside the professional model a customized version written to dynamically load according to a viewpoint in Google Earth. Figure 9.5 illustrates a section of the model in Google Earth.

The Google Earth version is specifically developed for the non GIS user. In terms of professional use, the level of functionality is compromised but the ability to navigate and overlay other datasets is increased. This is a common trade-off for functionality vs cost and ease of use. As such the choice to roll out a Google Earth version is important as it allows any local government employee to view the model. This links in to our second level of

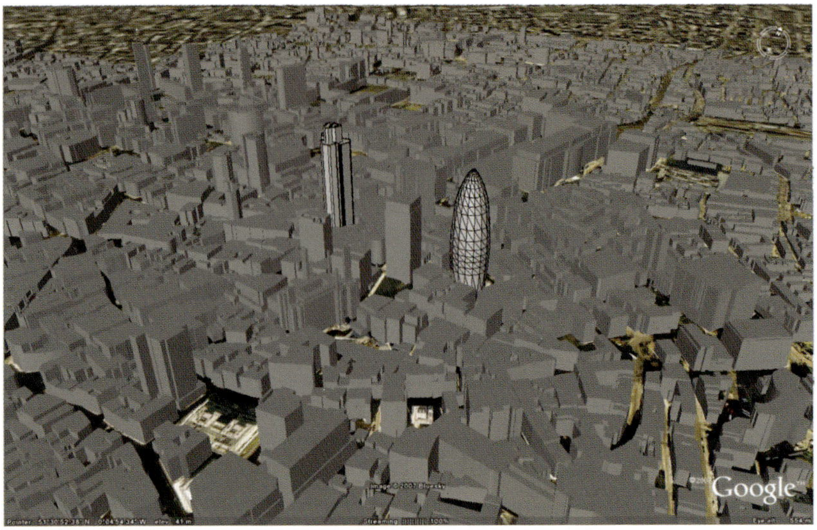

Figure 9.5 Virtual London in Google Earth

user – the concerned citizen for public participation. Initially this was seen as the main focus for using the model but has had to be restricted due to issues of copyright with the Ordnance Survey base data used to present this version. The restrictions on data use have been central to city visualization and GIS in general, especially outside of the academic community. In short, data costs money to collect and therefore licence to use the data is often restrictive in terms of further distribution. In terms of Virtual London, this led to the withdrawal of the public access version, illustrating the difficulty faced by Ordnance Survey in adapting its licensing policies for the new age (Cross, 2007).

The public face of Virtual London is therefore currently limited to movie files and as such indirect visualization of data within the city model. While this is restrictive in terms of public participation and allowing access to the data, it does result in improved visual output as with movie files one is not concerned with real-time visualization. A good example of this is how a three-dimensional city model can effectively communicate data in the visualization of air pollution where such levels of visualization are currently not possible in real time but possible offline as we show in Figure 9.6.

Figure 9.6 illustrates air pollution data from the Environmental Research Group at Kings College London, where a pollutant surface based on nitrogen dioxide in three dimensions is draped over the cityscape. The move to visualize data in three dimensions is controversial and often seen as mere 'eye candy' by some specialists in the field. Yet in terms of a communication tool, it illustrates the areas of intense air pollution arguably more effectively than any two-dimensional map. This may partly be due to the visual nature of the medium allowing a stronger sense of location and place to be obtained than a top-down two-dimensional view. As such any amount of data can be visualized with the model. Figure 9.7 illustrates how the city could flood as a result of sea level rise. With the animation file, it is possible to watch the water level rise and therefore identify which areas are more at risk according to the degree of rise. Again, this is a use of the Visual City in offline mode where we can sensibly embed data to visualize important outcomes.

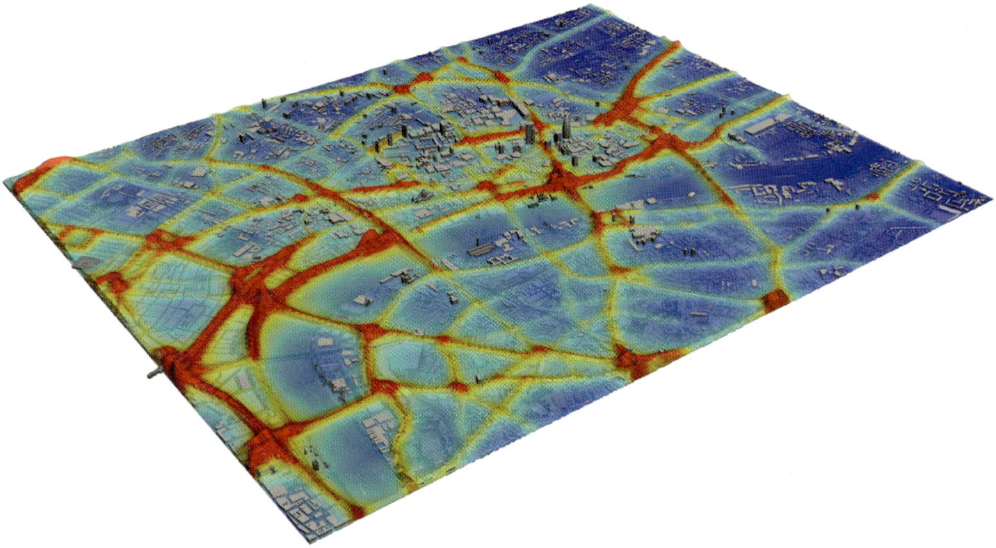

Figure 9.6 Air pollution rendering

The Visual City does not necessarily need to be three-dimensional. Indeed as we argue later, there are a number of emerging two-dimensional technologies that create a Visual City. Neither should a Visual City be seen as purely data or informational space, for social space is becoming increasingly important in its development, as we will now show.

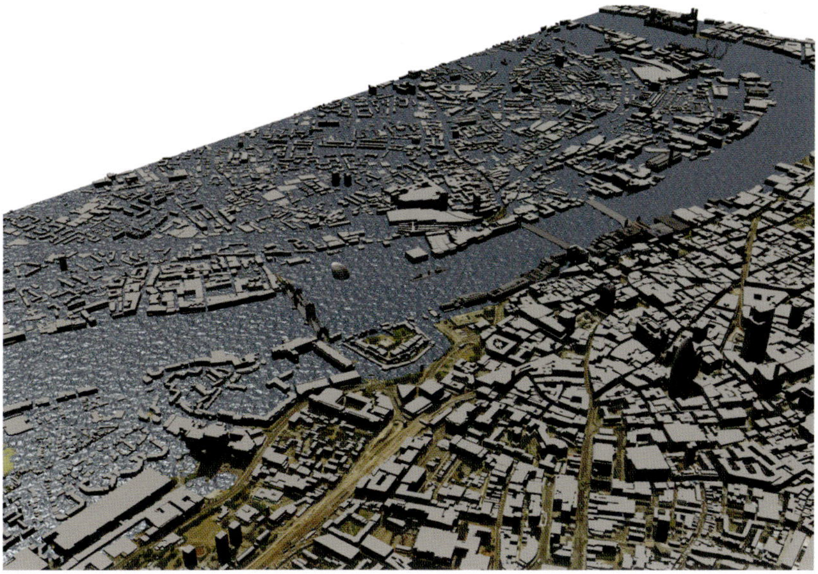

Figure 9.7 London flooding

9.4 The development of virtual social space

Terms and phases come into and out of fashion. Cyberspace, a once common term for describing the Internet, is now passé, as is the term Metaverse for the description of multi-user worlds. Yet it is Stephenson's (1992) textual definition of the Metaverse which is closest to today's visual virtual cities. Stephenson's novel *Snowcrash* depicts life in the Metaverse:

> As Hiro approaches the Street, he sees two young couples, probably using their parents' computer for a double date in the Metaverse, climbing down out of Port Zero, which is the local port of entry and monorail stop. He is not seeing real people of course. This is all part of the moving illustration drawn by his computer according to the specifications coming down the fiber-optic cable. The people are pieces of software called avatars. (Stephenson, 1992, p. 35)

Avatars are an individual's embodiment in the Visual City, providing the all-important visual and social presence in the digital environment. They are the citizens, the occupants and the commuters of the digital realm; indeed they are the inhabitants of the Visual City in all but a real physical presence. The term avatar – for use in terms of digital environments – was first used by Chip Morningstar, the creator of Habitat, the first networked graphical virtual environment, developed on the Internet in 1985. The term 'avatar' originates from the Hindu religion as an incarnation of a deity, hence an embodiment or manifestation of an idea or greater reality. Figure 9.8 illustrates typical designs for avatars in a virtual world, in this case in Second Life.

Second Life, launched in 2003, currently represents the most successful social/visual space on the Internet. It differs from other more game-based systems such as the popular World of Warcraft as it does not have any quests or goals. The system is purely a social geographic

Figure 9.8 Avatars in Second Life

space within which its users are able to construct the environment entirely themselves. From the elevation of the landscape to the scale of a city, every part of Second Life's visual space is editable. It is as close to the Metaverse that current technology allows and provides a unique insight into the future of the Visual City. Benedikt (1996) states that, as virtual worlds are not real in the material sense, many of the axioms of topology and geometry so compellingly observed to be an integral part of nature can therefore be violated or reinvented as can many of the laws of physics. It is this reinvention that allows attributes to be enhanced and emphasized and the laws of gravity, density and mass to be excluded, allowing buildings to be moved or deleted with the click of a mouse and allowing the user to fly above or anywhere within the environment.

As such Second Life is a Visual City which does not collate to the cities in Google Earth. It is a landscape of fictional space existing only on one of the 3000 servers that power Second Life. The lack of gravity and the ability of avatars to fly or teleport to locations creates a cityscape which differs considerably from the real world. With the limited design control – there are no planners or architects – any user can create a virtual sprawl of spiralling urbanity mixed with eccentric retail areas and recreational land use parcels.

In terms of the Visual City, you would not necessarily expect textual information to allow the creation of a cityscape. Yet combined with a social network, text-based communication can provide a uniquely visual view of the city as a whole. Text-based messages via mobile phones are now part of everyday life. The first text message was sent in December 1992, while SMS (short messaging service) was launched commercially for the first time in 1995 (Wilson, 2005). Text-based messaging is, in general, a one-to-one communication system. To create a social space, the SMS needs to be shared via a wider network and thus it becomes one-to-many in its communicative potential through newly emerging services such as Twitter.

Twitter is representative of the recent trend in social networking sites allowing people to connect and communicate. Where it differs from sites such as MySpace is that it is purely based on the SMS format of 140 maximum characters with the text entry box via Twitter asking the simple question 'What are you doing?' As such, the system is applicable to short and often pithy updates on a person's activity sent via mobile phone, instant messaging device or via the Twitter website. The question is how can this text-based information source create a Visual City? The answer is partly due to the shear number of users on Twitter (in excess of 200 000) and the ability to include a user's location in the messages. Combining the location of Twitter posts, known as Tweets, with a Google Maps Mashup generates the ability to visualize what people are doing at different locations in a city in real time. We illustrate the location of Tweets in Central London in Figure 9.9.

New visualizations of Tweets are currently emerging on an almost daily basis, allowing the concept to scale to the global level with the ability to visualize in real time feeds of people's thoughts and 'what they are doing'. Using Microsoft Live (Microsoft's web-based mapping service), it is possible to visualize these one-way conversation flows updated every 5 seconds with either a global or street level view. Developed using a system known as Atlas and GeoRss, a startup company, 'Freshlogic', have developed a mapping system that updates these feeds geographically. Of interest in terms of the city is the overview which is gained when viewing from above. By simply letting the system run, it will zoom into each new location, complete with address, users' photographs and Tweet every 5 seconds. The system also works with geotagged photographs via Flickr. Using the same Atlas system, the map will

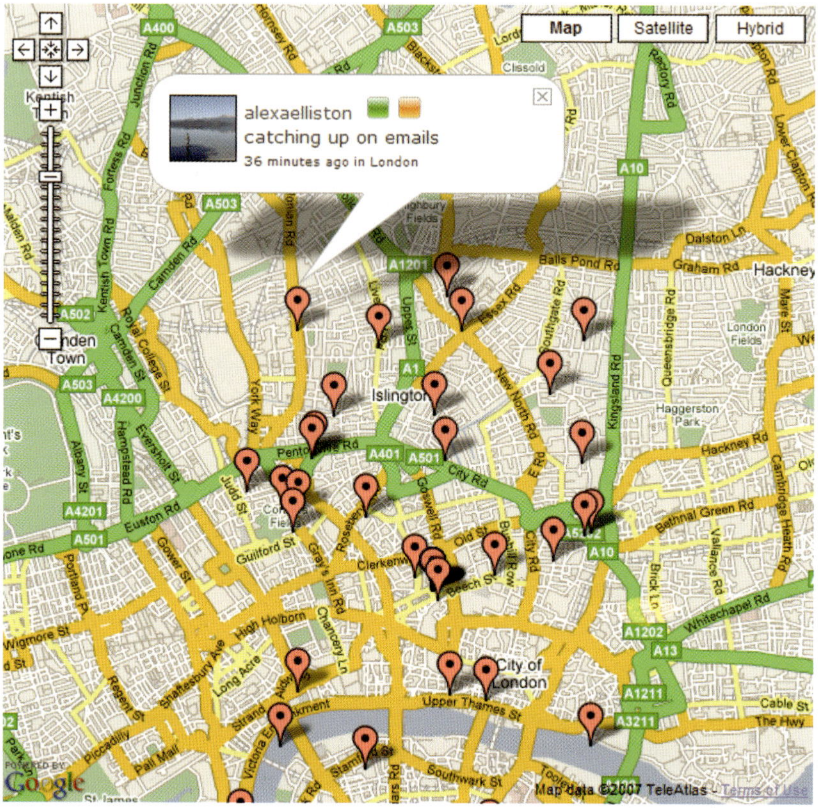

Figure 9.9 Tweets in Central London

update with new photography of places at each predetermined time interval. These again are live feeds into a web-based geographical visualization system, something that was unheard of and hardly imaginable a mere 18 months ago.

The key to this rise in geographical information is data, but not data as we would traditionally view it in large information sets from a central repository, often government-based, but personally gathered data. The move towards low-cost, yet powerful, software such as SketchUp creates a geographically tagged database of three-dimensional objects, mainly in relation to our built environment. Yet this is only one aspect, as we have seen. The move towards the increasing miniaturization of hardware and the demand for remote access to information is pushing forward hand-held personal digital assistants (PDAs) and more importantly the mobile phone market. These hardware innovations come in waves with each new wave adding increasingly complex functionality within increasingly easy-to-use interfaces. In the late 1990s, PDAs were the 'must have' gadget for remote access to information. Functionality was limited to email and internet access, firstly via slow modem connections linked to mobile phones and then later via wi-fi hotspots. Such devices allow access to information but not in the geographic sense *per se*. As with all waves of innovation, PDAs fell out of favour and are only just re-emerging, this time integrated within mobile phones, making available a portable digital tool kit for the data capture of Visual Cities available to the public at large.

Figure 9.10 Personal GPS tracking data

The latest of these devices is the Nokia N95, a phone which features a 5 mega pixel camera, wi-fi and more importantly a built-in GPS. As such it makes the perfect tool for both capturing and communicating within the built environment – a portable tool to create the Visual City. The camera has a high enough resolution for use in photomodelling and SketchUp as well as holding up the possibilities of panoramic capture. The GPS unit allows tracking of routes and the uploading of data to Google Earth.

Figure 9.10 Illustrates my route into Waterloo Station, London, tracked using the N95. The height of the route represents speed, providing a unique insight into my own travel into the city. The integration of GPS into devices such as mobile phones allows them to be used outside of the traditional car-based environment and thus they become part of our navigational abilities on foot. The ability to navigate through the physical city while capturing digital data in real time or sending Tweets or geotagged photographs to Flickr represents a key development in the Visual City. People generate data, data which up until now has generally not been logged, let alone sent to a digital Earth for visualization.

In terms of the Visual City, it should not be assumed that there is one Visual City for each urban area. Indeed we can identify numbers from one or two full three-dimensional city models to hundreds of thousands for individual city visualizations. There is not as such a single platform or database for the increasing amount of information that can be captured. Google Earth provides a good basis with its Community Layer, which provides information gathered by the public-at-large. The shear amount of information can, however, be overwhelming and in general this layer is left switched off and therefore unseen by the majority of users. The shear density of population in a city, and thus the amount of information that could be input into system such as Google Earth, is resulting in vast amounts of data of varying quality. While such data is of interest on a number of levels for display, the move seems to be towards one of the personal yet shared Visual City rather than a single collaborative database.

9.5 The future: the personal city

A familiar theme is the decrease in knowledge required to create and present geographical information, which is leading to a direct increase in the amount of information available. As we have seen, user-created data can be visualized within a global system such as Tweets and Flickr via Atlas and Microsoft Live or as personal tracks via mobile devices within Google Earth. While all these data streams can be built into one Visual City, such as our Virtual London, there is also a move to more personalized geographic data. The editing of Google Maps to create a location previously involved the manual editing of code and a moderate knowledge of XML, but with the release of Google's My Maps it is now possible to create one's own map in a matter of minutes. The My Map's application is a web-based service which allows the user to add points, lines and polygons as an overlay to Google Maps. This again is a significant addition to the visualization of the cityscape, both in two and three dimensions, as the overlays created can be exported to Google Earth or indeed any KML viewer. In addition to the ability to add points, polygons and lines to the map is the integration of video via either Google Video or YouTube.

In essence, we are but at the beginning of what will be a revolution in social, visual and informational data plotted geographically by general users. The ability to create one's own map of the cityscape is of prime importance as these maps can be either public or private. If the user chooses the public option which is the default, the map becomes searchable within the Google general search engine. Information embedded in the map thus, if searched for, directly links to the map. As such the map, be it city-based or otherwise, becomes the key interface to informational space.

The rise of social networks provides us with the ability to look down on the city and view the activities that its citizens are involved in. This ability provides unique social data and an insight into how the citizens are thinking, working and socializing. At the moment, Twitters are two-dimensional, but it is a short step to move these data streams into a three-dimensional world such as Google Earth. If you then combine this with avatars as in Second Life, then you not only have a Visual City with visual and informational space, you also introduce perceptual space into the context. This is more than a Visual City, for we now stand at the threshold of a Visual Earth.

References

Bell, J. (2006) Virtual spaces and places; cyberspace; space and place in computing. *Presence Research.* Available at: http://pegasus.cc.ucf.edu/~janzb/place/virtual.htm (accessed 3 December 2007).

Benedikt, M. (1996) Information in space is space in information. In *Images form Afar. Scientific Visualisation – An Anthology*, Michelson, A. and Stjernfelt, F. (eds). Copenhagen, Akademisk Forlaf, pp. 161–171.

Cross, M. (2007) Copyright sinks virtual planning. *The Guardian* 24 January 2007.

Lynch, K. (1960) *The Image of the City.* Cambridge, MA, MIT Press.

Mitchell, W. J. (1994) *Picture Theory.* Chicago, IL, University of Chicago Press.

Rigg, J. (2007) What is a Panorama, PanoGuide. Available at: www.panoguide.com/reference/panorama.html (accessed 3 December 2007)

Stephenson, N. (1992) *Snowcrash*. New York, Bantram Spectra.
Wilson, F. R. (2005) A history of SMS. Available at: http://prismspectrum.blogspot.com/2005_11_01_archive.html (accessed 3 December 2007).
Wyeld, G. T. and Andrew, A. (2006) The virtual city: perspectives on the dystopic cybercity. *The Journal of Architecture* 11(5): 613–620.

10

Travails in the Third Dimension: A Critical Evaluation of Three-dimensional Geographical Visualization

Ifan D. H. Shepherd

Middlesex University Business School, Middlesex University, London

10.1 Introduction

Interactive three-dimensional computer graphics are not new. Almost half a century ago, Ivan Sutherland developed the Sketchpad system (Sutherland, 1963b), which introduced the first interactive CAD-like toolkit that he felt would 'generalize nicely to three dimensional drawing' (Sutherland, 1963a, p. 138). Just a few years later, he was experimenting with a head-mounted three-dimensional display device (Sutherland, 1969). Even the use of three-dimensional computer graphics to visualize data is far from novel. In the 1980s, for example, a special report (McCormick, DeFanti and Brown, 1987) revealed a burgeoning use of the third dimension in what was then termed 'scientific visualization', and by the end of the decade significant progress had been made in the development of three-dimensional GIS (Raper, 1989).

Interactive three-dimensional computer graphics became more widely available in the 1990s due to the spread of relatively inexpensive graphics display technology, and domestic users of computer games were soon flying, running, driving and shooting their way around extensive and increasingly realistic three-dimensional worlds.[1] More recently still,

[1] It is generally agreed that *Wolfenstein 3D*, a first-person shooter game, which was released by id Software in 1992, triggered the initial mass appeal of 3D computer games. This success was followed in 1993 by *Doom*, and in 1996 by *Quake*, with its full 3D engine (id Software, 2007).

Geographic Visualization Edited by Martin Dodge, Mary McDerby and Martin Turner

augmented reality applications of three-dimensional displays have begun to appear in applications as diverse as telemedicine, travel guides and geocacheing (Güven and Feiner, 2003). In view of this considerable history and widespread contemporary deployment, it might appear to be rather late in the day to reconsider the use of the third dimension for data visualization, and particularly to suggest that there may be troubles in this particular virtual paradise. However, there are several good reasons for taking stock at this point in time.

The first is that we may be in danger of repeating the 'technology first' mistake that occurred in the 1980s when colour displays began to displace monochrome displays (Hopkin, 1983). In some quarters, a '3D for 3D's sake' tendency appears to be repeating the 'colour for colour's sake' trend of a couple of decades ago, among developers and users alike. It is perhaps timely, therefore, to reconsider the circumstances in which 3D is more appropriate than 2D, and when it is not. A second reason for undertaking a reappraisal is that some seasoned researchers – and many college students – who use current software to visualize their data do not fully appreciate the principles of effective data visualization, whether it be 1D, 2D, 3D or 4D. The growing popularity of 3D graphics in the media (from films and games to virtual web spaces and satnav gadgets), together with its appearance in applications software and operating systems (and perhaps most notably in the Vista release of the Windows operating system), suggest that a new era of 3D popularity may be about to dawn. If data analysts are not to be swamped by inappropriate 3D technology, and are to know how to use it effectively, then they need to develop a better understanding of the principles, roles and limitations of 3D data visualization.

A third reason for undertaking a critical review at this time is that, somewhat paradoxically, 3D data visualization techniques have only recently been included in mainstream GIS and desktop mapping systems. Despite the depth of innovation in 3D data visualization over the past quarter of a century, most current practice in geographical data analysis is still rooted in the 2D era, supported by a technology that reveals a significant paper-based legacy. It might therefore be an appropriate time to consider the potential future role of 3D in mainstream GIS and mapping software, and how this might best support effective visual data analysis.

A final reason for the current appraisal lies in the danger that 3D might be seen as a final step in the linear evolution of data visualization technology. It will be argued that data visualization is only part of a broader framework of representational technologies, in which 2D representations continue to provide powerful insight into data. In addition, multi-sensory representation (or perceptualization) not only offers solutions to some of the limitations of 3D data visualization, but it also provides much-needed alternatives for those who are visually impaired.

One further introductory observation is in order, which concerns the scope of this chapter. A great deal of what is termed geovisualization concerns the creation of (photo-realistic) views of actual or proposed world features, usually based on digital spatial data. The current chapter, by contrast, explores ways in which suitable data of any kind, whether spatially referenced or not, may be visualized in 3D scenes. Such scenes may be based on real-world objects or locations (as when thematic layers are draped or otherwise superimposed over terrain models), but they may also consist of recognizable but artificial landscapes, or even more abstract spatial scenes. For this reason, the term geographical data visualization better indicates the concerns of this chapter.

Several broad questions will be posed about the role of the third dimension in data visualization. First, how far have we come in developing effective 3D displays for the analysis of spatial and other data? Second, when is it appropriate to use 3D techniques in visualizing data, which 3D techniques are most appropriate for particular applications, and when might 2D approaches be more appropriate? (Indeed, is 3D always better than 2D?) Third, what can we learn from other communities in which 3D graphics and visualization technologies have been developed? And finally, what are the key R&D challenges in making effective use of the third dimension for visualizing data across the spatial and related sciences? Answers to these questions will be based on several lines of evidence: the extensive literature on data and information visualization; visual perception research; computer games technology; and the author's experiments with a prototype 3D data visualization system.[2]

10.2 What is gained by going from 2D to 3D?

There are several good reasons for using the third dimension when visualizing data. Some of these are briefly reviewed below.

10.2.1 Additional display space

However large the monitor, and however high the screen resolution, users will always reach limits to the amount of data that can be reasonably displayed in their data visualizations. Research in the field of information visualization (or infovis) indicates that the volume of data objects that can be comfortably displayed on screen is considerably larger when using a 3D representation than when using a 2D representation (Card, Robertson and Mackinlay, 1991). Where information exists in tabular form, then conventional 2D box displays may be converted to 3D solids to increase the information shown on screen; and where the information involved is hierarchically organized, such as the folders and files stored on a computer, or the documents available in an online repository, then conventional 2D trees may be upgraded to higher capacity 3D cone trees (Robertson, Mackinlay and Card, 1991). Finally, where data are organized as a network, then 2D connectivity graphs may be replaced by 3D node-and-link graphs (Hendley *et al.*, 1995). A major advantage of this increase in information for the user is that it permits a larger amount of contextual information to be seen while focusing on particular objects of interest. A significant drawback is that 3D visualizations often tend to impose severe interaction demands on users.

The conversion of existing 2D data visualization techniques into 3D equivalents is an attractive development strategy, and many conversions have proved successful, as is the case with 3D fish-eye distortion displays (Carpendale, Cowperthwaite and Fracchia, 1997), 3D beamtrees (van Ham and van Wijk, 2003) and 3D distribution glyphs (Chlan and Rheingans, 2005). However, other kinds of 2D visualization techniques may not deliver equally significant benefits when converted into the third dimension. For example, although software has been developed to generate 3D versions of parallel coordinates and star glyphs

[2] The experimental 3D data visualization software used to generate the illustrations that accompany this chapter was devised by the author, based on data visualization principles and the published results of experimental research in visual perception and human–computer interaction. The software was programmed by his son Iestyn Bleasdale-Shepherd, who is a software engineer specializing in real-time computer graphics at Valve Corporation in Seattle.

(Fanea, Carpendale and Isenberg, 2005), the added complexity of the adapted versions means that the upgrade from 2D to 3D may not provide as much benefit as expected. The pseudo-3D versions of bar and pie graphs which appear increasingly in research publications introduce apparently unnoticed distortions into the messages conveyed to readers.

10.2.2 Displaying additional data variables

An immediate gift of the third display axis is that it enables at least one additional data variable to be mapped during visualization. At the simplest level, this means that the height axis may be used for showing the values of any interval-scaled variable. This technique has been used (arguably to mixed effect) in the prism map, which overcomes the inability to apply the size visual variable to the surface area of polygons without distorting the area of real-world features. (2D cartograms solve this problem in another way, but at the expense of introducing spatial distortions to well-known maps that are often confusing to data analysts.)

However, it should be noted that the third display axis is not restricted to displaying a single additional data variable, as in the prism map. One way of maximizing the potential of the *z*-axis is to map several data variables onto more complex 3D symbols or glyphs. Several examples have been discussed in the literature, and they generally fall into two groups. The first group involves the construction of complex glyphs from multiple data variables, which are typically distributed across the *x*–*y* plane of the display space. These glyphs are most commonly used in information and scientific visualization, a representative form being a tree whose branches are sized, angled and coloured to represent particular data variables. Analysts interprets the distribution of such glyphs in much the same way that they would explore the distribution of any set of point symbols.

Where data are available for points on an equally spaced grid, an alternative approach has been adopted. In the iconographic display (Levkowitz, 1991; Pickett and Grinstein, 1988), data values control the display of an array of articulated icons across the data grid, whose varying orientations, thicknesses and colours result in spatial textures that vary visually across the grid. The interpretation of these textures is undertaken differently from the more discrete display of 3D glyphs, because the iconographic display is designed to be perceived preattentively rather than cognitively. Although this display was first devised using 2D icons, later versions have been developed for both 3D and 4D visualizations. For example, the 3D approach has been used to display multiple environmental variables in North America by Healey (1998). As we will show later, however, even imaginative use of multivariate 3D glyphs and icons is insufficient for the visualization of datasets containing large numbers of variables.

10.2.3 Providing a familiar view of the world

Two-dimensional representations of three-dimensional features in the real world have significant interpretive drawbacks. By flattening topography and using abstract symbols to represent surface features, the 2D paper map imposes a significant learning burden on occasional map users. (In a similar way, the traditional architectural drawing, with its attempt at capturing 3D reality in a multiple view – plan, side elevation and front elevation – divides visual attention and makes spatial integration more difficult.) By contrast, it is often claimed that viewers find it easier to interpret data visualizations that are constructed as 'natural' or 'familiar' scenes (Gee *et al.*, 1998; Robertson, 1990), which more closely represent the real world. One basis for this argument is that the vertical viewpoint used in the traditional printed map is unnatural for humans who spend most of their lives moving around a world that is seen largely in oblique view from ground level. Another is that the symbolic forms of representation widely adopted on paper maps to represent familiar surface features are too divorced from the everyday experience of buildings, roads or trees.

There are many recent examples where 2.5D and 3D data visualizations have been used to communicate environmental problems and issues to the general public more effectively than conventional 2D maps. The extent and impact of flooding, both in the case of actual floods (e.g. the impact of hurricane Katrina in New Orleans in 2005) and also predicted floods (e.g. the effect of global warming on central London), are often given a greater impact by being visualized using oblique views that show surface variations of the areas involved. However, some caution is needed here, because although Sweet and Ware (2004) suggest that an oblique viewpoint enables better judgements to be made about surface orientation, other interpretive benefits of adding perspective are questioned by Cockburn (2004).

Even where data are spatially referenced, it may be more appropriate to construct metaphoric spatial scenes to visualize such data rather than reproduce the actual 3D environments from which these data are derived. Over 20 years ago, for example, the rooms metaphor was proposed for a general computer interface (Henderson and Card, 1986), and the BBC Domesday system adopted a museum metaphor for the display of images stored on videodisc (Rhind, Armstrong and Openshaw, 1988). More recently, e-commerce web sites have introduced the virtual shopping mall as a context for browsing purchasable products, collaborative learning websites have adopted 3D virtual worlds (e.g. Börner *et al.*, 2003; Penumarthy and Börner, 2006), while some 3D social networking web sites have adopted familiar 3D environments such as landscapes (e.g. Second Life) or hotels (e.g. Hotel Habbro) for their avatars to inhabit.

In experiments undertaken to visualize telecommunications network data, Abel *et al.* (2000) suggest that specific tasks may be better supported with 3D scenes built using visual metaphors such as buildings, cities and the solar system, rather than on the inherent geographical structure of the network infrastructure. Their research suggests that an analysis should be undertaken of task requirements in order to determine the utility of adopting natural rather than unconventional spatial representations. Similar considerations also apply to the choice between 3D interfaces that mimic natural human actions and those that overturn conventional principles. Pierce (2000), for example, argues for the need to break existing assumptions about how people interact in the real world when designing interaction methods for virtual 3D worlds, suggesting that '3D worlds are governed by more complex laws than 2D worlds, and in a virtual world we can define those laws as we see fit'.

10.2.4 Resolving the hidden-symbol problem

One of the largely unrecognized problems of displaying large numbers of point symbols on a conventional 2D map is that they often obscure one another. Where this results from overlaps in the extent of proportional or graduated symbols, then various techniques may be used to resolve the problem, including transparency and symbol cut-outs (e.g. Rase, 1987). However, where symbols obscure one another because the objects they represent share identical locations, an alternative solution is required. An example of this problem is illustrated in Figure 10.1(a), in which the social classes of individual families living in part of the East End of London in the late 1880s are displayed on a conventional 2D map, using data from the manuscript notebooks of Charles Booth's poverty enquiry (Shepherd, 2000). Unfortunately, because of the widespread incidence of dwelling multi-occupancy in this area, many of the point symbols visible on this map conceal others beneath them. A solution to this problem (Shepherd, 2002) is to use the third display dimension in a 3D visualization to display stacks of point symbols at each location, as shown in Figure 10.1(b). Because of the widespread occurrence of locationally coincident phenomena, 2D point symbol maps should be used extremely carefully, and 3D stacked symbol visualizations should be used wherever appropriate.

10.3 Some problems with 3D views

Although 3D data visualization has undoubted benefits when compared with some 2D approaches, a considerable body of user experience and experimental research reveals significant problems in using the third dimension to create effective data visualizations. A number of these are explored below. However, space prevents detailed consideration of the highly significant problem of user interaction within 3D virtual worlds, which are the subject of a separate study (currently under preparation).

10.3.1 Scale variation across 3D scenes

A perspective view tends to be adopted for most 2.5D and 3D visualizations of spatial data. Unfortunately, because of the foreshortening effect in such views, which increases with distance from the observer, it is difficult to make accurate visual comparisons of objects within a 3D scene. This problem was recognized early on in the history of 3D digital mapping in an evaluation of the problems of interpreting graphical output from the SYMVU software (Phillips and Noyes, 1978). More recently, evidence from perception research reveals that not only is depth consistently underestimated (Swan *et al.*, 2007), but that the human visual system perceives relationships in each of the three directions separately, and that the relationship between physical and perceived space is non-Euclidean (Todd, Tittle and Norman, 1995). A number of solutions have been proposed to assist users in making effective distance and measurement estimates in 3D worlds, four of which are briefly reviewed here. It should be noted, however, that no single technique is entirely effective across the range of tasks that need to be performed in 3D scenes.

Reference frames

Most 3D histograms and bar charts generated from non-spatial data provide a set of 3D axes as an integral part of the display, and this assists across-scene position and size estimation by the viewer. However, when spatially referenced data are visualized in 3D, a standard reference framework of this kind is usually absent. In such cases, a user-defined bounding box may be drawn around some or all of the objects within the scene in order to provide some sense of scale. This may be a simple wireframe box, or a more elaborate set of calibrated axes, and the geographical extent of the reference framework may be chosen by the software and/or the

(a)

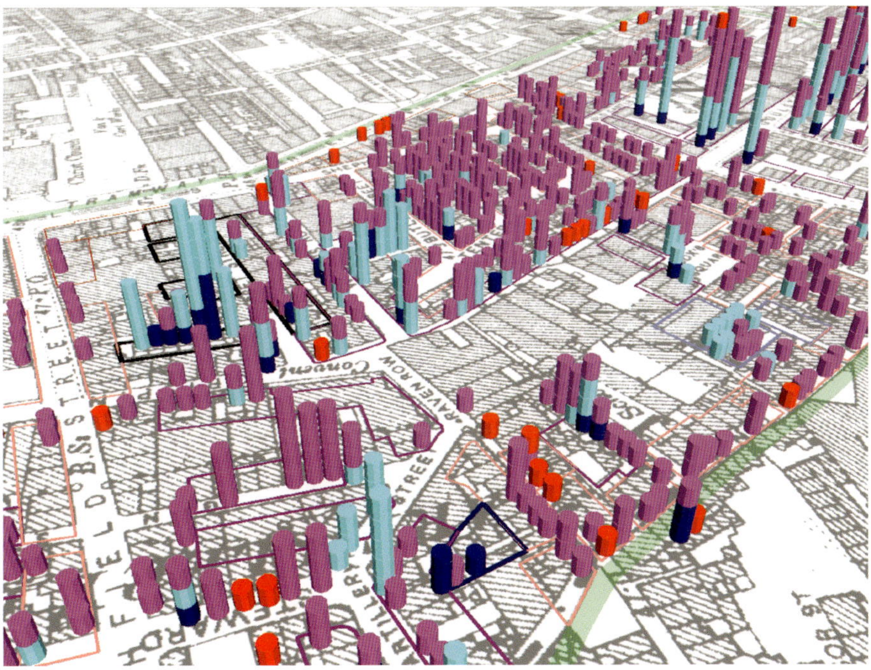

(b)

Figure 10.1 (a) A 2D vertical view of family classes in the East End of London in the late 1880s (source: author). (b) A 3D oblique view of family classes in part of the East End of London in the late 1880s (source: author) user. (Several examples may be seen in the images created using the OpenDX software, which are available at www.opendx.org/highlights.php.) Where large parts of the globe are shown in a 3D scene, the bounding box may need to be more complex, and where thematic layers are stacked above the globe, each layer may also require its own local reference frame. Clearly, the number of variables being displayed places limits on the effectiveness of this approach, which will be considered again later. For the moment, it is worth noting that perceptual research has revealed a possible distortion effect in the use of an enclosing frame, which may flatten 3D scenes, and reduce perception of depth within them (Eby and Braunstein, 1995).

Reference and slicing planes

In this solution, a plane is drawn at a specific level within the view, thus acting as a visual plane of reference for the analyst. The plane may be opaque or transparent, it may have grid lines drawn across it, and it may be drawn on any of the three axes, though in most geographical visualizations it will be positioned along the z-axis. Cutting planes are commonplace in 'slice and dice' voxel medical models, which have become more widely known through the well-known 'Virtual Body' and 'Visible Human' projects. User-adjustable cutting planes are also widely used for geophysical exploration (e.g. Fröhlich *et al.*, 1999).

A simpler example is illustrated in Figure 10.2, in which a semi-transparent reference plane is used to enable users to visually identify locations in part of north London where more

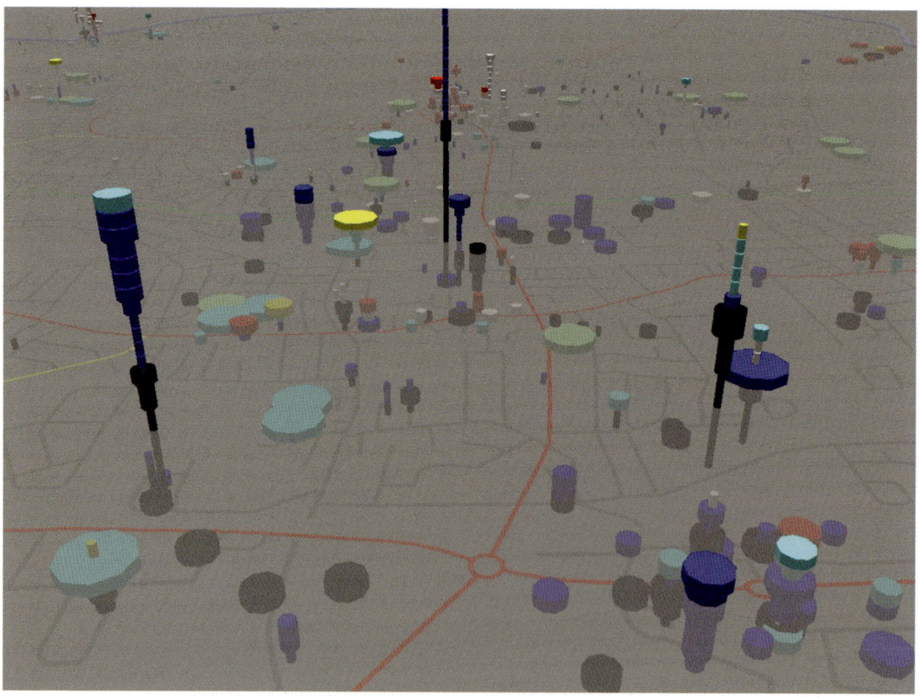

Figure 10.2 Three-dimensional view showing various types of commercial property in north London, with a semi-transparent horizontal reference plane used to highlight locations with more than five properties (source: author)

than five businesses are located at a single address. However, there are design problems with this kind of visualization technique. First, it is only effective where the stacked symbols are drawn with equal heights. (This is an example of the kind of conflict that frequently occurs between visual variables and other visual effects in 3D data visualizations.) Second, where the individual stack symbols are sorted by business attributes, as in this example, then the kinds of businesses visible above the reference plane will not necessarily be representative of the range of businesses within the entire stack. Despite these and other limitations, however, reference planes afford numerous insights into value relationships across 3D scenes, such as which symbols lie above or below a particular threshold value, or which land is lower/higher than other land.

Divided symbol stacks

In 3D data visualizations which display stacks of locationally coincident point symbols, there are usually no visual cues to indicate the vertical extent of each symbol. This prevents the analyst from making comparative visual estimates of the number of items of particular colours, forms or sizes in each stack within the view. Where stacks contain several symbols using the same set of visual variables (e.g. size, type, colour), it becomes even more difficult to make comparative estimates of the number of symbols in stacks across the entire scene. One way of providing an additional visual cue is to insert small gaps between the symbols in each stack, automatically adjusted by the software to remain proportional to the standard symbol height. An example is provided in Figure 10.3, which uses stacks of coloured cylindrical solid symbols to indicate the genders of all individuals living in residential properties in part of the East End of London, using data from the 1881 census of population. Not only do the inter-symbol gaps enable the analyst to make rough visual estimates of the number of individuals resident at each location, but the software also assists the making of broader comparative judgements by automatically sorting the symbols into common value groups (males vs females in this example) within each symbol stack.

Non-perspective projections

Although non-perspective projections have been widely used in architecture and engineering, and the isometric grid became popular with computer games in the 1990s (e.g. SimCity, Civilisation, A-Train and Theme Park), the perspective projection is almost universally used for 3D data visualization. Wyeld (2005a, b) suggests that this is part of a visual tradition that has dominated art and other visual media since the Renaissance, and that a deliberate effort is needed to wean people away from this expected form of representation. For some data interpretation tasks which involve the comparison of objects across a scene, a case can be made for setting aside the perspective view, and using instead a non-perspective projection. In Figure 10.4, for example, the view shown in Figure 10.3 is redisplayed using an orthogonal projection, which makes it easier to make cross-scene comparisons of the sizes of symbol stacks. (With this software, the viewer is able rapidly to toggle between the two projected views with a single keystroke to assist their interpretation of the data.)

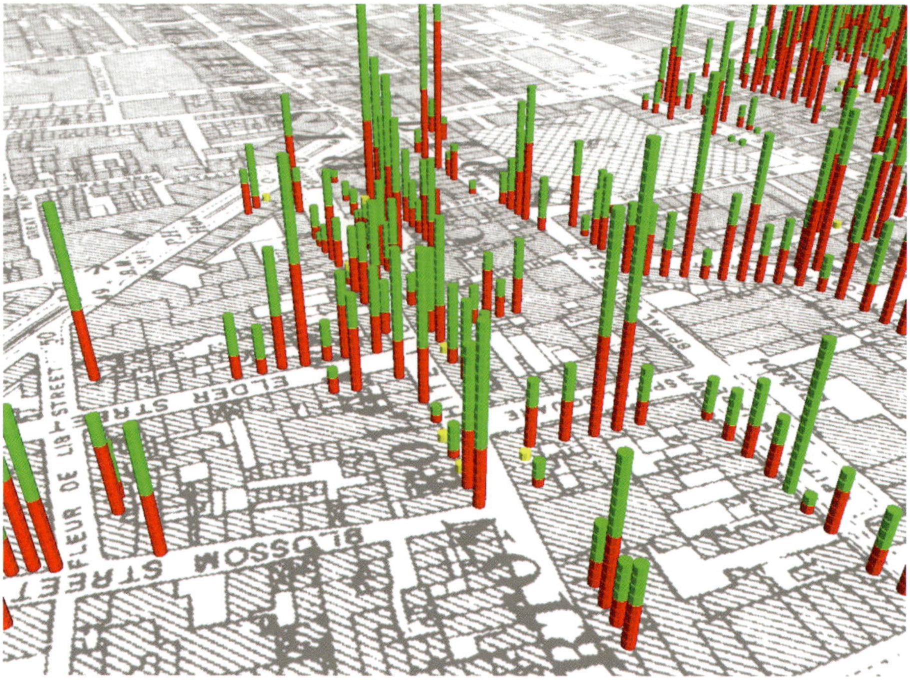

Figure 10.3 Distribution of males (red symbols) and females (green symbols) in part of the East End of London in 1881 (source: author)

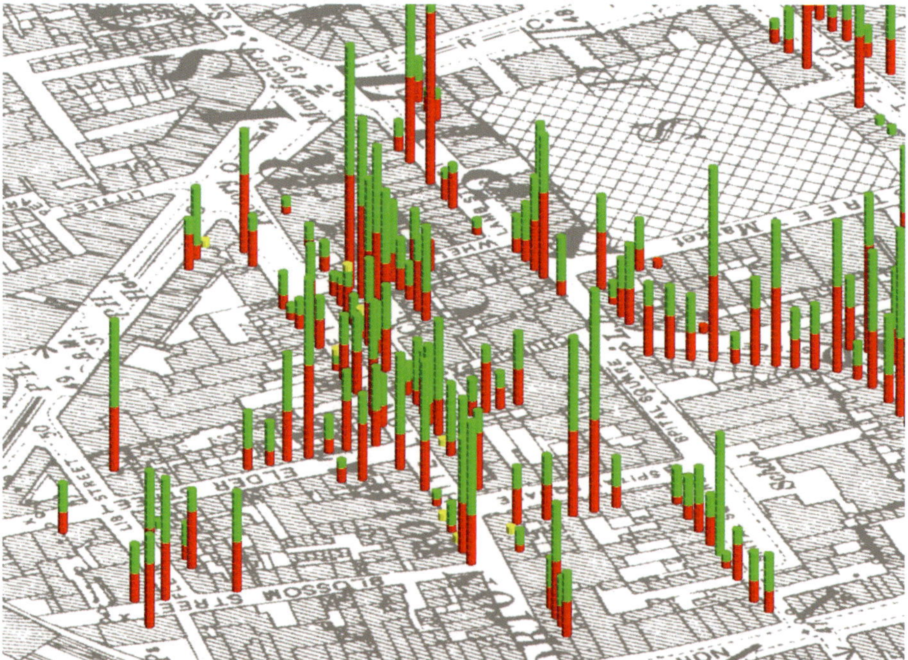

Figure 10.4 Orthogonal projection of the view shown in Figure 10.3 (source: author)

10.3.2 Symbol occlusion within 3D scenes

Although 3D data visualization may be used to resolve the problem of hidden symbols in 2D maps, as discussed above, most 3D visualizations still suffer from symbol occlusion, due to the alignment of objects within a scene in relation to the user's viewpoint. Some of the more common solutions to this problem are discussed below. The first four involve often significant modifications of the scene contents to address the problem, while the last four involve less radical surgery. As with solutions to the problem of scale variation, none of the proposed solutions is entirely satisfactory.

Object culling

This is perhaps the most obvious, but also the most draconian, solution to the occlusion problem. The chief drawback is that the removal of occluding objects reduces the ability of the analyst to view objects of interest in their natural context. Because much of the power of both 2D and 3D data visualization derives from the user's ability to simultaneously appreciate both detail and context in a single view, object culling considerably reduces the benefits of being able to see contextual data during data exploration. Various approaches have been taken to resolve this problem. For example, in some experimental visualizations of 3D node-and-link graphs (e.g. Hypergraph, 2007), the analyst is able to select the focal point of interest interactively, and selectively expand and collapse sub-trees. A more rigorous solution is proposed by Shen *et al.* (2006), who suggest the use of semantic and structural abstraction to declutter selected nodes and links from such graphs.

Object minimization

An alternative approach is temporarily to redraw inessential objects so as to reduce their visual interference with the point(s) of interest. For example, in experiments with 3D point symbols on a planar base map, Chuah *et al.* (1995) reduced the sizes of unimportant, occluded symbols, leaving the sizes of the focal symbols unchanged. Although reductions to both the heights and widths of occluding symbols is possible, the study found that width reduction was more useful where the analyst needed to compare the heights of selected symbols with all other symbols within a scene.

Object displacement

By moving selected objects away from one another in a cluttered scene, it is possible to reveal objects of interest without entirely removing or rescaling the selected objects as in the previous two methods. Three approaches have been proposed. In the first, a subset of objects of interest, together with a reference plane, is raised above the full set of objects in the original visualization. This technique is adopted in experimental data visualization software by Chuah *et al.* (1995) and Schmidt *et al.* (2004). The second method involves the differential movement of objects, or parts of objects, away from one another within a scene. A 2D example is provided by the point symbol dispersion routine provided with the MapInfo desktop mapping system. A variation of this method is more widely used in the 3D display of anatomical and other medical illustrations, creating an 'exploded view' (e.g. Bruckner and Gröller, 2006). However, the use of exploded views in statistical data visualization, as in the case of 3D exploded pie graphs, includes potential interpretative errors into the display. The third object displacement method involves the repositioning and spatial distortion of contiguous (usually zonal) objects in geographical space, as in 2D area cartograms. A danger with this family of techniques is that they render the user less likely to be able to interpret the exact spatial relationships between objects that have been moved (often by some arbitrary amount) from one another.

View distortion

By differentially distorting the overall geometry of a scene, objects near the user viewpoint may be more clearly seen. Perhaps the best-known technique is the fish-eye view (Furnas, 1991), which selectively distorts a scene so as to enlarge objects near the point of interest. One of the problems of this technique is that users may be unable to obtain a proper sense of the spatial relationships between objects, though this is more likely to affect displays of inherently spatial data than displays of purely statistical data, and it should be noted that many geographers will be used to the spatial deformations used in map projections and cartograms.

Rotation or viewer movement

One of the biggest advantages of interactive 3D data visualization over printed 3D images is that users have the ability to resolve some of the occlusion problems interactively, either by rotating objects within the scene or by moving their viewpoint with respect to the scene. This induces the kinetic depth effect, which not only reduces symbol occlusion, but also enhances the viewer's appreciation of the depth relations between objects in a scene (Ware and Franck, 1996). However, this may be accompanied by unpleasant user side effects, especially in an immersive 3D environment and may also required adeptness in navigating within 3D scenes which many analysts will not possess.

Symbol transparency

By displaying selected symbols in reduced opacity, occluded symbols may be seen through foreground symbols. Several options are available: symbols in a scene may be drawn with equally reduced opacity, or the transparency of selected occluding symbols near the focus of the user's attention may be increased, leaving those further away at full or increasing opacity. A somewhat different role for symbol transparency is to enable the comparison of two datasets in a single scene by applying transparency to one set of symbols and rendering the other set opaquely. An example is illustrated in Figure 10.5, which shows the comparative distribution of cigar makers in the East End of London from two datasets: the population census of 1881 (the green transparent symbols) and Booth's poverty survey of the late 1880s (the red opaque symbols) (Shepherd, 2000).

Figure 10.5 Use of opaque and semi-transparent symbols to compare the distribution of cigar makers in the East End of London in 1881 (transparent) and 1887 (opaque) (source: author)

Symbol shadows

When objects displayed in 3D space are viewed from a particular direction, symbol self-occlusion often makes it difficult to perceive their spatial distribution within the three axes of that space. One way of providing information about the spatial distribution is to project object shadows onto one or more planes of a bounding box surrounding the objects. (Some of the key design issues are discussed in a medical illustration context by Ritter *et al.*, 2003.) An example of this technique is illustrated in Figure 10.6, which shows the distribution of earthquakes below the Big Island of Hawaii, as viewed from a medium-oblique angle. The grey base plane displayed below the scene shows the shadows of the 3D earthquake symbols projected from above, and helps to clarify their distribution in the x–y plane. Some studies show, however, that not all 3D tasks are equally enhanced by including object shadows (Hubona *et al.*, 2000).

Multiple linked views

The occlusion of objects caused by scene complexity sometimes prevents users from comprehending the entire visualized scene from a single viewpoint. One potential solution to this problem is to display the virtual world simultaneously as seen from two or more alternative

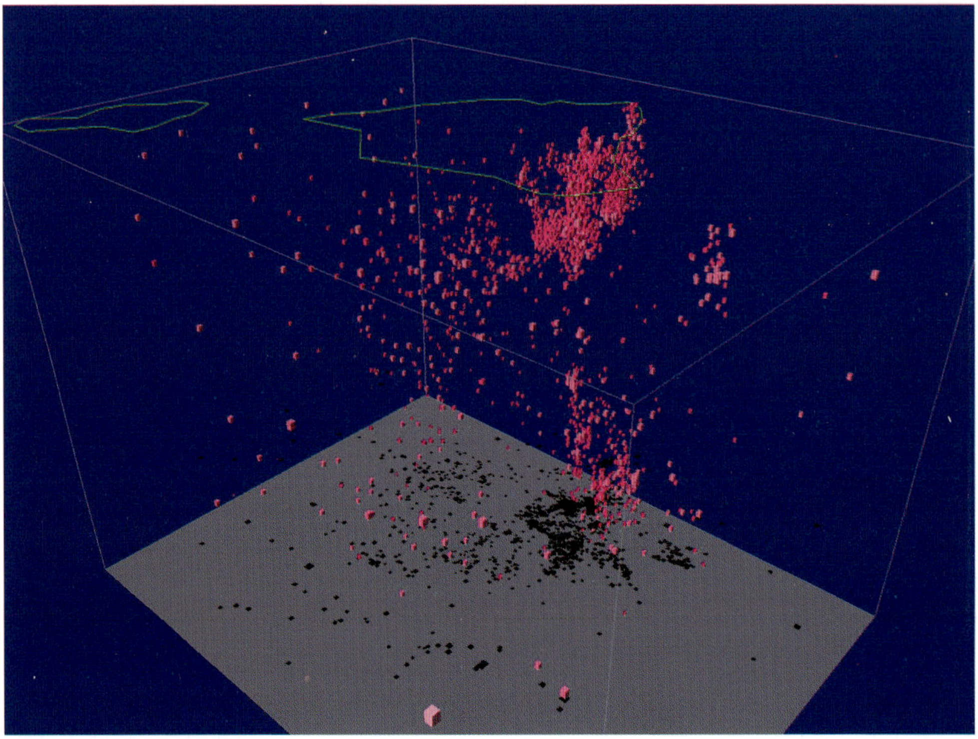

Figure 10.6 Use of projected symbol shadows to reveal the distribution of Hawaiian earthquakes in the x–y plane of a 3D visualization (source: author earthquake data from CNSS)

viewpoints in linked views. In one experiment with a multi-view 3D visualization, Plumlee and Ware (2003) found that the display of a view proxy (i.e. a triangular symbol in a vertical view indicating the field-of-view in an oblique view) and view coupling (i.e. keeping vertical and oblique views oriented in the same direction) were both beneficial for undertaking certain tasks in 3D worlds. In an evaluation of alternative methods of presenting route instructions on mobile devices, Kray *et al.* (2003) revealed that, in order to find their current location on a 2D map, users often found that a 3D view was useful in identifying their location in the real world. This study also confirmed a finding of several other studies, which is that, although the 3D display did not improve task performance, it was found to be 'fun' to use.

10.3.3 Symbol viewpoint dependencies

A normal feature of navigating through 3D scenes is that the perceived dimensions and/or shapes of objects in the scene vary according to their orientation with respect to the viewer. This becomes problematic in those data visualizations where the appearance of the 3D symbols is meant to represent data variables. An example occurs where the dimensions of 3D bar symbols (i.e. height, width, and thickness or depth) are used to encode data variables. The problem in doing this is that the perceived bar widths will not only vary according to the assigned data values, but will also vary with the observer's viewing angle. [An example occurs in a study of retail businesses in Toronto (Hernandez, 2007), in which 3D bar symbols representing businesses in Toronto are aligned with the streets on which they are located.] In extreme circumstances, where bars symbols with minimal thickness are viewed from the side, they may all but disappear. This problem was identified by the authors of a study involving the visualization of ocean-bed characteristics by Schmidt *et al.* (2004) and was resolved by using spherical 3D glyphs whose perceived sizes were independent of their orientation.

The viewpoint dependency problem not only affects the ability of symbol dimensions to carry data information, but it may also undermine the use of other symbol variations to represent data. For example, where symbol shapes are varied to reflect nominal scale data, then the analyst's viewpoint may make it difficult to perceive these shape differences across the scene (Lind, Bingham and Forsell, 2003). There are several other reasons why the 3D equivalents of 2D symbols may not work effectively in 3D scenes. As Kraak (1988, 1989) has indicated, the surface shadowing applied to 3D point symbols to enhance their realism may conflict with the visual variations in the lightness and texture applied to their surfaces by 3D rendering. Krisp (2006) illustrates a similar problem with 3D density surfaces, in which the colours used to encode height are locally modulated by the hill-shading used to enhance viewing realism. The general conclusion seems to be that, although it is relatively trivial to convert the standard 2D geometrical symbols used in conventional thematic mapping into 3D equivalents for data visualization, in which a square becomes a cube, a circle becomes a sphere, a line becomes a ribbon or wall, and a region becomes a prism, variations in the size and shape of these symbols may not be accurately perceived in a 3D data visualization because of viewpoint dependencies.

10.3.4 Stereo 3D: pretty useful, or just pretty?

Viewing 3D data visualizations in full stereo is assumed to provide the benefits of binocular viewing enjoyed by human primates in their natural environments. Although stereo images possess a certain 'wow' factor, as indicated by the entertainment value of the iMAX cinematic experience, effective stereo data visualizations are not always easy to create, and require that close attention be paid to known perceptual principles. In Figure 10.7, for example, which shows the distribution of earthquakes below the Big Island of Hawaii in 2003, use is made of the principle that variations in the lightness visual variable are better able to heighten the viewer's appreciation of depth than variations in hue (Ware, 2004).

An even more important decision-making factor concerns the evidence from visual perception research, which suggests that stereo is less effective than several other techniques for indicating depth and scene layout to the viewer when undertaking tasks in 3D visualizations. Indeed, stereo is only one of nine major visual depth cues and, in many circumstances, is not the most important (Cutting and Vishton, 1995). A question worth asking is whether the visualization in Figure 10.7 provides a better indication of the three-dimensional distribution of the earthquakes than, say, a monographic view that includes symbol shadows (as in Figure 10.6). One of the more significant alternatives to stereo is the kinetic depth effect, in which an awareness of the depth relations among objects in a scene is induced by the relative motion of foreground and background objects, either as the viewer moves or as the objects

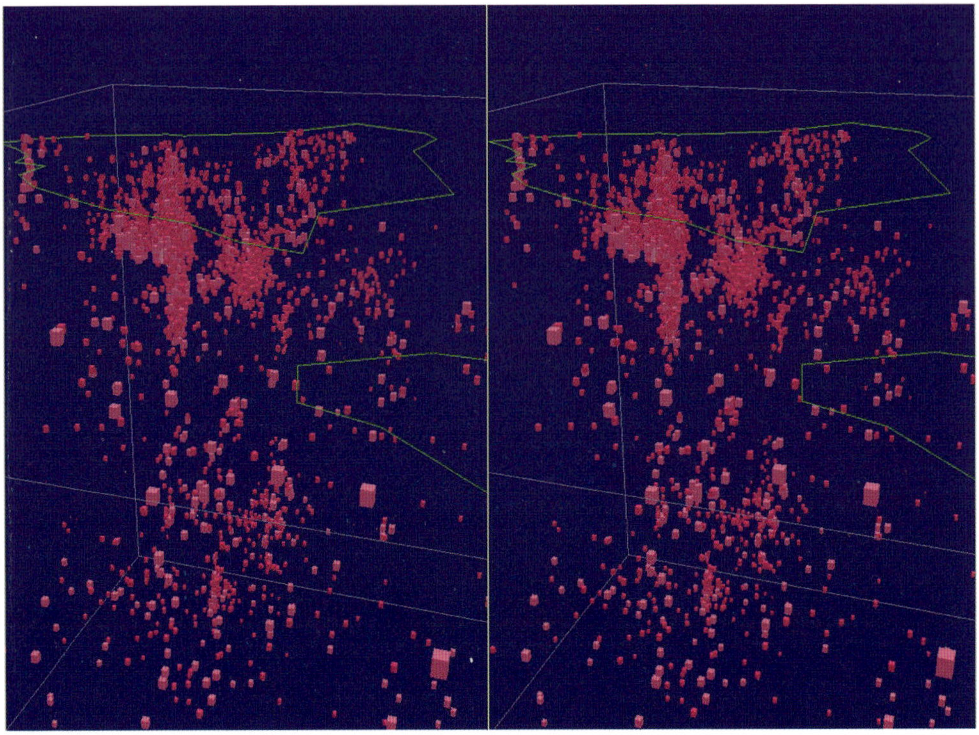

Figure 10.7 Stereo view of earthquakes below Big Island, Hawaii, using luminescence/lightness variations to enhance the depth effect (source: author; data from CNSS)

are rotated. In experiments using a head-mounted device to induce 3D perception, Ware and Franck (1996) found not only that motion parallax alone is better than stereo alone for tasks involving full 3D awareness, but also that motion parallax in combination with stereo provides the best depth cues for the analyst. Stereo should not therefore be considered a 'must have' facility but, as Ware and Franck (1996) suggest, individual data visualizations should use particular combinations of 3D-inducing effects. Few hard and fast rules are available, because the relevant combinations need to be task-specific, and further research is needed to evaluate which 3D effects are best suited to particular data visualization tasks.

10.3.5 *z*-Axis contention

With the increasing availability of spatial tracking data, several attempts have been made to automate the production of the space–time cube (Oculus, 2006; Kraak, 2007), first introduced in the 1960s (Hagerstrand, 1970). In the resulting visualizations, the third dimension is used to display time, with the *x*- and *y*-axes retaining their traditional role of displaying ground surface features. Unfortunately, when spatial data are visualized in a space–time cube, the *z*-axis is not really available as a spare dimension to be used exclusively to show time, because the movement of people and objects also takes place within the vertical spatial dimension. Although it is tempting to 'overload' the *z*-axis by using it to display both the three-dimensional landscape surface and the space–time lines above it, doing so can lead to difficulties in visual interpretation. This is because the space–time lines attached to different points on the surface are no longer visually comparable, because they are offset in the vertical plane by different amounts depending on the height of the landscape surface to which they are attached

In some low-lying study areas, as in parts of the Netherlands (Kraak 2003), this problem may be largely ignored. In other areas, where it is known that topography has little influence on space–time patterns, this problem may be solved by adopting the age-old fiction of the traditional 2D map: a flat world. However, in many hillier and mountainous areas, where it is important to understand how the landscape surface impacts on people's movement patterns, terrain height must also be shown in the cube. In such cases, *z*-axis contention becomes a serious problem that threatens to undermine the benefits of this particular 3D visualization technique.

One possible solution may be to attach the bases of all space–time lines to an arbitrary plane above the highest point of the landscape surface, although this may make it difficult for the interpreter to relate the two sets of information. As previously mentioned, this stacking technique is used in several visualizations of spatial point-located data (e.g. Chuah *et al.*, 1995; Schmidt *et al.*, 2004), and also in numerous 3D meteorological visualizations where a single surface layer is displayed some way above a globe. In one example (Aoyama *et al.*, 2007), the suspended isosurface showing land-surface temperatures is visually related to the plane base map of the USA below it by having the outlines of regions of interest in the former projected down onto the latter. However, where the *z*-axis is used to display data for several variables, this results in a stack of multiple 3D thematic layers displayed above a reference surface. Even more complex techniques are needed to relate the locations of objects on one layer to those on other layers. In oil reservoir visualizations (e.g. Calomeni and Celes, 2006), the inclusion of vertical lines representing boreholes partly reduces this problem. In general, however, and despite the undeniable artistry of some of the complex visual models, one begins to wonder whether one has reached the limits of 3D data visualization as an exploratory and interpretive tool. Rather than overload the *z*-axis, it might be more effective for the analyst to resort to map overlay analysis or, as discussed in the next section, to adopt some form of dimension reduction techniques in order to simplify the data before or during the visualization process.

10.3.6 The 'dimensionality curse'

The challenge of what Robertson, Mackinlay and Card (1991, p. 189) refer to as 'intellectually large data collections' is a problem that is only partly solved by moving from 2D to 3D visualizations. Although it is possible, as previously discussed, to display very large numbers of objects in 3D scenes with powerful hardware and clever algorithms, visualization limits are rapidly reached as the number of dimensions (i.e. variables, fields or attributes) increases above a relatively small number. As Bertin (1967, 1977) fully appreciated, the graphical sign system cannot be used to display more than a handful of variables in a single scene, and this remains the case even after it has been augmented with additional visual variables (e.g. transparency, blur, and specular highlights), or extended into the third dimension through complex articulated glyphs and icons.

Even when 3D symbols are used to encode up to half a dozen data variables, this is still a long way away from being able to satisfy the needs of many analysts, whose datasets commonly include scores or even hundreds of variables. In the innovative 3D seafloor visualizations created by Schmidt *et al.* (2004), for example, only five variables were simultaneously displayed, and in the iconographic visualizations created by Gee *et al.* (1998) and Healey (1998), only a similar number of data variables were encoded by means of icon geometry and colour. Claims (e.g. by Wright, 1997) that certain 3D data visualization techniques are capable of displaying hundreds of variables are therefore little more than marketing hype.

Given this problem, it becomes necessary to think the unthinkable, and to consider setting aside the representational spatial framework in which much of our data is gathered. Indeed, relevant 2D and 3D visualization techniques that are stripped of a geographical frame of reference may provide powerful insights into complex data in ways that cannot be provided by conventional spatial visualizations.

Three broad strategies have been adopted. The first is to retain the geographical semantics of the real world in the 3D display space, but to switch between visual representations of selected (small) subsets of variables. Most of the discussion in this chapter so far has focused on this strategy. A second strategy is to abandon geographical semantics in the 3D display space, and to use visual data mining or visual analytics techniques (de Olivera and Levkowitz, 2003) to display information for multiple variables in 2D displays. Such visualizations include parallel coordinates (Inselberg and Dimsdale, 1989), pixel-based displays (Keim, 2000), 'dimensional anchors' (Hoffman, Grinstein and Pinkney, 2000), and an increasingly large number of other multivariate data visualization techniques (Wong and Bergeron, 1997). Where the number of variables or dimensions is extremely large, 2D projections may be made from multidimensional data into 2D spaces, or components may be derived from the original data by various data reduction techniques, and displayed in 2D display space (e.g., Yang *et al.*, 2004, 2007). By coupling data reduction techniques to data visualization, visualization continues its tradition of helping the analyst to steer the data mining in potentially fruitful directions (Keim and Kriegel, 1994; Keim, 2002), but operating within statistical space rather than geographical space.

A third strategy is to combine the spatial and non-spatial approaches in a multiple linked views environment. For example, parallel coordinates displays have been incorporated into several software systems designed for the analysis of spatial data, along with 2D maps and other forms of tabular and graphical display (e.g. Stolte, Tang and Hanrahan, 2002; Andrienko and Andrienko, 2003; Guo, 2003; Guo, Chen and MacEachren, 2006; Marsh, Dykes and Attilakou, 2006). (This multiple linked views approach has also been developed in other research fields, e.g. Gresh *et al.*, 2000.) A notable feature of these hybrid systems is that 3D spatial representations are notable by their absence; almost without exception, they only incorporate 2D maps and 2D statistical graphics.

This suggests two interesting possibilities. The first is that 2D display techniques may be more effective than 3D techniques for routine data interpretation purposes, even for spatial data. The second is that analysts looking for comprehensive data visualization software should not expect to find them solely among the offerings of current GIS and desktop mapping software vendors, whose focus is mainly on spatial data management and analysis. Most of the innovative data visualization software of the past two decades has emerged from the non-spatial sciences, and especially the information visualization community. By interfacing effective modular software from these sources to standard 2D mapping tools developed within geography, it may be possible to acquire the most effective toolkit for both spatial and non-spatial data analysis. This has been the motivation for the author's own experimental 3D data visualization software, which imports data from MapInfo and

ArcView, in the spirit of his earlier advocacy of DIY GIS (Shepherd, 1990) and software federations (Shepherd, 1991). However, thorough evaluation of the usefulness of such software federations for routine analytical tasks is necessary if we are to determine whether they outperform currently available GIS and mapping software, and whether they render eye-catching 3D visualizations an unnecessary luxury.

10.4 Conclusions

Over the past quarter of a century, a wealth of experimentation in 3D data visualization has taken place, so that just about anything one wants to see in 3D can now be produced almost automatically. However, our review suggests that 3D is not always as useful for data visualization as it has sometimes been made to appear. Each of the advantages and benefits claimed for 3D has its caveats, and not all of the known problems with 3D have completely satisfactory solutions. As graphics hardware gets increasingly powerful, and visualization software gets ever more sophisticated, it becomes increasingly important to step back from the compelling visual image on the screen and ask some relatively simple questions: is the visualized scene free from distortion, bias or other visual error? Are the display methods used appropriate for the task in hand? Would any patterns hidden in the data be more evident if 2D visualization methods were used? Are the visualization techniques being used best suited to the current user?

Just because it can be done does not meant that it should be done; some 3D effects are of questionable analytical value, and 3D is not always better than 2D for visualizing data. Indeed, Lind, Bingham and Forsell (2003) suggest that, because of the distortions introduced by human space perception, 'the general usefulness of a 3D visual representation may be limited', particularly in situations where analysts are meant to discover relations based on Euclidean distances or shapes. They suggest that the primary role for 3D may be in providing users with a general overview of object relationships in a scene, and especially for spatial data. Others (e.g. Kray *et al.*, 2003) have suggested that a large part of the appeal of 3D displays for users undertaking particular spatial tasks lies in their entertainment or 'fun' value. For his part, the guru of web usability has thrown the following provocative claim into the ring: '3D is for demos. 2D is for work' (Nielsen, 2006; see also Nielsen, 1998).

A great deal of evaluation remains to be undertaken to identify which, when and how currently available 3D data visualization tools and techniques should be used. Despite the considerable progress made in recent years, the technology of data representation is still in its formative stages, and developers, researchers, educators and users alike have major contributions to play in improving the technology, and its effective use. Developers need to bridge the gap between what is currently available and what is desirable; researchers need to undertake rigorous evaluations of alternative approaches to visualization and interaction, in order to identify the fitness for purpose of existing and emerging technologies; educators and trainers have a responsibility to help users understand the principles and limitations of 3D visualization, as well as teaching them how to make effective choices in harnessing the power of available tools for their needs; and individual users face a continuous learning challenge in making effective use of the many dimensions available to them in making sense of their data. We may all have been born into a 3D world, but we need to be continually aware that our virtual 3D worlds are sometimes more challenging than the real thing.

Acknowledgements

It is a pleasure to thank my son, Iestyn, for the many fruitful discussions on 3D computer graphics we have had over the years, for helping me to understand the potential contributions that modern computer games technology can make to geographical data visualization, and also for cutting the code. Many of the ideas in this chapter could not have been developed without his continuing partnership in my research.

References

Abel, P., Gros, P., Loisel, D., Russo Dos Santos, C. and Paris, J. P. (2000) CyberNet: a framework for managing networks using 3D metaphoric worlds. *Annales des Télécommunications*, 55(3–4): 131–142.

Andrienko, N. and Andrienko, G. (2003) Informed spatial decisions through coordinated views. *Information Visualization*, 2: 270–285.

Aoyama, D. A., Hsiao, J.-T. T., Cárdenas, A. F. and Pon, R. K. (2007) TimeLine and visualization of multiple-data sets and the visualization of querying challenges. *Journal of Visual Languages and Computing*, 18: 1–21.

Bertin, J. (1967) *Sémiologie Graphique: les diagrammes, les reseaux, les cartes.* Paris, Editions de l'Ecole des Hautes Etudes en Sciences. [Second edition, 1979, Paris–La Haye, Mouton. Translated by Berg, W. J. as: *Semiology of Graphics.* Madison, WI: The University of Wisconsin Press.]

Bertin, J. (1977) *La Graphique et le Traitement Graphique de l'Information.* Paris, Flammarion. [Translated by Berg, W. J. and Scott, P. (1981) as *Graphics and Graphic Information-Processing.* Berlin, Walter de Gruyter.

Börner, K., Penumarthy, S., DeVarco, B. J. and Kearnety, C. (2003) *Visualizing Social Patterns in Virtual Environments on a Local and Global Scale.* Lecture Notes in Computer Science. Berlin, Springer, pp. 1–14

Bruckner, S. and Gröller, M. E. (2006) Exploded views for volume data. *IEEE Transactions on Visualization and Computer Graphics*, 12(5): 1077–1084.

Calomeni, A. and Celes, W. (2006) Assisted and automatic navigation on black oil reservoir models based on probabilistic roadmaps, *Proceedings of the 2006 Symposium on Interactive 3D Graphics and Games*, pp. 175–182, 232.

Card, S., Robertson, G. G. and Mackinlay, J. (1991) The Information Visualizer, an information workspace, *CHI Proceedings*, pp. 181–188.

Carpendale, M. S., Cowperthwaite, D. J. and Fracchia, F. D. (1997) Extending distortion viewing from 2D to 3D, *IEEE Computer Graphics and Applications*, July/August: 42–51.

Chlan, E. B. and Rheingans, P. (2005) Multivariate glyphs for multi-object clusters, *Proceedings of the 2005 IEEE Symposium on Information Visualization (INFOVIS '05)*, pp. 141–148.

Chuah, M. C., Roth, S. F., Mattis, J. and Kolojejchick J. (1995) SDM: selective dynamic manipulation of visualizations, *Proceedings of the ACM Symposium on User Interface Software and Technology, UIST 95*, pp. 61–70.

Cockburn, A. (2004) Revisiting 2D versus 3D implications on spatial memory, *Proceedings of the Fifth Conference on Australasian User Interface*, Vol. 28, pp. 25–31.

Cutting, J. E. and Vishton, P. M. (1995) Perceiving layout and knowing distances: the integration, relative potency, and contextual use of different information about depth. In *Handbook of*

Perception and Cognition, Vol. 5: Perception of Space and Motion, Epstein, W. and Rogers, S. (eds). San Diego, CA, Academic Press, pp. 69–117.

de Olivera, F. M. C. and Levkowitz, H. (2003) From visual data exploration to visual data mining: a survey, *IEEE Transactions on Visualization and Computer Graphics*, 9(3): 378–394.

Döllner, J. (2005) Geovisualization and real-time 3D computer graphics. In *Exploring Geovisualization*, MacEachren, A. M. and Kraak, M.-J. (eds). London, Pergamon, pp. 325–344.

Eby, D. W. and Braunstein, M. L. (1995) The perceptual flattening of three-dimensional scenes enclosed by a frame. *Perception*, 24(9): 981–993.

Fanea, E., Carpendale, S. and Isenberg, T. (2005) An interactive 3D integration of parallel coordinates and star glyphs, available at: pages.cpsc.ucalgary.ca/~isenberg/papers/Fanea_2005_I3I2.pdf (last accessed 14 February 2007).

Fröhlich, B., Barrass, S., Zehner, B., Plate, J. and Göbel, M. (1999) Exploring geo-scientific data in virtual environments, *IEEE Visualization, Proceedings of the Conference on Visualization '99*, pp. 169–173.

Furnas, G. W. (1991) Generalized fisheye views, *Proceedings CHI '91 Human Factors in Computing Systems*, pp. 16–23.

Gee, A., Pinckney, S., Pickett, R. and Grinstein, G. (1998) Data presentation through natural scenes, available at: www.cs.uml.edu/~agee/papers/Gee1998.html (last accessed 14 January 2007).

Gresh, D. L., Rogowitz, B. E., Winslow, R. L., Scollan, D. F. and Yung, C. K. (2000) WEAVE: a system for visually linking 3-D and statistical visualizations, applied to cardiac simulation and measurement data, *IEEE Visualization, Proceedings of the Conference on Visualization '00*, pp. 489–495.

Guo, D. (2003) Coordinating computational and visual approaches for interactive feature selection and multivariate clustering. *Information Visualization*, 2: 232–246.

Guo, D., Chen, J. and MacEachren, A. M. (2006) A visualization system for space–time and multivariate patterns (VIS-STAMP), *IEEE Transactions on Visualization and Computer Graphics*, 12(6): 1461–1474.

Güven, S. and Feiner, S. (2003) Authoring 3D hypermedia for wearable augmented and virtual reality, *Proceedings of the ISWC '03 (Seventh International Symposium on Wearable Computers)*, White Plains, NY, pp. 118–226

Hagerstrand, T. (1970) What about people in regional science?, *Papers of the Regional Science Association*, 24: 7–21.

Healey, C. G. (1998) On the use of perceptual cues and data mining for effective visualization of scientific datasets, *Proceedings Graphics Interface '98*, pp. 177–184.

Henderson, D. A. and Card, S. K. (1986) Rooms: the use of multiple virtual workspaces to reduce space contention in a window-based graphical user interface, *ACM Transactions on Graphics*, 5(3): 211–243.

Hendley, R. J., Drew, N. S., Wood, A. M. and Beale, R. (1995) Narcissus: visualizing information, *Proceedings of the International Symposium on Information Visualization*, Atlanta, GA, pp. 90–94.

Hernandez, T. (2007) Enhancing retail location decision support: the development and application of geovisualization, *Journal of Retailing and Consumer Services*, 14(4): 249–258.

Hoffman, P., Grinstein, G. and Pinkney, D. (2000) Dimensional anchors: a graphic primitive for multidimensional information visualizations, *Proceedings of the Workshop on New Paradigms in Information Visualization and Manipulation: 8th ACM International Conference on Information and Knowledge Management*, pp. 9–16.

Hopkin, D. (1983) Use and abuse of colour, *Proceedings of the '83 International Computer Graphics Conference*, Pinner Green House, Middlesex, pp. 101–110.

Hubona, G. S., Wheeler, P. N., Shirah, G. W. and Brandt, M. (2000) The relative contributions of stereo, lighting and background scenes in promoting 3D depth visualization, *ACM Transactions on Computer–Human Interaction*, 6(3): 214–242.

Hypergraph (2007) Hypergraph software website, hypergraph.sourceforge.net (last accessed 15 February 2007).

id Software (2007) Official id Software history web page, www.idsoftware.com/business/history/ (last accessed 2 February 2007).

Inselberg, A. and Dimsdale, B. (1989) Parallel coordinates: a tool for visualizing multidimensional geometry, *Proceedings of the 1st conference on Visualization '90*, pp. 361–378.

Keim, D. A. (2000) Designing pixel-oriented visualization techniques: theory and applications, *IEEE Transactions on Visualization and Computer Graphics*, 6(1): 59–78.

Keim, D. A. (2002) Information visualization and visual data mining, *IEEE Transactions on Visualization and Computer Graphics*, 7(1): 100–107.

Keim, D. A. and Kriegel, H.-P. (1994) VisDB: database exploration using multidimensional visualization, *IEEE Computer Graphics and Applications*, 6: 40–49.

Kraak, M.-J. (1988) *Computer-assisted Cartographical Three-dimensional Imaging Techniques.* Delft, Delft University Press.

Kraak, M.-J. (1989) *Computer-assisted cartographical 3D imaging techniques.* In *Three-dimensional Applications in Geographical Information Systems*, Raper, J. (ed.). London, Taylor & Francis, pp. 99–114.

Kraak, M.-J. (2003) The space–time cube revisited from a geovisualization perspective, *Proceedings of the 21st International Cartographic Conference (ICC)*, pp. 1988–1996.

Kraak, M.-J. (2007) Geovisualization and time – new opportunities for the space–time cube, Chapter 15 in this volume.

Kray, C., Laakso, K., Elting, C. and Coors, V. (2003) Presenting route instructions on mobile devices, *Proceedings of the 8th International Conference on Intelligent User Interfaces*, pp. 117–124.

Krisp, J. M. (2006) *Geovisualization and Knowledge Discovery for Decision-Making in Ecological Network Planning.* Helsinki, Publications in Cartography and GeoInformatics, Helsinki University of Technology.

Levkowitz, H. (1991) Color icons: merging color and texture perception for integrated visualization of multiple parameters, *IEEE Visualization, Proceedings of the 2nd Conference on Visualization '91*, pp. 164–170.

Lind, M., Bingham, G. P. and Forsell, C. (2003) Metric 3D structure in visualizations. *Information Visualization*, 2: 51–57.

Marsh, S. L., Dykes, J. and Attilakou, F. (2006) Evaluating a visualization prototype with two approaches: remote instructional vs. face-to-face exploratory, *Proceedings of the Information Visualization (IV '06).*

McCormick, B. H., DeFanti, T. A. and Brown, M. D. (eds) (1987) *Visualization in Scientific Computing.* New York, ACM SIGGRAPH.

Nielsen, J. (1998) 2D is better than 3D, Jacob Nielsen's Alertbox, 15 November 1998, available at: www.useit.com/alertbox/981115.html (last accessed 19 February 2007).

Nielsen, J. (2006) Usability in the movies – top 10 bloopers, Jacob Nielsen's Alertbox, 18 December 2006, available at: www.useit.com/alertbox/film-ui-bloopers.html (last accessed 19 February 2007).

Oculus (2007) GeoTime, www.oculusinfo.com/SoftwareProducts/GeoTime.html (last accessed 15 February 2007).

Penumarthy, S. and Börner, K. (2006) Analysis and visualization of social diffusion patterns in three-dimensional virtual worlds. In *Avatars at Work and Play: Collaboration and Interaction in Shared Virtual Environments*, Schroeder, R. and Axelsson, A.-S. (eds). London, Springer, pp. 39–61.

Phillips, R. J. and Noyes, L. (1978) An objective comparison of relief maps produced with the SYMAP and SYMVU programs. *Bulletin of the Society of University Cartographers*, 12: 13–25.

Pickett, R. M. and Grinstein, G. G. (1988) Iconographic displays for visualizing multidimensional data, *Proceedings of the IEEE Conference on Systems, Man, and Cybernetics '88*, Piscataway, NJ, pp. 361–370.

Pierce, J. (2000) Expanding the Interaction Lexicon for 3D Graphics. PhD thesis, Carnegie Mellon University, Pittsburgh. Available at: www.cs.cmu.edu/~jpierce/publications/publications.html (last accessed 1st February 2007).

Plumlee, M. and Ware, C. (2003) An evaluation of methods for linking 3D views, *Proceedings ACM SIGGRAPH 2003, Symposium on Interactive 3D Graphics*, pp. 193–201. New York, ACM Press.

Raper, J. F. (ed.) (1989) *Three-dimensional Applications in Geographical Information Systems.* London, Taylor & Francis.

Rase, W. D. (1987) The evolution of a graduated symbol software package in a changing graphics environment. *International Journal of Geographical Information Systems*, 1(1): 51–65.

Rhind, D., Armstrong, P. and Openshaw, S. (1988) The Domesday machine: a nationwide geographical information system, *The Geographical Journal*, 154(1): 56–68.

Ritter, F., Sonnet, H., Hartmann, K. and Strothotte, T. (2003) Illustrative shadows: integrating 3D and 2D information displays, *Proceedings of the 8th International Conference on Intelligent User Interfaces*, pp. 166–173.

Robertson, G. G., Mackinlay, J. D. and Card, S. K. (1991) Cone trees: animated visualizations of hierarchical information, *Proceedings of the ACM CHI 91 Human Factors in Computing Systems Conference.* New York, ACM Press, pp. 189–194.

Robertson, P. K. (1990) A methodology for scientific data visualization: choosing representations based on a natural scene paradigm, *Proceedings of the First IEEE Conference on Visualization (Visualization '90)*, pp. 114–123.

Schmidt, G. S., Chen, S.-L., Bryden, A. N., Livingston, M. A., Osborn, B. R. and Rosenblum, L. J. (2004) Multi-dimensional visual representations for underwater environmental uncertainty. *IEEE Computer Graphics and Applications*, September/October: 56–65.

Shen, Z., Ma, K.-L. and Eliassi-Rad, T. (2006) Visual analysis of large heterogeneous social networks by semantic and structural abstraction.*IEEE Transactions on Visualization and Computer Graphics*, 12(6): 1427–1439.

Shepherd, I. D. H. (1990) Build your own GIS? *Land and Minerals Surveying*, 8(4): 176–183.

Shepherd, I. D. H. (1991) Information integration and GIS. In *Geographical Information Systems: Principles and Applications*, Maguire, D., Rhind, D. and Goodchild, M. (eds), Vol. 1. London, Longmans, pp. 337–360.

Shepherd, I. D. H. (2000) Mapping the poor in late-Victorian London: a multi-scale approach. In *Getting the Measure of Poverty: the Early Legacy of Seebohm Rowntree*, Bradshaw, J. and Sainsbury, R. (eds). Aldershot, Ashgate, Vol. 2, pp. 148–176.

Shepherd, I. D. H. (2002) It's only a game: using interactive graphics middleware to visualise historical data. *Society of Cartographers Bulletin*, 36(2): 51–55.

Stolte, C., Tang, D. and Hanrahan, P. (2002) Query, analysis, and visualization of hierarchically structured data using Polaris, *Proceedings of the Eighth ACM SIGKDD International Conference on Knowledge Discovery and Data Mining*, pp. 112–122.

Sutherland, I. E. (1963a) SketchPad: a man-machine graphical communication system, PhD thesis, available at www.cl.cam.ac.uk/techreports/UCAM-CL-TR-574.pdf, Massachusetts Institute of Technology.

Sutherland, I. E. (1963b) SketchPad: a man–machine graphical communication system. *AFIPS Conference Proceedings*, 23: 323–328.

Sutherland, I. E. (1969) A head-mounted three dimensional display, Fall Joint Computer Conference. *AFIPS Conference Proceedings*, 33: 757–764.

Swan, J. E., Jones, A., Kolstad, E., Livingston, A. and Smallman, S. (2007) Egocentric depth judgments in optical, see-through augmented reality. *IEEE Transactions on Visualization and Computer Graphics*, 13(3): 429–442.

Sweet, G. and Ware, C. (2004) View direction, surface orientation and texture orientation form perception of surface shape, *Proceedings of Graphics Interface*, pp. 97–106.

Todd, J. T., Tittle, J. S. and Norman, J. F. (1995) Distortions of three-dimensional space in the perceptual analysis of motion and stereo. *Perception*, 24(1): 75–86.

van Ham, F. and van Wijk, J. J. (2003) Beamtrees: compact visualization of large hierarchies. *Information Visualization*, 2: 31–39.

Ware, C. (2004) *Information Visualization: Perception for Design*, 2nd edn. San Fransisco, CA: Morgan Kaufmann.

Ware, C. and Franck, G. (1996) Evaluating stereo and motion cues for visualizing information nets in three dimensions. *ACM Transactions on Graphics*, 15(2): 121–140.

Ware, C. and Plumlee, M. (2005) 3D geovisualization and the structure of visual space. In *Exploring Geovisualization*, Dykes, J., MacEachren, A. M. and Kraak, M.-J. (eds). London, Pergamon, pp. 567–576.

Wong, P. C. and Bergeron, R. D. (1997) 30 years of multidimensional multivariate visualization. In *Scientific Visualization: Overviews, Methodologies, and Techniques*, Nielson, G. M., Hagen, H. and Muller, H. (eds). New York, IEEE Computer Society Press, pp. 361–370.

Wright, W. (1997) Multi-dimensional representations – how many dimensions? Cluster stack visualization for market segmentation analysis, available at www.oculusinfo.com/expertise.html (last accessed 10 February 2007).

Wyeld, T. G. (2005a) 3D Information Visualisation: An Historical Perspective, *Proceedings of the Ninth International Conference on Information Visualisation (IV '05)*, pp. 593–598.

Wyeld, T. G. (2005b) The pedagogical benefits of stepping outside the perspective paradigm: challenging the ubiquity of Western visual culture, *ETCC '05*, Taiwan.

Yang, J., Patro, A., Huang, S., Mehta, N., Ward, M. O. and Rundensteiner, E. A. (2004) Value and relation display for interactive exploration of high dimensional datasets, *IEEE Symposium on Information Visualization*, pp. 73–80.

Yang, J., Hubball, D., Ward, M., Rundensteiner, E. A. and Ribarsky, W. (2007) Value and relation display: interactive visual exploration of large datasets with hundreds of dimensions, *IEEE Transactions on Visualization and Computer Graphics*, 13(3): 494–507.

11

Experiences of Using State of the Art Immersive Technologies for Geographic Visualization

Martin Turner and Mary McDerby

Research Computing Services, University of Manchester

11.1 Introduction

Over the past couple of decades we have seen a rapid decrease in the price of hardware components for visualization technology, including graphics card, projectors and display screens. As a result, within the academic scientific community in the UK there are now a few hundred high-end visualization centres offering various levels of immersive or semi-immersive experiences. Immersive systems try to make you unaware of your physical self by presenting visual and at times non-visual stimuli through devices including headsets, large-scale screens and high-resolution displays. This expansion has, therefore, opened the door for many new researchers, including those in geographic studies, to explore new forms of visualization experiences and different modes of interactive graphical investigation.

Scientific and information visualizations have immense power to convince and illustrate, and at times enable users to gain a higher level of insight and inspiration within their geographic, spatial or statistical data. Spence (2006, p5) described the process of gaining insight from data through visualization as the 'ah-ha – *now* I understand' exclamation from possibly just a single glance. For a variety of data types some of the most famous good examples of visualizations have been collated by Tufte in his series of books (Tufte, 1991, 1997, 2001, 2006), and specifically for the field of cartography and statistical visualization, Friendly and Davis (2006) have compiled an extensive milestone timeline. When applied within visualization centres, the experience has been termed creating a heightened level of 'presence' (Slater, Steed and Chrysanthou, 2002), achieved by combining good visualization techniques

Geographic Visualization Edited by Martin Dodge, Mary McDerby and Martin Turner

with well-specified and maintained immersive and semi-immersive environments. So with the expansion in number of these centres there has also been an attempt to extend and control the level of 'presence' achieved, and thus aid a user to effectively explore and understand their own data sets.

Visualizations also have the unfortunate ability to optically trick and fool users, which neurobiologists and even artists have known and used for a long time. We will explore in this chapter a few examples of how increasing awareness in a controlled manner can and has created better ways to describe space and localization with minimal visual confusion. Most people have heard the saying that 'a picture describes a thousand words', but to quote Hewitt *et al.* (2005), to countless students, what a good visualization requires is a 'thousand words to describe it'.

To illustrate some best practice in displaying visualizations within large purpose-built centres we will first describe the human visual system and some of the illusions that are inherent. Then, illustrated with relevant examples, we will consider only a few of the different visualization systems that are available and importantly some of the modes they can be operated in. Remembering the limitations of the human visual system, we wish to finally explain how textural or aural descriptions are required, in fact we believe essential, for effective geographic visualization.

11.2 The human visual system

In order to understand visualization and immersive technologies that try to cover the complete visual experience, we will first step back and analyse the human visual system. Depending on the reference, approximately one-third of the human brain is involved with the processing of visual input (Oyster, 2006). Therefore, to transmit a large amount of information, using visual stimuli can be one of the most efficient methods. Schroeder, Martin and Lorensen (2006, chapter 1) summarized this as:

> visualisation is the transformation of data or information into pictures. Visualisation engages the primal human sensory apparatus, vision, as well as the processing power of the human mind. The result is a simple and effective medium for communicating complex and/or voluminous information.

With the continual increase in resolution of capturing devices, the flow of raw data has become larger and larger and the role of appropriate visualization more important. In fact the quantities of data being produced by simulations of physical, natural and theoretical problems are, for example, frequently so large that graphical representations offer the only viable way for researchers to assimilate them. Immersive systems therefore must be considered, as well as the way the visual system will perceive them. The psychologist Alcock (Alcock, Burns and Freeman, 2003) described our brain as a 'belief engine' – constantly processing information from our senses and then creating an ever-changing belief system about the world we live in. We will next consider the route a photon of light takes to become a known signal, which is used to help create a specific belief, as well as how this simple transformation process from light to data can lead to misinformation and a false belief.

11.2.1 Photons of light turned into information

Neurobiologists have probed deeper within real vision systems to gain a better understanding of the biological and organizational arrangement of this visual 'belief engine' (Kandel, Schwartz and Jessell, 1991; Thompson, 2000). Once light is absorbed by the retina it is translated into neural information, to become a set of synaptic responses. The retina at the back of the eye consists of about 125 million photoreceptors – one-third are cones that are used for colour vision and two-thirds are rods that are used for grey-scale vision. Probabilistically it is possible for a single photon to excite one rhodopsin molecule in a rod, causing about 500 molecules of transducin to be released, resulting in the blocking of about 1 million sodium ions in the rod that then results in a change of synaptic response (Kingsley, Gable and Kingsley, 1996). So in the right conditions, that is, a dark room and no stimulus for 20 plus minutes, a single photon on the retina can be visible – which shows one of the many ways the evolution of the human visual system is at both the biological and physical limits. Unfortunately, knowing the limits of the system do not tell us about the way we see whole objects and explore in our case geographic structures and maps.

Although there are about 125 million receptors, there are only about one million ganglion cells that are used to transmit this information from the retina to the brain along the optic nerve. These signals also flow relatively slowly, resulting in a communication bottle-neck that requires about a 250:1 compression ratio. This means a lot of data will be lost. In the human eye this is achieved by intermediary cells, illustrated in Figure 11.1, that are positioned between the retina and the optic nerve. These cells consider differences in intensity value

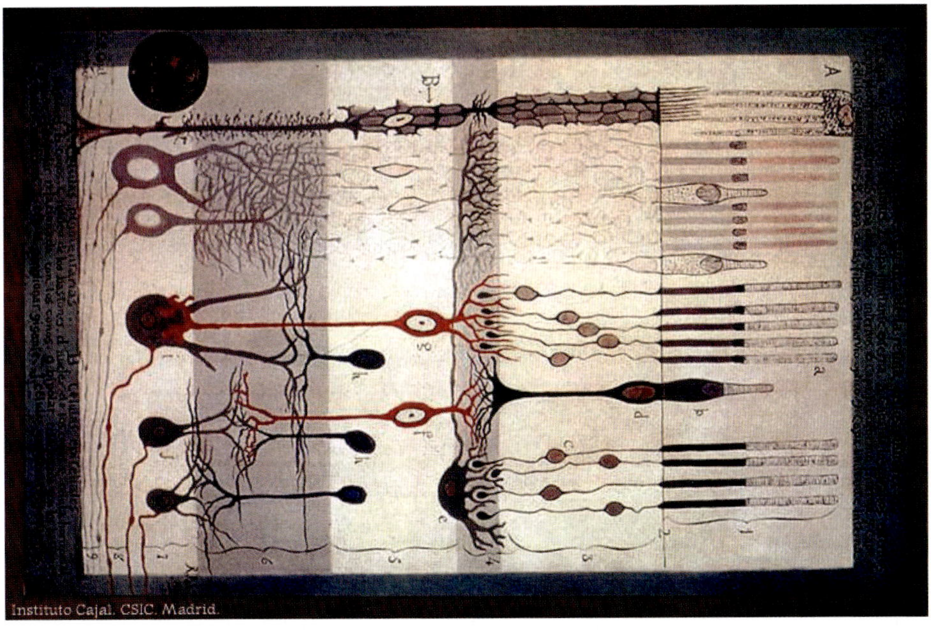

Figure 11.1 Structure of the mammalian retina (source *ca* 1900 by Santiago Ramon y Cajal). Light enters from the left and the rods and cones are shown on the far right. In the middle are the intermediary cells with connectors going to the optic nerve on the far left

between the summations of large neighbouring regions, so we have the situation of a huge but controlled information loss (Gregory, 1998).

The visual pathway also splits the right and left sides of the two retinas (information travelling along the optic nerve) to the different halves of the brain within the visual cortex. This is vitally important, especially for stereoscopic vision, as the two images of an object focused on the two retinas will be found right next to each other, within the visual cortex of the brain. We therefore have two stages of large information processing being carried out: first at the early parts of the visual system in the intermediary cells and then later reconstructed in the brain, which is especially useful for considering intra-eye analysis for stereoscopic processing.

In the visual cortex of the brain a series of further interactions occur to create more complicated combined signals. These resulting neurons have been termed simple, complex, hyper-complex and contour-detecting, amongst others. For example, a specific complex neuron could detect a bar of light at any angle and position but of fixed length and width. In the 1990s it was postulated that every recognized object may have a specific individual neuron in the brain (proposing the existence of the famous 'grandmother cell' that would detect and fire specifically only when you saw your grandmother), but this was demonstrated numerically to be impossible [although recent work does show the brain has the ability to trigger familiar scenes with single cell responses (Quian Quiroga *et al.* 2005)], and the brain has shown itself to be far more malleable and plastic in behaviour. This was summed up by D. H. Hubel and T. N. Wiesel in their 1981 Nobel Prize acceptance speech:

> what happens beyond the primary visual area, and how is the information on orientation exploited at later stages? Is one to imagine ultimately finding a cell that responds specifically to some very particular item? (Usually one's grandmother is selected as the particular item, for reasons that escape us.) Our answer is that we doubt there is such a cell, but we have no good alternative to offer. To speculate broadly on how the brain may work is fortunately not the only course open to navigators. To explore the brain is more fun and seems to be more profitable. [Full details of their 25 years of collaboration are available in Hubel and Wiesel (2004)]

As we cannot have cells for all types of objects that we see, it is now known that by training and concentration the human brain can become hyper-sensitive to certain stimuli. This means that, owing to a life time of training, both in the short term and in the long term, everyone sees objects differently relating to age, sex, extrachemical stimulus, living conditions and all previous experiences (Ludel, 1978). Thus what you see in a geographic visualization is almost certainly not exactly what someone else sees in the same image.

The signal from the eye, which is a fraction of the total raw information, is then combined with experience and training within your brain to constitute the 'belief engine'. Therefore, when a person gains insight from a specific visualization, this may not be apparent to all viewers, and the final requirement that we recommend, is for there always to be a detailed description, which 'tells a story' alongside the visualization.

11.2.2 Illusions – creating too much information

The human brain's 'belief engine' means that, at times, we do not actually believe the right thing, and this can be incredibly convincing, even when logic tells us otherwise.

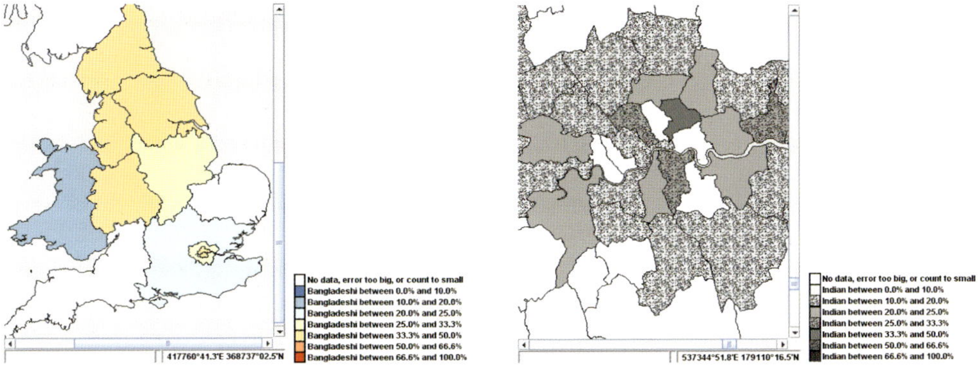

Figure 11.2 A good example of legend selection for geospatial data, carefully selecting both colour and grey-scale values with respect to the human visual system [data source from Peters *et al.* (2007), colour choice adapted from Stone (2002) courtesy of Blackwell Publishing]

We list here four key features that can be observed directly within the brain through visualizations, which then specify some of the limiting features of the human-visual system.

Edges and change

The intermediate cells convert absolute numbers for colour and grey-scale values into relative values over different sized areas. This means we effectively see only edges and local changes in value over neighbourhoods and not precise values. It is, therefore, impossible to recognize a specific intensity value except in relation to neighbouring intensity values. This has a significant impact in the choice of values, for example, in legend colours (see an example in Figure 11.2).

A simple illusion, based upon this principle and described by Ernst Mach, is illustrated in Figure 11.3. It is impossible for a viewer to correctly estimate that the central small squares are in fact the same intensity values. The exact reasons for these illusions are, as always, contended by different researchers, but the effects in human response are statistically significant and agreed upon.

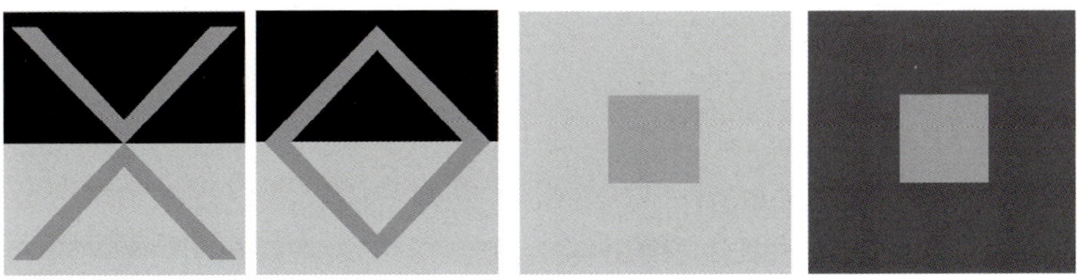

Figure 11.3 Two versions of the famous Mach illusion (Adapted from Kaiser, 2007) www.yorku.ca

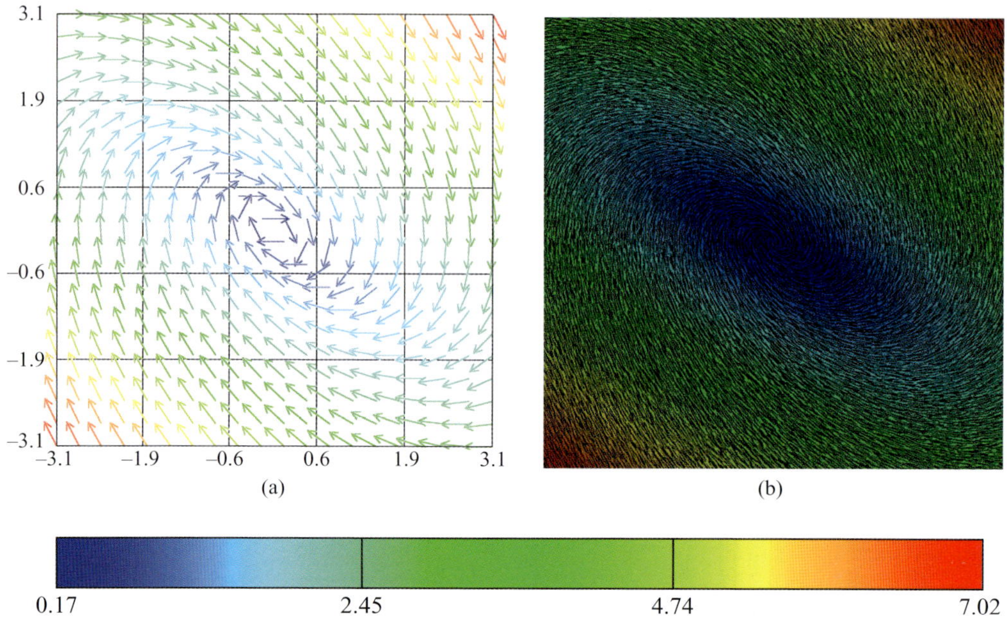

Figure 11.4 Visualization and reducing enhancing clutter: (a) arrows represent flow textures, whereas (b) translates this flow by convolving it with a noise function that reduces distractions. In both cases colour maps the numerical value of the flow magnitude [*source:* these images were created by Jia Liu at the University of Manchester, to show new visualizations texture modes, for vortex flow analysis (Liu *et al.*, 2005)]

Angles and shapes

Levels of cells within the brain by combination create more and more complicated neural results. This shows that it is not just value changes of data that are important, but as these more complex cells respond specifically to particular angles, widths and sizes of stimuli, certain shapes will therefore be easier to see and thus more obvious to a user. A simple kind of artefact based on visual clutter is shown in Figure 11.4. This illustrates a vector flow field using arrows that in their mass visually distract as there are many highly visible edges. The edges of the lines are all sharp and are thus highlighted artificially within the human visual system, which is not likely to be the required response. The second image (b) tries to address this issue by convolving a vector field with a texture map. This texture field illustrates both continuous flow characteristics as well as using frequencies that minimize distractions. The frequencies chosen within the texture field are those with a high response and visual significance to the human visual system.

Misconnected parts

Higher visual elements allow for unconnected parts to be automatically associated with each other with these neurons being termed contour cells. This means that some areas that

Figure 11.5 Two versions of the Kanizsa triangle illusions (Adapted from Kaiser, 2007) www.yorku.ca

are meant to be disconnected can appear to be very strongly connected. There are many examples of optical illusion that illustrate this and one of the simplest is the Kanizsa triangle (Figure 11.5). The white triangle is seen even with drastic miss-alignments of the angles and large separations of the end points. The way the human visual system is able to interpret a triangle is understood through these higher-level 'contour' cells that signal a continuous line even when there are disconnected edges.

Other illusions are now resolvable with better understanding of the neurological wiring. Figure 11.6 illustrates the spatial location variation within our visual system, with a device called the Herman grid illusion. There is a higher degree of accuracy for our vision at the central area of focus. When you focus on a small part of the grid of squares, the edges of the neighbouring squares are very sharp, whereas at the periphery grey dots appear between the squares. The intermediary cells have smaller and larger receptive fields at the focus and periphery of vision, respectively, and the process of convolution involved is a plausible explanation.

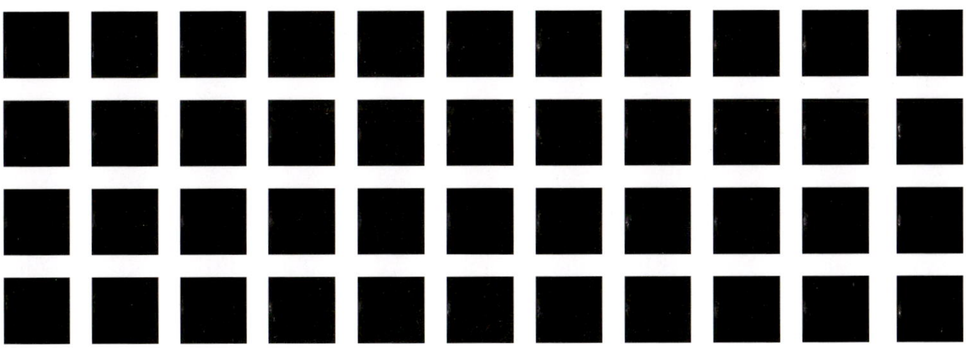

Figure 11.6 Version of the Herman grid illusion (Adapted from Kaiser, 2007) www.yorku.ca

Stereoscopic vision

Owing to specific focus depending on where we look, we are able to closely observe objects from both eyes directly with their neural response signals occurring at the same neighbouring regions of the brain. This gives us fine-grained and very useful stereoscopic vision for very specific points in space.

In an immersive projected system, to maintain a similar quality of visual image, it is often necessary to have multiple projectors that are stereoscopically enabled. To create the illusion each eye needs to see a slightly different view of the world and the visual system fuses these images into one three-dimensional stereo view (Slater, Steed and Chrysanthou, 2002). The difference in the two images is called binocular disparity. In stereoscopy special glasses are often used to fuse the images – fooling the mind's vision into believing it is seeing three dimensions. Most stereoscopic systems, excluding holographic images, have the disadvantage of requiring the user to focus on the screen plane, thus causing slight discomfort if the stimulus is uncomfortable or mis-aligned.

Perfect matching of colours and intensities between displays is very difficult even with professional colour-matching solutions. As we have shown, in the human visual system if the same colour presented on two displays is fractionally different, this may be very noticeable (see illusion in Figure 11.3). Large visualization systems virtually always use multiple projectors, and this can cause an immediate issue regarding quality with regard to matching colours and intensity values. A common solution has been to go for an edge blended system. Blending using interpolation can be handled by extra hardware, built into the projects or occasionally via software. This can both solve the edge issues and match colour and brightness values. Using multiple projectors has the advantages of increasing resolution and providing stereoscopic cues (one projector per eye), as well as providing a method to increase brightness.

These examples show how we can see things that are not there, see things differently from the periphery as opposed to the central focus regions of our vision and indicate the way we fail to consider absolute variations in values, as well as creating the illusion of stereoscopic effects.

The significance of illusion has also been considered by art. This is because the science of art has already considered many of these illusions, and discussed some of them in great detail – many a long time before the neurological investigation of the vision system (Gombrich, 2002; Gregory and Gombrich, 1998). For example, the movement in optical art has been presenting work for many years, most notably within the 1950s to the 1970s. Three of the main terms used by the Op-Art movement have direct neurological explanations (Parola, 1996; images in Figures 11.7–11.9 have been adapted from illustrations within this text).

Assimilation

Assimilation is our tendency to minimize stimuli and create uniformity. It is a simplifying process, sometimes termed grouping with respect to proximity or similarity. Figure 11.7 illustrates how it is difficult to differentiate the lines into groups so, although they are thinner on the left, the whole looks very similar as a single group.

Figure 11.7 Illustration of *assimilation* – all the lines progressively becoming thicker from the left to the right

Figure 11.8 Illustration of *contrast* – adding a dividing block in Figure 11.7 causes the two groups to appear distinct

Figure 11.9 Illustration of *negative and positive* – in this image the white and black compete to become the positive part of the illustration. The white stripes do not physically exist as a feature, as they are the background, but in a local focused region have the appearance of a feature

Contrast

Contrast is the antithesis of assimilation and may be described as the accentuation of differences. These two terms indicate how our brain, taking only a subset of the information, assumes similarity or difference until there is enough visual evidence to the contrary. By placing a dividing barrier into the same illustration as the one shown in Figure 11.7, the two halves now look distinctly different, but in fact they have internally the same *assimilation* tendency as the whole (Figure 11.8).

Negative and positive

Artists define the figure as positive space and the ground as negative space. Negative space surrounds positive space and appears to extend continuously behind the figure. Although, it has less importance, it is usually larger in size. Positive space appears more clearly defined. A line or a contour is by definition a very positive feature. The white stripes in the Figure 11.9 illustration definitely appear to be there, but in reality are just an illusion.

The human visual system is a wonderful structure built on evolution, but due to limits in information flow, there are times when it fails. We need to know when and how these failures can affect visualizations. Presenting data is a balance from creating the most appropriate visual experience, and avoiding the pitfalls of visual illusions, to finally create a meaningful geographic representation.

An important part of this equation is the physical space, which the viewer is enclosed in while observing the visualization. This needs to provide both a stereoscopic and a wide viewing angle display. We will consider next just a small selection of visualization centres that have aimed at presenting data at the correct resolution and colour balance, but also try to understand how the user will experience and interact with the visualization.

11.3 Constructing large-scale visualization systems

There are a number of specialized stereoscopic enabled visualization centres used by earth scientists and other researchers for geographic representations in the UK (this includes all of the oil companies, as well as the British Geological Survey). These try to increase the level of presence with all the visual cues, and to avoid issues associated with confusing the human visual system. We also illustrate various modes in which these centres have been used.

In 1999 Manchester Visualization Centre constructed one of the first large-scale scientific visualization spaces in the UK, consisting of an active stereo 7 m curved wall similar to a flight simulator (Figure 11.10). This allowed scientific users to achieve a higher understanding of their data, but its cost, including building work, was over £1 million, which meant only a few institutions could afford to build such a centre for their research. Figure 11.10 shows views of different earth scientists exploring a three-dimensional geological data field consisting of underground surface levels of rock, as well as a volume of ultrasound readings. The user operates on her own, viewing the data through active stereoscopic glasses, providing near perfect three-dimensional vision as the system tracks both her head and the interactive wand in her right hand. This enables the intuitive exploration in three dimensions of this large data

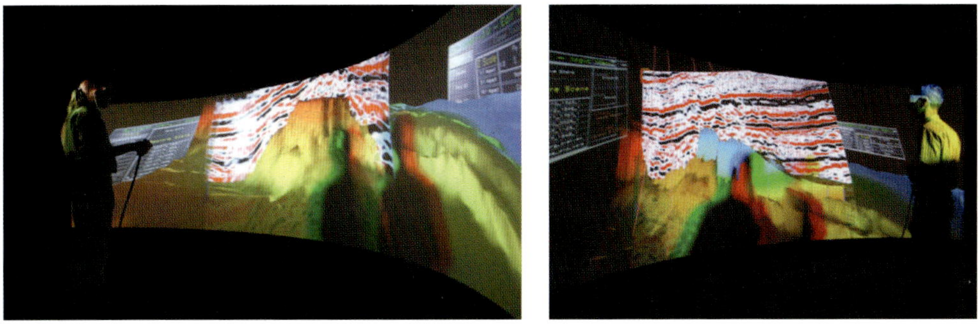

Figure 11.10 Earth scientists using Schlumberger's Inside Reality; Virtual Geological Exploration software. The data shown is the geological subsurface areas found in a region of the North Sea (software from Schlumberger 2007) courtesy of Schlumberger Ltd.

set and also the ability to interactively cut away layers of ultrasound data or reveal different intensity levels of the rock. This is an important mode of interactive visualization, involving just a single individual exploring a data set in order to discover for themselves new insights. It is postulated, although far from proved, that filling the user's visual senses with data aids the process of discovery. Later in 2003, one of the authors, while working at De Montfort University, Leicester, UK, specified and built a similar visualization centre. More modern components were used, which decreased the cost, and a generally more frugal reconstruction brought the overall price to just under £250 000. Reducing the cost had the effect of extending the user base from architects to product designers. It is seen that using curved screens at high resolutions allows a visualization to cover both the focus and periphery of the user's field of view (so-called semi-immersion). More extensive environments have been constructed that allow for full immersion, including the ReaCTor or CAVE™. These consist of a cubic display system often with four active stereo screens, three walls and the floor, although it is also possible to project onto all six sides of a cube. This increases the cost and some users who have experienced this type of system refer to it as a pre-cursor of the Holodeck from Star Trek. The authors have worked in many systems and links to resources are available through the UK visualization support network (vizNET, 2007).

In 2004 the authors oversaw the construction of the first Access Grid node, the next generation of video conferencing system (Access Grid, 2007), which was equipped with passive stereoscopic visualization projection equipment, and cost only £80 000. This enabled stereoscopic visualization to augment or take over a normal remote video conferencing meeting. It should be noted that this new video conferencing system also embraces the idea of using large screens offering high resolution, creating semi-immersive presentations. Figures 11.11–11.13 show visualizations presented to a research audience within this node. The first, Figure 11.11, shows a representation of geological borehole samples, positioned in their correct three-dimensional geographic location, but abstracted with a pseudo-colour histogram of the elements superimposed along the borehole length. The audience can interactively explore the three-dimensional space and understand spatial locations; for example detecting areas where specific heavy metals exist (GeoExpress software product range from Oxford Visual Geosciences Ltd, 2007). Figure 11.12 presents Google Earth stereoscopically projected onto part of the screen describing a virtual mapping tour through a three-dimensional

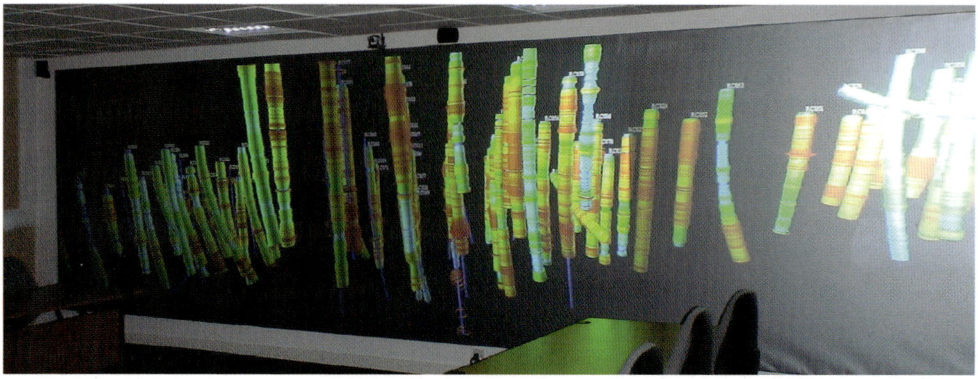

Figure 11.11 Geographical visualizations in the combined Access Grid conference node using a stereoscopic three-dimensional display. This shows the GeoExpress visualization software, projecting boreholes presented as a histogram displayed along their length and placed in their correct spatial drill position

University of Manchester combined with curved map overlays (Google Earth, 2007). It is proposed that using stereoscopic presentations with a large screen allows an environment where understanding of the space is more intuitive. With a speaker and an audience these examples show a second mode of visualization, one-way presentation and demonstration. The important part here is that there is a presenter who can 'tell a story' to the audience.

The final mode of visualization is team interrogation, where a small group of experts interact collaboratively with data sets to discover new patterns and understandings. Figure 11.13 shows LiDAR (light detecting and ranging) data of a detailed scan from the 'badlands' of South Dakota, captured during fossil gathering by university palaeontologists. This was presented for open discussion to a group of earth scientists, including geologists and seismologists. They were encouraged, and able, to discuss openly contentious issues with the

Figure 11.12 Google Earth projected stereoscopically within the conference node with added layers of three-dimensional objects superimposed at the correct position

Figure 11.13 A joint research group discussion where palaeontologists interacted with geologists and seismologists (public engagement event for a National Geographic Channel science documentary)

palaeontologists regarding the spatial layout, in this case understanding why certain fossil finds should occur at specific locations and where to explore in the next field trip. This visual approach to analysis was found to have great savings in both time and financial resources. To enable this visually, the fossil find locations are superimposed as sets of markers positioned on top of the three-dimensional geological features of the surrounding terrain. This mode of interactive visualization enabled a virtual field trip to occur that could never happen, or would be very expensive, and to be repeated again and again.

Virtual environments becoming affordable

Unfortunately, all these systems suffer from being relatively expensive, in both resources and management time, and are not portable. In recent years it has become possible to build cheaply a small, portable version based upon a design concept product called a GeoWall (Turner and LeBlanc, 2006), using commercially available components, for under £10 000. Quality has been sacrificed at this stage, but these systems are now affordable across disciplines for a variety of purposes. A basic projection and recording system is shown in

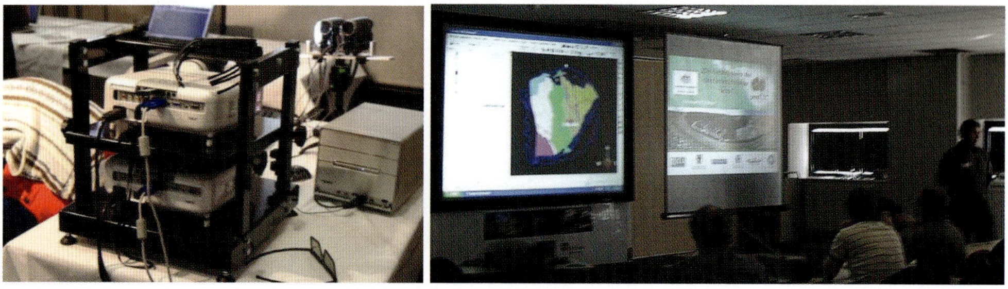

Figure 11.14 Portable system being exhibited, consisting of two projectors powered by a compact PC shuttle system as well as a combined two- and three-dimensional presentation at the GSA Penrose conference (*source:* N. Holliman and J. Imber for the image; details of construction are available at Turner and LeBlanc, 2006)

Figure 11.14 that can be easily taken to meetings, lecture rooms and conferences with minimal complexity and still produce high-quality, semi-immersive presentations.

The whole unit, including computer, projectors and screen, fits comfortably within the back of a car and it is just possible for a single user to carry all the components at once. This system has, over the last 18 months, been to over 40 locations and presented to thousands of users, and versions of the system are available at various universities across the UK. Set-up time by non-experienced operators can be under half an hour and a minimal amount of training is required. Importantly it does still allow all three of the modes of interaction to occur – solitary detailed interrogation, audience presentation and discovery – as well as complex group interaction.

11.4 Rules and recommendations

Best practice should always be considered for any geographic, information and statistical information visualization, whether on the small screen or in a large presentation system. Section 11.1 briefly mentioned some examples and references to generic guidelines for best visualization techniques. At all stages the user and their ability to interpret both as a methodology, in terms of interaction, and as a simple ability to visually understand, through knowledge of the human visual system, need to be always kept in mind. When considering large immersive technologies there are three extra considerations that need to be addressed: size, interactivity and narrative.

11.4.1 Increase the size

As multiple projection systems are used, display size correspondingly increases, and there is an additional requirement to increase the resolution needs to match. This can require the use of more and higher specification graphics cards as well as needing synchronized computers within clusters. As the human visual system is still being used, good quality visuals and comfortable viewing positions are still important, which act as guidance values for the required resolution.

Also, with the increase in presentation space one must be aware of usability issues. No longer are the users in the familiar territory of the single desktop monitor. They are immersed in a large environment which might consist of a 15 m screen. This, as Robertson *et al.* (2005) reports, results in myriad issues, such as 'losing the cursor': users tend to accelerate their mouse movements to overcompensate for the larger 'monitor' and end up 'lost in space'. Other issues include accessing menus across large distances, the number of windows being opened and overlapping issues within blend areas. An awareness of such issues must be brought to the attention of the users, if not compensated for (e.g. in the case of 'losing the cursor', an auto-locator cursor tool can be used).

With a defined physical space dedicated as a visualization centre, it is also useful that it is both accessible and welcoming, especially if users need to access it often or for a long period of time. Care needs to be taken that this becomes a useful place to work, and as appealing for the user as their office workspace is at present. Integration within a multi-purpose space, for example combined with video conferencing or as a project management room, also adds to the features available and helps to enable real work to occur. A final concern is with ease of use for day-to-day operation, and making sure that any system does not require extensive training or continual use by specialist operators.

11.4.2 Interactivity

During this chapter we have described three different modes of interactive visualization that through experience often occur within large spaces, which we will briefly summarize here.

Single interactive journey of discovery

It has been shown from experience that deep insight may not be instantaneous but may require an individual to interact for long periods of time with a visualization. This allows them to explore and understand their data in a sense that any demonstration or showcase cannot offer. For this reason visualizations need to be comfortable for extended use, and this implies correct calibration to avoid any physical discomfort.

Presentation and group explanation

The showcase and demonstration use is one of the most successful and exploited modes for any visualization centre, partially due to the instantaneous experience achieved. This mode is also one that requires the most care in order for the narrated story, which may involve unknown group interaction, to be carefully presented. What this mode should not be used for is to present the technology itself rather than the data.

Group discussion to collaborate and explore disparate ideas

If the power of a single visualization is the third of the brain processing it, when you have 10 people in a group, the argument would proceed that there is 10 times the visual processing occurring. Having multiple 'belief engines' operating at the same time means there is a greater chance of interaction and group insight. The visualizations do need to be presented correctly as optical illusions seen by one member of the group are likely to be seen be them all.

11.4.3 Telling a 'story'

An important point is that all of the modes should start with some hypothesis and create a 'story' that describes the visualization. There should always be a story to tell even if it is one of explorations in content, and this needs to be incorporated within the visualization. We started saying that 'visualizations need a 1000 words to describe them', and the original reference by Hewitt *et al.* (2006) described within a long paper a complete product range using seven different examples, each having a visualization of the problem, as well as a complete textual description defining both the methods and the findings. It is likely that without these descriptions we are just left with inspirational visualizations of nothing.

The final stage, which is possibly the most important one, is how to fully document a visualization which will include all parts of the process including the 'story' and the interpretations that were involved. This we are not going to cover in detail here, but there are some experiences and standards from other communities that should be considered. Documentation of visualizations from the scientific viewpoint has been described through meta-data inspired by Berners-Lee, Hendler and Lassila's (2001) Semantic Web and De Roure and Sure's (2006) Semantic Grid ideas. Both of these processes aim to record each step of the scientific data capture and workflow analysis that a user takes. This would allow systems and experiments to be repeated and reinterpretations made. An alternative set of standards for interpreted data for architectural visualization has been considered within The London Charter (Beacham, Denard and Niccolucci, 2006). This defines, as well as the meta-data involved, the subjective decisions that have been taken to create visualizations and has been termed para-data. The charter covers eight internationally recognized principles for disseminating three-dimensional data to the community.

11.5 The future – a better and cheaper place

The current generation of video conferencing system, called the Access Grid, is based upon the idea of a grid of people interacting and assimilating information in a similar manner to a Grid of computers. The success of this has come from available increased networking bandwidth, meaning that in the UK alone there are over 200 specialist large purpose-built rooms (similar to the one in Figure 11.10), and worldwide over 20 000 licences for the desktop software have been issued (Access Grid Support Centre, 2006 and 2007).

We have shown in this chapter how visualization is not just about the equipment and presentation devices but, importantly, about the human visual system, the modes of interaction and a reminder that it is a way of telling a story. As computer components and networking have become cheaper, the technical and presentational issues are slowly being addressed. The future should consider a new form of grid visualization systems that constitute a network of human visual systems, and provides a convenient means for them to interact and debate. This should adhere to continual best practice, but then new stories may be told and come from large global interactions.

References

Access Grid, video conferencing system; available at: www.accessgrid.org/ (accessed 2007).

Access Grid Support Centre, User survey results and statistics 2006 and 2007; available at: www.agsc.ja.net/.

Alcock, J. A, Burns, J. and Freeman, A. (eds) (2003) *PSI Wars: Getting to Grips with the Paranormal.* London, Academic Press.

Beacham, R., Denard, H. and Niccolucci, F. (2006) An introduction to the London Charter. In Ioannides, M. *et al.* (eds), *The E-volution of Information Communication Technology in Cultural Heritage: Where Hi-tech Touches the Past: Risks and Challenges for the 21st Century.* Budapest, Archaeolingua; available at: www.londoncharter.org/.

Berners-Lee, T., Hendler, J. and Lassila, O. (2001) The semantic web. *Scientific American*, May: 35–43.

De Roure, D. and Sure, Y. (2006) Semantic grid – the convergence of technologies. *Journal of Web Semantics*, 4(2): 82–83.

Friendly, M. and Davis, D. (2006) www.math.yorku.ca/SCS/Gallery/milestone/ visualization milestone timeline website.

Gombrich, E. H. (2002) *Art and Illusion: a Study in the Psychology of Pictorial Representation*, 6th edn. London, Phaidon Press.

Google Earth (2007) http://earth.google.com/ version 4 including three dimensional plug-ins.

Gregory, R. L. (1998) *Eye and Brain: The Psychology of Seeing*, 5th edn. Oxford, Oxford Press.

Gregory, R. L. and Gombrich, E. H. (eds) (1998) *Illusion in Nature and Art.* Gerald Duckworth.

Hewitt, W. T., Cooper, M., John, N. W., Kwok, Y., Leaver, G. W., Leng, J. W., Lever, P. G., McDerby, M. J., Perrin, J. P., Riding, M., Sadarjoen, I. A., Schiebeck, T. M. and Venters, C. C. (2005) Visualization with AVS, Chapter 25. In Johnson, C. and Hansen, C. (eds), *Visualization Handbook.* Oxford, Elsevier.

Hubel, D. H. and Wiesel, T. N. (2004) *Brain and Visual Perception: The Story of a 25-Year Collaboration.* New York, Oxford University Press.

Kaiser, P. K. (2007) *The Joy of Visual Perception: A Web Book*; available at: www.yorku.ca/eye/thejoy.htm

Kandel, E. R., Schwartz, J. H. and Jessell, T. M. (1991) *Principles of Neural Science*, 3rd edn. Oxford, Elsevier.

Kingsley, R., Gable, S. and Kingsley, T. (1996) *Concise Text of Neuroscience.* Philadelphia, PA, Williams and Wilkins.

Liu, J., Perrin, J., Turner, M. and Hewitt, W. T. (2005) Perlin Noise and 2D second-order tensor field visualization. In *Theory and Practice of Computer Graphics.* Eurographics UK/Chapter UK Conference, pp. 113–118.

Ludel, J. (1978) *Introduction to Sensory Processes.* New York, W.H. Freeman.

Oxford Visual Geosciences Ltd. www.geoexpress.co.uk/ (accessed 2007).

Oyster, C. W. (2006) *The Human Eye – Structure and Function.* New York, Sinauer.

Parola, R. (1996) *Optical Art; Theory and Practice.* New York, Dover.

Peters, S., Clark, K., Ekin, P., Le Blanc, A. and Pickles, S. (2007) Grid enabling empirical economics: a Microdata application. In *Computational Economics.* Berlin, Springer.

Quian Quiroga, R., Reddy, L., Kreiman, G., Koch, C. and Fried, I. (2005) Invariant visual representation by single neurons in the human brain. *Nature*, 435: 1102–1107.

Robertson, G., Czerwinski, M., Baudisch, P., Meyers, B., Robbins, D., Smith, G. and Tan, D. (2005) The large-display user experience. *IEEE Computer Graphics and Applications*, July/August: 44–51.

Schlumberger. Inside reality remote collaboration product range; available at: www.slb.com (accessed 2007).

Schroeder, W., Martin, K. and Lorensen, B. (2006) *The Visualization Toolkit: an Object-oriented Approach to 3D Graphics*, 4th edn. New York, Kitware.

Slater, M., Steed, A. and Chrysanthou, Y. (2002) *Computer Graphics and Virtual Environments; From Realism to Real-time.* Reading, MA, Addison Wesley.

Spence, R. (2006) *Information Visualization: Design for Interaction*, 2nd edn. Englewood Cliffs, NJ, Prentice Hall.

Stone, M. (2002) *A Field Guide to Digital Color.* Warriewood, NSW, A K Peters.

Thompson, R. F. (2000) *The Brain: A Neuroscience Primer*, 3rd edn. New YorkWorth.

Tufte, E. R. (1991) *Envisioning Information.* Cheshire, CT, Graphics Press.

Tufte, E. R. (1997) *Visual Explanations: Images and Quantities, Evidence and Narrative.* Cheshire, CT, Graphics Press.

Tufte, E. R. (2001) *The Visual Display of Quantitative Information*, 2nd edn. Cheshire, CT, Graphics Press.

Tufte, E. R. (2006) *Beautiful Evidence.* Cheshire, CT, Graphics Press.

Turner, M. J. and Le Blanc, A. (2006) JISC Programme VRE: CSAGE (Collaborative Stereoscopic Access Grid Environment) Final Project Report; available at www.kato.mvc.mcc.ac.uk/rss-wiki/SAGE and www.geowall.org.

vizNET; UK Visualization Support Network, www.viznet.ac.uk/ (accessed 2007).

12

Landscape Visualization: Science and Art

Gary Priestnall and Derek Hampson

School of Geography, University of Nottingham

School of Fine Art, University College for the Creative Arts, Canterbury

12.1 Landscape visualization: contexts of use

The launch of Google Earth in June 2005 boosted public awareness of both digital geographic information and visualization techniques in a way probably not seen before. Although other web-based 'virtual globes' existed, the level of detail, ease of interaction and wide acceptance of the search engine helped to propel Google Earth into millions of homes around the world. The arrival of Microsoft Virtual Earth 3D late in 2006 gave users alternative data coverage and representations. Aside from the clear potential for posting and sharing spatially tagged information set in a global context, virtual globes are raising both awareness and expectations of what is possible in terms of interactive photorealistic visualization. Big steps have also been taken by the computer games industry along a similar timeframe with the launch of Microsoft Xbox 360 and Sony PlayStation 3, offering a new level of detail and photorealism. Large development teams will ensure rapid progress in both virtual globes and games technologies, underpinned by continued developments in computer hardware. Any attempt to create and visualize a virtual landscape in any other context will inevitably run the risk of being compared with these widely available representations. Landscape visualization is used in both education and public consultation to communicate existing landscapes but also alternative scenarios, past and present. There is a tendency to strive for photo-realism, in part a response to the developments described above, and in so doing make the visualization believable. With this often comes an acceptance from the observer that the visualization represents truth, a surrogate for the real landscape. In this chapter we seek to explore how landscape visualizations relate to our experience of the real landscapes concerned. First we

Geographic Visualization Edited by Martin Dodge, Mary McDerby and Martin Turner

address this from the perspective of field validation of computer-generated visualization, considering the factors which affect how faithful a visualization is to the real landscape. We then discuss factors contributing to what could be termed the 'essence' of a landscape and present some experiences from 'Chat Moss' a project that examined an artistic representation of a case study site.

When photorealistic landscape visualization is used for environmental decision-making, the audience may often be unfamiliar with the nature of the digital data and processing lineage used to create the models. It has been suggested that some kind of code of ethics is required to encourage best practice in the use of computer visualization in such contexts (Sheppard, 2001). Visualization software and digital data commodities are becoming more widely available and easier to work with, so the range of contexts in which virtual landscapes are used is set to increase. Sources of digital elevation data to form the base of landscape models include direct ground survey, contours and spot heights, photogrammetry, radar and LIDAR (light detection and ranging). The data chosen should be 'fit for purpose', as with any form of analysis within geographical information science, but it is all too easy to produce virtual landscapes from sparse or erroneous elevation data. This data can be embellished with aerial photography and surface features such as trees, roads and buildings from CAD packages to produce quite compelling and realistic looking virtual worlds, but which may bear little relationship to the actual landscape.

This striving for realism in many contexts seems an inevitability, although some researchers question whether it is always required and whether there may be some intermediate 'sufficient' level of realism (Appleton and Lovett, 2003). The simple process of draping high-resolution aerial photography over a digital terrain surface is a very common technique and can play a critical role in creating an effect of realism but can also mask limitations in the surface model. Aerial photography often contains strong shadowing and, when the direction of this is coincident with the lighting direction within the three-dimensional rendering software, then a level of surface topographic detail is suggested which can far exceed the actual surface detail given by the terrain data. This can fool the observer into thinking that the elevation data forming the basic structure of the virtual model is at a similar high level of precision, and indeed accuracy, when it is invariably not. An issue we will return to later in this chapter is the need to consider how explanatory 'metadata', describing a particular dataset or in this case visual artefact derived from various digital datasets and processing stages, could be utilized more effectively to inform the observer of what they are looking at. As computer-generated landscape visualizations become more like photographs, then the adage 'the camera never lies' can hold true in the eyes of many observers. It is with this danger in mind that it is considered important to foster a critical awareness of the nature of various forms of digital surface models and visualization techniques and to explore broader notions of how landscape can be represented.

12.2 The need for ground truth

The term 'ground truth' in the context of geographical information science is most commonly applied to the field calibration of satellite-borne sensing devices. Measuring reflectance properties of the Earth's surface at known locations in the field allows a satellite image of that area to be processed in such as way as to allow land cover to be mapped with reasonable confidence. Here we use ground truth in a more qualitative sense to mean the observation of the field locations of computer-generated visualizations with a view to assess the degree to which they offer faithful representations of the landscape. From an educational perspective fieldwork offers an ideal opportunity to engage students in such ground truth exercises and in so doing encourage them to look more critically at both digital datasets and visualization techniques in their ability to mimic reality.

Some work has been done in assessing what could be termed the 'representational fidelity' of interactive virtual landscapes in providing a sense of presence when simulating fieldwork in a laboratory environment (Whitelock and Jelfs, 2005). An alternative approach is to develop techniques for taking landscape visualization out into the field to engage students in a direct comparison of the models with the actual landscapes they represent. These exercises form part of a first-year undergraduate geography field course near Keswick, Cumbria, North West England, described in more detail in Priestnall (2004). In advance of the field-trip digital terrain data, aerial photography, geological data and a model of glaciation in the area were prepared and integrated into a three-dimensional modelling package called Bryce. This would allow students to quickly locate and render a number of landscape visualizations from known positions, and then locate them in the field to undertake a direct comparison.

Data were acquired which represent widely available commercial base data often used in the creation of virtual models in a range of environmental visualization contexts, namely:

- *digital surface model* (DSM) – 5 m resolution gridded terrain model from airborne radar;
- *aerial photography* – 25 cm resolution colour aerial photography;
- *geology* – solid and drift geology polygons and fault line vectors;
- *glacial model* – created from limited field evidence and distribution of glacial till deposits given in geological data.

The field exercise occupied a full day, beginning with a briefing where students were organized into groups of four and the concept of a grid-based digital surface model derived from airborne radar was explained, as were the various contexts in which landscape visualization could be used. The aim was described as the development of a schema for assessing computer-generated three-dimensional landscape visualization against real scenes. Groups were asked to select at least three locations which they believe represent a range of landscape types suitable for exploring the strengths and weaknesses of the digital data and visualization techniques for representing reality. The 'observer' points for each of the chosen views were located within the three-dimensional modelling package and a perspective view rendered and printed out for each. A printout of each view with the glacial model superimposed was also produced, this time onto transparency to experiment with simple augmentation of the user's view in the field.

Out in the field each of the chosen views was visited in turn, the printouts of the computer-generated scenes were annotated, additional field sketches made as appropriate, and observations made as to the extent to which each image conveyed a true sense of the landscape in front of the students. Figure 12.1 gives an example of one such view point looking towards

Figure 12.1 Landscape visualization in the Newlands Valley, Cumbria. Photographic representation (left), computer-generated visualization from radar terrain model (centre); and a digital reconstruction of a retreating glacier in the valley around 15,000 years ago (right)

the southwest into the Newlands Valley, Cumbria. The reporting stage back in the field centre involved groups developing a structured schema by which landscape visualizations could be assessed against the real scenes, illustrating this with evidence gathered in the field alongside the computer-generated imagery.

12.3 Outcomes from fieldwork exercises

The general approach has been found useful in engaging students in the nature of digital surface modelling and raising awareness of visualization techniques. Having run similar exercises for over eight years, it is possible to summarize some of the elements common to the schemata developed by students in attempting to assess computer visualization against reality. It should be remembered that the students were dealing with static views and that landscape visualization can include a range of animated, interactive and immersive techniques in addition to the photo-like snapshot. A conscious effort was made to maintain focus on the static viewpoints in an attempt to manage the complexity of the problem and separate representational fidelity issues from those concerning modes of interaction. It should be acknowledged therefore that the following observations are not intended to be a complete list, more an attempt to categorize some general issues which may be worthy of further investigation.

- *Scale and complexity* – it is only when faced with real objects in the field that the implications of data scale or resolution become clear. The terrain model used here, for example, was a 5 m regular grid of elevation values, so features of width 5–10 m had little chance of being represented and would at best be small undulations in the surface model. Objects of this size were relatively unimportant when views contained predominantly distant terrain characterized by broader topography of hills and valleys, where the scale of such features became dominant over the resolution of the data.

- *Landscape composition and landmarks* – following on from the issue of scale, the relative importance of key landmark features in the foreground or middle ground of a view can be seen as affecting the type or composition of view. Clearly where foreground objects are prominent, then the relative position of the observer to these is crucial and the view changes markedly over a very small area. The nature of the landscape in Cumbria is such that views are often dominated by, indeed generated by, hills or fells, which are often referred to by name and used for orientation and navigation.

- *'Camera' properties* – upland landscapes such as Cumbria are invariably more impressive in the field than when represented on a photo-like image. In part this is because the focal length of the human eye (equating approximately to a 50 mm camera lens) has wider peripheral vision. Even ignoring the effect of depth cues through stereo vision, it is difficult to replicate this effect when rendering computer-generated imagery. The use of the transparency to augment the centre of the view with the glacial scene emphasized the importance of human peripheral vision as being a major contributing factor to the overall impression of a landscape. Different modes of interaction, such as interactive stereo widescreen formats or more immersive techniques, may go some way to addressing these issues.

- *Spatial context and 'a-priori' knowledge* – the ability of observers to recognize viewpoints in the field is affected to a degree by their familiarity with the area, through *a-priori* knowledge gained through previous visits, by studying maps beforehand or through the spatial context gained by walking to the viewpoint, aided perhaps by a map or positioning device. In this exercise the student groups familiarized themselves with the area by generating the viewpoints themselves, which inevitably involves exploration of the three-dimensional model of the area. This is significant as landscape visualization is often used in a context where the audience is not privy to some or all of the knowledge associated with spatial context and therefore the viewer may find it difficult to orientate themselves when presented with a new visualization in isolation.

- *The 'essence' of a view* – many non-visual factors were important in contributing to the 'essence' of the view, such as weather conditions, sound and sense of exposure or the conditions under foot. This may be significant when considering whether people orientate and navigate in virtual environments in a similar way as they do in the real environments.

- *Evidence for model building* – the reconstruction of a past landscape in the form of a retreating glacier raised an interesting issue in terms of communicating the evidence used to create such a model. As soon as historical or hidden landscape components, such as glaciers and geology, respectively, are added, we must put our trust in the way the nature of evidence and processing to build such models is communicated. Despite careful communication of the sparse and uncertain nature of the evidence to support the glacial formation shown in Figure 12.1, many students regarded this reconstruction as 'truth'.

Clearly many of the observations above relate to the scale and complexity of the landscape scene being modelled. Cumbria is generally regarded as a picturesque landscape and the frames of reference for many visitors would be distant mountainous views. As soon as the landscape scene becomes dominated by closer surface objects such as trees and buildings, the types of digital representation offered by the radar model become inadequate for portraying a recognizable scene, as shown in Figure 12.2.

Replicating scenes like this requires finer resolution terrain data and a huge amount of time and effort capturing and modelling individual surface features, often photogrammetrically. Typically the undulations on the radar model, or higher resolution LIDAR models, would be stripped off and CAD-like building and tree objects would be placed on the surface. When we begin to consider what could be termed 'non-picturesque' landscapes, then scenes dominated by surface features become more common. We should ask at this point what exactly we should be attempting to model in a certain landscape given the magnitude of

Figure 12.2 Scale and complexity of landscape: a valley floor scene represented in a photograph (left) and a computer-generated visualization from a radar terrain model (right)

the task if attempting full photorealism. Also, we should look beyond the purely visual comparison of virtual and real landscape scenes and consider the role of spatial context, *a-priori* knowledge and what was termed above the 'essence' of a landscape scene. It is in this general area of interest that several exploratory collaborations have taken place between the authors over a number of years which explore issues of representation and understanding as encountered in the fields of geographic information science (GIScience) and fine art, in particular figurative painting.

12.4 Broadening the context

The question of 'what we should be representing' is an important issue that has faced geographical information scientists for many years (Fonseca *et al.*, 2000; Hart, 2004) and certainly too broad in scope to do justice to in this work. It is, however, the intention to identify certain interdisciplinary approaches to landscape representation which may be relevant in this broader context and which may contribute to the GIScience research agenda more specifically. As geographic information becomes used more in a 'first person' context on mobile devices then the immediate and direct relationship between virtual representation and the real world around the user becomes ever more important. The design of more egocentric representations (Meng, 2005) required to work on small screens in a range of outdoor contexts may need to address the issues of spatial context, *a-priori* knowledge and the 'essence' of the user's immediate surroundings as introduced earlier. A good example of developments in this field is the mobile phone application Viewranger developed by Augmentra Ltd (2007), which offers recreational walkers a dynamically updated and annotated skyline given by the user's orientation from the recent GPS track. Such representations, in this case inspired by pictorial guides to the Cumbria fells (Wainwright, 1964), seek to identify a suitable frame of reference for a given type of user in a given situation. What will continue to be a huge challenge is to explore landscape representations which attempt to serve many different types of users with different backgrounds, levels of expertise and motives.

The authors have developed an ongoing dialogue around themes of representation through collaboration on a number of projects driven by a desire to explore regimes of representation that are of relevance to both painting and computer-based landscape visualization and GIS. The decision of what to represent and how to represent it is faced in both disciplines. Rather than expect to find a comfortable common ground, the intention was to share common experiences with a view to possibly offering new perspectives on our own practices. This engagement has had a number of outcomes to date including;

- 'Hawley Square' (Hampson and Priestnall, 2001) paired five artists with five geographers around themes of representation, focussing on Margate in Kent, England, culminating in an exhibition, CD, symposium and web site.
- 'Real World Mapping' investigated techniques for capturing and re-presenting information about an inner city neighbourhood to derive a collective 'vision' of a place from local residents to contribute to the planning of regeneration schemes by the local authority.

The recently completed Arts and Humanities Research Council (AHRC)-funded project 'Representations of Embodiment' continued to explore these issues through the development of collaborative working practices which allowed the authors to approach these challenges in new ways. The challenge was to consider the ways a landscape can be represented as it is experienced through a range of encounters with data as well as with the landscape itself, using Chat Moss in Lancashire as the case study site.

12.5 The Chat Moss case study

A challenge facing both fine art and GIScience is to represent landscapes as they are experienced through the complexity of direct and indirect encounters; i.e. the understanding that comes from being in the landscape bodily and the understanding that comes through less immediate encounters, for example when we imagine a landscape through reading, thinking, or looking at photographs.

The general aim of the project 'Representations of Embodiment: Implications for Fine Art and Geographical Information Science' was to create and analyse an art work which embodied such complexity and to explore the implications for GIScience. The method we employed to do this was to collaboratively focus on a specific case study site, Chat Moss in Lancashire, and to work towards creating an artwork that would incorporate data representations of the site from a variety of sources, and then to analyse the process through which this data contributed to the site's representation in the final artwork – in this case a ceiling painting.

Chat Moss in Lancashire (pictured in Figure 12.3), is a non-picturesque landscape between Manchester and Liverpool, dominated by a peat bog and once the site of a significant engineering endeavour to route the first passenger railway across the landscape in the early nineteenth century. Many of the outputs from the project, including the web site (www.chatmoss.co.uk), the exhibition, and the accompanying publication, are named 'Chat Moss' to reflect the geographical focus of the project.

During the development of the work, the nature of the collaboration proved hugely effective in engaging both Priestnall and Hampson in research questions centred around issues of the nature of data, geographic data representation, and methodologies for their

Figure 12.3 The present day landscape of Chat Moss, Lancashire

definition, collection and visualization. The theoretical underpinnings to Derek Hampson's practice developed dramatically during an intense period of research, and the artwork itself took on the unexpected form of a ceiling painting. The main question we identified was the relationship between seeing and experience, where seeing is thought beyond mere optical sensing and includes the notion of a more cognitive, 'conceptual' vision. This is a broad and significant issue that has faced researchers in GIScience for many years, and not one that this project can realistically hope to offer a final answer to; but we are able to propose certain interdisciplinary understandings, underpinned by readings in phenomenology which may contribute to the GIScience research agenda more specifically.

Underlying the project was the desire to test the concept of embodiment, in particular the idea that our existence as embodied beings plays a crucial role in the way visual representations are made and understood, based on Michael Fried's writings on nineteenth century German artist Adolph Menzel (Fried, 2002). Embodiment here can be characterized as prioritizing of the body, as opposed to the mind, as the thing through which we understand the world. The theory is that artists translate their naturally embodied everyday experience into the way artworks are created. Viewers as embodied beings 'read' the representations as intuitively as they are created. Coming to an understanding of the phenomenological concept of embodiment was a central concern, initially through the writings of the French philosopher Maurice Merleau-Ponty which had influenced Fried. The search for a deeper understanding of these concepts led Hampson to the origins of phenomenology in the work of Edmund Husserl and Martin Heidegger.

During the development of the artwork, the potential of phenomenology became increasingly significant to Hampson. Phenomenology, as instigated by Edmund Husserl (1859–1938) and developed by Martin Heidegger (1889–1976), creates a method through which a profound analysis of data is possible. Putting to one side the 'natural attitude' of our everyday understanding of the world, and engaging in a presupposition-less analysis of what is given in our encounter with things in the world, leads to understandings that uncover profound connections between the viewer and the viewed. In many ways phenomenology should be thought of as a method of investigation rather than a belief system. This 'phenomenological method' offered us a methodology that allowed us a way of reaching a complexity of understanding about the historical and contemporary nature of the site culminating in the ceiling painting.

The major methodological problem that we faced was how to analyse the process through which the separate elements of data that contributed to the composition of the artwork were combined in the process of art-making. Artists are famously reluctant to analyse the procedure through which their works are made. The common concept of artistic creativity is that it is instinctive, and that any process of clarification will lead to its disappearance.

There is some justification in this as any self-conscious analysis by the artist of their creative process in the moment of its employment will inevitably become part of that practice, which would amount to an unwarranted intrusion into the creative process. Therefore we had to find an approach that would allow the process through which the Chat Moss artwork was to be made to be analysed in terms of its engagement with data without intruding into the creative process.

This was achieved through an approach that built upon phenomenology's own methods of investigation. Phenomenology can in some ways be seen as the antithesis of psychology. If for psychology meaning resides internally in states of mind, for phenomenology meaning is derived through an engagement with objects, things in the world outside us. To understand the world in this way a particular methodology was needed that allowed questions to be asked about the nature of our relationship to objects without the asking affecting the answer. Phenomenology developed a number of approaches that enabled this, leading to a series of discoveries that were instrumental in allowing us to scrutinize more effectively how different types of data contributed towards the artwork's creation.

A particular implication for GIScience in this area is the way in which methodologies for describing and communicating the nature of the real world as experienced by different people can be developed. Phenomenology offers a different way of thinking from the largely computer science-based ontological approach to knowledge representation which has received most attention in recent years. Knowledge representation schemes in GIScience are dominated by visible 'mappable' objects and their properties, and rarely accommodate more ill-defined experiences.

If we are aiming at a more complete representation of a place, then these ill-defined experiences and their contribution to our understanding need to be clarified. We are talking here about the relationship between us and things, whether an artwork, a photograph, a landscape or a fleeting thought – and the way in which the relationship, through a data exchange between us and it, allows an intuition about the site to be constructed. Intuition is central to the phenomenological account of our structures of perception. Heidegger defines intuition as 'simple apprehension of what is itself bodily found just as it shows itself' (Heidegger, 1985). Intuition is a deep understanding, derived from an encounter with things, allowing us to see something as it actually is in itself, without any unwarranted conceptual constructions being placed on it.

12.5.1 Phenomenological analysis

In order to get to an understanding of what phenomenology teaches us about the structures of perception, we will now offer an analysis of four of the experiences, encounters with things, which led to the creation of the Chat Moss artwork. These experiences involved an engagement with three forms of data: spoken, visual and direct 'bodily' encounters. On the surface there is a disparity between these, yet if we examine them in detail it becomes clear that they are linked in unexpected ways.

The first experience, which began Hampson's fascination with the place, was that of a spoken account of the story of Chat Moss. This occurred in childhood when, in a bright Victorian classroom, he was told the story of Chat Moss. Told with no visual aids, the teller described the conquering of the Moss by George Stephenson, his struggles to survey the route in the face of recalcitrant landowners and suspicious locals, the subsequent labours to build a railway line across what was then an undrained bog into which everything and everyone, including the line itself, sank, leading to the final triumphant solution, the floating of the railway across it on a bed of gorse and heather.

If we understand the story objectively we can say that it delivered some pieces of information; it gave the bare factual bones – names, dates and an understanding of the place's historical importance. Yet it also gave something more, it gave an intuition of the Victorian world of commerce and enterprise, a world of historical overcoming which seemed to find fulfilment in the name 'Chat Moss', at once familiar and mysterious, in the process giving a particular understanding of this place in the landscape.

Hampson listened to the story, but it was more than mere listening; in the process of listening he engaged with the story as he was told it through thinking and imagining through which process the intuition was constructed. How can we account for this? Part of the answer can be found in the concept of intentionality the elucidation of which was crucial in the development of phenomenology. First developed by Edmund Husserl at the beginning of the twentieth century, in his *Logical Investigations* (Husserl, 2001) and then further clarified by Martin Heidegger, intentionality enabled an attack to be made on the Cartesian world view. It discovers that we do not simply see things, or in this case hear things; instead, in that hearing we have feelings for the thing that is heard. In our every act of perceptual or imaginative engagement we direct ourselves towards the object of our attention. These acts are inescapable.

Yet the intuitions are not automatically given: everyone listening to the Chat Moss story would not have had the same insight. It would depend on the manner in which listeners engaged with the data given in the story, leading to different degrees of 'givenness'. Phenomenology discovered three levels of 'givenness': empty intended, the self-given and the bodily given.

Empty intended refers to the manner in which we talk about something without bringing it to mind, it is the way in which we converse and listen in everyday conversation. In normal conversation we do not analyse each word or each reference to a person or thing; we do not bring them to mind – we accept the reference and pass on.

The self-given allows us to construct an intuition of the subject of conversation by giving us much more fully that towards which we direct our attention, in this case the importance of Chat Moss as a foundational act for Victorian Society intuitively fulfilled from the story.

Bodily given is the next level of understanding, which involves intuitive fulfilment complemented by the physical thing being before us. As such Heidegger calls it the 'superlative mode of the self-givenness of an entity' (Heidegger, 1985, p. 41). To experience something as it is in the world with us is to experience it in its full truthfulness. From this it would seem to follow that, if the story enabled a 'self-given' intuition of the historical importance of the Chat Moss, then all that was needed to develop an even deeper insight would be to go to the place itself in order for it to be 'bodily given'.

This leads to the second experience that contributed towards the project's development, a visit to the site itself. Yet when this happened, rather than making the understanding more concrete, it only served to increase the complexity of understanding. All we could say of the flat nondescript landscape of stumpy trees and expanses of grass before us, across which the railway still ran, was that it was just there. It appeared to have an intractable nature that was highly resistant to insight; it did not connect with the story in any way. By visiting Chat Moss we confronted the place itself, which gave no inkling of the historic story that had taken us there.

This would seem to suggest that Chat Moss as a place was not a thing of history, and if we wanted to discern its historical standing then we would have to look elsewhere. This we did with more research into the history of the building of the railway across Chat Moss. The main source of information on the construction of the Liverpool to Manchester railway is Samuel Smiles' 1864 publication, *The Lives of the Engineers* (re-published as Smiles, 1975). The book contains information on the vicissitudes of the line's construction, yet this written information did little to advance the project. Rather like the objective facts of the original story, it did little to excite the reader's imagination.

Instead the third insight was provided by a single small image from the book, a vignette line engraving that imagined the scene at the surveying of Chat Moss in the 1820s, which once more unlocked a door leading to intuitive understanding. Our visit to Chat Moss itself had been a quest to associate our historical understandings with a place in nature, which had failed. Here, through this image on a page the intuition of Chat Moss's historical importance became attached to an extant entity, a printed image on the page. Thus understanding is developed through intentional acts of imagining based on a bodily given entity, the print. Here we think it would be useful to analyse the nature of the print as an image and as an extant thing, again through phenomenologically based research, in order to clarify the levels at which they 'operate'.

The print as an image allows us to see something; seeing here is thought in two senses of the word to see as optical sensing and to see in its phenomenological sense as 'simple cognizance of what is found' (Heidegger, 1985, p. 39). This means being able to see what the image is, in addition to its pictorial aspect, what it is as a thing. Examining the vignette as a thing enables us to study it using categories that Heidegger develops in his *History of the Concept of Time.*

Now we can understand it as: an environmental thing; a natural thing; and a picture thing. Everything is both an environmental and a natural thing. The environmental thing is the thing as it exists in the environmental contexture of our world, in this case as an image on paper in a book by Samuel Smiles on page 129 at the end of one chapter and prior to the next that will deal with Chat Moss. The image is oval in shape with no frame; it fades off at its edges off into the surrounding page. The thing's shape, size and pictorial nature tell us that it is a vignette (Figure 12.4).

SURVEYING ON CHAT MOSS.

Figure 12.4 The vignette (Smiles, 1975, p. 129)

We can then look at the way in which the vignette is constructed as a natural thing, i.e. the type of thing it is; immediately we say it is a print using clues from the thing itself. At the bottom of the vignette there is a raised bank indicated in the foreground by a drawn line that wanders freely over the surface. In contrast the moss on which the surveyors stand is represented though a series of mechanistic horizontal lines. This representational schema continues into the background with the delineation of the distant landscape. This insistent horizontal hatching continues into the sky, where the clouds echo the foreground bank in their winding construction, and yet on closer inspection can be seen to be constructed from the same horizontal lines.

The image is still not exhausted by this analysis. Now we can examine it as a picture thing, a thing through which something is pictured or represented. The vignette is the thing through which we see Chat Moss. In the foreground there is a raised bank and beyond it the moss on which we see a group of four men to the left; their dress indicates their status, the top-hatted overseer and the flat-capped white-coated workers who operate the theodolite. To the right a lone figure of a young boy holds the measuring staff, allowing a reading to be made. Immediately we are given an intimation of the organizational aspect of Victorian society through the dress and the different roles of the characters sketchily indicated. This idea is also carried by the woods and hills indicated at the edge of the moss and the smoking chimneys of encroaching industrialization beyond. The landscape scene is dwarfed by the sky.

Let us return to an analysis of the type of image the print is – a vignette. Developed in this Victorian form by Thomas Bewick, the vignette functions as a contemplative temporal image. This is how the Chat Moss vignette works; its position in the book is prior to the written description of the scene it imagines; it calls on the viewer to think ahead to what is to come in the following chapter, a written description of the surveying and construction of the railway over Chat Moss. It also calls on us to think ahead in terms of the scene displayed; following the survey comes the construction and the subsequent triumph of the Liverpool to Manchester railway – a major milestone in the establishment of the Industrial Revolution. In addition, the scene itself is one of temporal looking, looking into the future. The survey is part of a process of planning; to plan is to look to the future.

As such the vignette has a relationship to our intrinsically *a-priori* understanding that is our capacity to know things in advance of empirical experience. Within the Cartesian ordering of the world the *a-priori* is located purely in the subject. Phenomenology discovers that it can be detected in both the subjective and in the objective. A further potential understanding to be drawn from this analysis is that images operate in different ways. The 'temporal looking' that appears to be a function of the vignette may not be found in, say, a photograph, where a different type of looking might be engendered. What that looking might entail in relation to a photograph will be explored briefly in our final example of understandings derived from data, in this case from an aerial photograph.

Early in the project Priestnall created an aerial photograph of Chat Moss (Figure 12.5), made up of a mosaic of many individual tiles. As we said earlier, images allow us to see things through them; through this photograph, with a single glance, we could see the whole of Chat Moss as if from above. It also enabled us to see Chat Moss more perceptually, as an entity over which we had a kind of dominion because of our ability to 'take it in', to view it at one go. The photograph as such gave Chat Moss in its immediate visual totality, but it

Figure 12.5 Aerial photograph of Chat Moss (Getmapping UK)

also gave a kind of ownership to us the viewers. Thus we can speculate that the relationship between the aerial photograph and the viewer might be a proprietary one.

In an effort to undercut this proprietary viewing, we carried out a mind experiment in which we reversed the relationship between viewer and the image by raising the aerial photograph into the air and flipping it 180°. Rather than being looked down upon, the landscape now looked down upon us, undercutting the seeming authoritative view that the aerial photograph appeared to offer. This experience led directly to the decision to construct a ceiling painting (shown in Figure 12.6) as the visual mode through which the understandings of Chat Moss would be embodied. One of the primary differences between a ceiling painting and an easel painting is the relationship between image and viewer. The standard Renaissance understanding of this relationship is a stable projective one, in which the image in the painting is projected onto its picture plane; from there it is projected into the eye of the beholder. Ceiling paintings undercut this by having no one static picture plane; in addition the observer is encouraged, by the composition of the work, to take up a variety of viewpoints, in the process creating viewing conditions which have a greater correspondence with the way in which we naturally engage with the world around us.

There are a number of research implications of this, including speculation on how an image can be made to create different types of perceptual engagements between it and the viewers.

Figure 12.6 The ceiling painting and a computer-generated mock-up showing the gallery context

12.6 Discussion

The use of a case study site was critical to focus the collaboration and to allow a 're-survey' in terms of digital data and historical, textual and image data. This analysis in combination with several field visits allowed the discussion of the nature and meaning of digital data in relation to their real-world counterparts. Also, studies of the relative importance of various data sources to our understanding of the landscape and how ideas of 'spatial context' were affected by varying degrees of *a-priori* information were possible. A careful consideration of different data sources was only possible through focus on a study site and Chat Moss proved rich in this regard despite being considered a rather 'nondescript' landscape by some.

The ceiling painting by Hampson was exhibited in November 2004 at the REDUX Gallery, London, and during the event a colloquium was held involving the authors, cultural geographers, artists and a senior researcher from the Ordnance Survey national mapping agency. The focus on the case study site as seen through the artwork, where some participants had visited the site and some not, proved another interesting aspect of the forum. The colloquium helped to broaden the debate and focus on the research agenda for the remainder of the project and beyond, and produced the edited volume *Chat Moss* (Hampson and Priestnall, 2005).

The nature of the interdisciplinary collaboration between the authors has been enlightening, in particular the process of verbalizing or visualizing certain aspects of their working 'practice', which will be a model for any future collaborative research undertaken. From the perspective of landscape representation and visualization, there seem to be three interrelated themes which constitute the research agenda for further collaboration: data and metadata, methodological and visualization.

12.6.1 Data and metadata

An important consideration to emerge from the colloquium was the difficulty in attempting to generate what may be termed 'general purpose' representations of a landscape. From a GIScience perspective, without a specific application and user community in mind it becomes almost impossible to define requirements and to identify data which might be deemed 'fit-for-purpose'. The actual meaning of data became an interesting focus, in particular how data are interpreted and lead to understanding. An examination of the critical instances where a dataset had significant impact on the development of the artwork allowed an unusually methodical and focussed decomposition of the meaning of particular data, for example the influence of the aerial photograph. This analysis of data and the thinking behind the development of the artwork could be likened to 'metadata' from a GIScience perspective. The data describing the data themselves, and their processing lineage, are important, or at least should be, in determining the way people gain understanding. That said, metadata is rarely used in the context of visualization almost to the extent an artwork typically functions without such supporting commentary. The colloquium gave an opportunity for GIS-based outputs to be presented and discussed in an unusual context and it was clear that these images were open to great misinterpretation without very careful explanation. The role of metadata in supporting landscape visualization will be an area of ongoing investigation.

12.6.2 Methodological

The problems with developing a 'general purpose' representation of a landscape relates to the broader issue of defining geographic ontologies which accommodate different personal perspectives on the same landscape. Although the development of an ontology was not the aim of the project discussed here, the nature of the collaboration certainly helped to avoid presuppositions of commonly used GIS data structures and readily available datasets and concentrate on the elements of the landscape that appear to be important to people. The phenomenological method used by Hampson may offer potential for exploring landscape descriptions as generated by a range of people, including mapping agency practitioners, and a joint exercise with members of the Ordnance Survey research unit has been suggested. A particular aim here would be to assess whether the processes through which people gain understanding can themselves be understood and can then contribute in some way to developing a transparent methodology for developing a more collective understanding.

12.6.3 Visualization

The exhibition and colloquium offered an opportunity to explore how people engage with geographical representations of different types. The ceiling painting itself proved a significant focus in terms of studying issues of viewpoint, scale and geographical frames of reference. There was no requirement for a single viewing position and orientation as there would be for a conventional wall-mounted work or indeed a map or aerial photograph as displayed via a computer monitor (with the assumption that 'North is up'). In some ways this relates to the way in which interactive, immersive and mobile forms of visualization offer less prescriptive views on the data but at the same time can require additional cues to provide spatial context and to help orientate the user within the digital representation. The role of mobile technology and visualization forms a part of SPLINT (spatial literacy in teaching), a Centre for Excellence in Teaching and Learning funded by the Higher Education Funding Council of England (HEFCE). Involving the University of Leicester (lead partner), the University of Nottingham and University College London, SPLINT aims to explore and enhance spatial literacy skills in Higher Education both within and beyond Geography. Through this initiative it is hoped that the effectiveness of various visual and non-visual, geographic representations are explored as alternative spatial frames of reference (for example Priestnall and Polmear, 2006). Lab-based stereo visualization facilities will offer a test bed for studying how groups gain understanding from landscape visualization scenarios, including the exploratory use of metadata in this context. The user's geographic location within the virtual model is also transmitted within the lab as a virtual GPS signal in order to test spatially aware applications being developed on mobile devices before being taken out into the field, following on from the work of Li and Longley (2006). This environment will also offer an opportunity to explore the key components of landscape visualization which prove influential in helping viewers orientate themselves and gain understanding from the presentation.

12.7 Conclusion

The desire to aim for photorealistic landscape visualization is understandable in public contexts given rising expectations and familiarity with cinematic and gaming graphics in the audiences. There do, however, appear to be some interesting avenues of exploration relating to how people gain understanding of a landscape in various contexts and given various forms of geographic representation. No specific common methodological approach for both artistic and GIScience approaches to landscape representation has yet been suggested. The mode of collaboration, however, has proved important in encouraging approaches less reliant on presuppositions. Great research challenges remain in terms of exploring the capture and representation of multiple 'world views' of a landscape and accommodating the varied historical and cultural dimensions which may be present. Less ambitious perhaps would be to consider how existing knowledge relating to the data and processing involved in creating a virtual landscape can be communicated effectively via metadata to better inform the viewer of what they are actually looking at.

Acknowledgements

The 'NEXTMap' Digital Surface Model, derived from Intermap's IFSAR (Interferometric Synthetic Aperture Radar), and colour aerial photography were supplied by Getmapping UK. Digital geology datasets were supplied by the British Geological Survey (BGS). Mastermap vector data used is courtesy of the Ordnance Survey, ©Crown Copyright Ordnance Survey. The Hawley Square project was funded by the European Union 'Interreg' fund and South East Arts. The 'Representations of embodiment' project was funded under the Arts and Humanities Research Council (AHRC) and Arts Council England (ACE) Science–Art Fellowship scheme where artists take inspiration from working within a scientific discipline.

References

Appleton, K. and Lovett, A. (2003) GIS-based visualization of rural landscapes: defining 'sufficient' realism for environmental decision-making. *Landscape and Urban Planning*, 65(3): 117–131.

Augmentra Ltd (2007) Viewranger. Available at: www.viewranger.com (accessed 15 March 2007).

Fonseca, F., Egenhofer, M., Davis, C., and Borges. K. (2000) Ontologies and knowledge sharing in urban GIS. *Computers, Environment and Urban Systems*, 24(3): 232–251.

Fried, M. (2002) *Art and Embodiment in Nineteenth Century Berlin*. New Haven, CT, Yale University Press.

Hampson, D. and Priestnall, G. (2001) *Hawley Square, Exhibition*. Exhibition Catalogue. Interactive CD-ROM and web site available at www.nottingham.ac.uk/~lgzwww/contacts/staffPages/gary/research/mappingchange/mappingchange.htm (accessed 15 March 2007).

Hampson, D. and Priestnall, G. (2005) *Chat Moss*. Nottingham, CMG.

Hart, G. (2004) Tales of the river bank: an overview of the first stages in the development of a topographic ontology, *Proceedings of Geographical Information Science Research UK (GISRUK '04)*, University of East Anglia, 28–30 April.

Heidegger, M. (1985) *History of the Concept of Time*. Translated by Kisiel, T. Bloomington, IN, Indiana University Press.

Husserl, E. (2001) *Logical Investigations*, Vol 2. London, Routledge.

Li, C. and Longley, P. (2006) A test environment for location-based services applications. *Transactions in GIS*, 10(1): 43–61.

Meng, L.Q. (2005) Egocentric design of map-based mobile services. *The Cartographic Journal*, 42(1): 5–13.

Priestnall, G. (2004) Augmenting reality? 3D modelling and visualization in geography fieldwork, *Proceedings of the 12th Annual Geographical Information Science Research UK Conference (GISRUK '04)*, University of East Anglia, 28–30, pp. 35–38.

Priestnall, G. and Polmear, G. (2006) Landscape visualization: from lab to field, *Proceedings of the First International Workshop on Mobile Geospatial Augmented Reality*, Banff, Alberta, 29–20 May.

Sheppard, S. R. J. (2001) Guidance for crystal ball gazers: developing a code of ethics for landscape visualization. *Landscape and Urban Planning* (special issue), 54: 183–199.

Smiles, S. (1975) *The Lives of George and Robert Stephenson.* London, The Folio Society/The Chaucer Press (originally published 1874).

Wainwright, A. (1964) *A Pictorial Guide to the Lakeland Fells, Book 6: The North Western Fells.* London, Frances Lincoln.

Whitelock, D. and Jelfs, A. (2005) Would you rather collect data in the rain or attend a virtual field trip? Findings from a series of virtual science field studies. *International Journal of Continuing Engineering Education and Lifelong Learning,* 15(1/2): 121–131.

13

Visualization, Data Sharing and Metadata

Humphrey Southall

Department of Geography, University of Portsmouth

13.1 Introduction

In the early years of statistical computing, file formats associated with particular program packages, and especially SPSS (Statistical Package for the Social Sciences), provided a *lingua franca* for researchers exchanging data. One consequence was that in its earliest years the UK Data Archive, and similar organizations elsewhere, saw their job as archiving SPSS files. Imagine if this simple situation had continued: you could take someone else's data and load it straight into visualization software, and the software would automatically identify variables, observations and associated labels and coding schemes; intelligent software would make sensible decisions about how to present data to non-experts; crucially, Google and its rivals would know how to scan the contents of such files, and searching for statistics on the internet would be trivial, and not the current nightmare.

Sadly, our closest current approximation to an omnipresent interchange format is Excel, but this defines almost no internal structure. This chapter reviews a new standard developed by the Data Documentation Initiative working with archives worldwide, and describes how it was used to directly control statistical visualization for inexperienced users in the web site *A Vision of Britain through Time*. Using it in this way proved to require significant extensions to the standard, and the chapter as a whole argues that an effective standard linking data gathering, statistical analysis and visualization will only emerge once data analysts become more involved in standards setting.

Geographic Visualization Edited by Martin Dodge, Mary McDerby and Martin Turner

13.2 The data documentation initiative and the aggregate data extension

The Data Documentation Initiative (DDI; www.ddialliance.org) is very much a collective enterprise of data archivists, not data analysts: the initiative has 25 members, all institutions and including the Inter-university Consortium for Political and Social Research (ICPSR) at the University of Michigan, the UK Data Archive at the University of Essex and the World Bank's Development Data Group. The DDI Alliance was formally established only in 2003, but work on the standard began in 1995, the first beta version appeared in 1999 and the DDI DTD Version 1.0 was published in March 2000 (Blank and Rasmussen, 2004). The work described here was based broadly on DDI Version 2.0, and especially on the aggregate data extension it introduced, but at the time of writing Version 3.0 is being finalized for release.

Version 1 of the DDI focused purely on survey data, meaning the results of directly computerizing the replies to questionnaires, ignoring time series and aggregate data. The highest level of documentation would be for a *collection*, meaning a complete data library, but this would be made up of *studies*, such as a particular questionnaire survey. The DDI document describing the data created by a study was to be divided into five parts: the first described the structure of the document itself; the second was a description of the study that created the data; the third was a description of the physical format of the data files; the fourth contained the variables, which for a questionnaire survey meant the questions asked; and, finally, any other materials. All this information was designed to be not just *machine-readable* but *machine-processable*, and was therefore represented using XML.

Version 2.0 of the DDI specification was released in 2003 and included an extension covering aggregate and tabular data, such as appears in census reports. This extension has provided a foundation for much of the work of the Minnesota Population Center on the US National Historical GIS project (www.nhgis.org), whose aims included converting all existing transcriptions of data from US census reports to a standard format. Any given census is, of course, a questionnaire survey which could be described using DDI 1.0. However, that description would cover only the individual-level microdata, which are usually confidential for a long period. Those censuses which are old enough to be no longer confidential were carried out long before the computer, and would be enormously expensive to computerize. Research is therefore largely limited to the tabulations created by the census offices from the confidential individual-level data, and published either in printed volumes or, more recently, machine-readable small area statistics. DDI 1.0 was used experimentally by the Minnesota project to describe aggregate data, but coverage of cross-tabulations was obviously inadequate.

The key innovation in the DDI aggregate data extension is the *nCube*, which is essentially a matrix. The dimensions of an nCube are defined by its component variables. The simplest possible nCube is a population count: just one variable and that with only one category. Two separate counts of the numbers of men and women in each area are only slightly more complex: a one-dimensional nCube, based on a single variable containing two categories. However, more complex structures are common. For example, the six *Decennial Supplements* published by the Registrar General for England and Wales between 1851 and 1910 consist mainly of sets of tables, one for each of around 600 Registration Districts, giving numbers of deaths in each combination of an age group and a cause of death, sometimes further sub-divided by gender. The report for 1861–1870 is mainly concerned with a 2 × 12 × 25 three-dimensional nCube: men or women; one of 12 different age groups; and 25 different cause of death categories. An even more extreme example is the first ever report on Britain's occupational structure, from the 1841 census. No occupational classification had been devised in advance, so the listings are a mixture of some ad hoc groupings and many individual job titles. The occupations listed vary between counties but, ignoring purely typographic differences, we found a total of 3647 different occupational categories. The tables divide the population into men and women, and into those aged over and under 20, so we have a 2 × 2 × 3647 nCube.

The three most important entities in the DDI aggregate data extension are *variables*, which combine into *nCubes* and are made up of *categories*. However, there are a number of other important concepts. First, while the original questionnaire may ask a simple question, like age, an almost infinite number of different categorizations can be imposed in the creation of aggregate statistics. For example, we have so far found 17 different sets of age groups used in British census and vital registration reports. These we group together into a single *variable group*, and we have similar collections of occupational and cause of death classifications. The DDI specification also defines *category groups* as collections of categories within a single variable that can be treated as one for analysis. nCubes can also be assembled into groups.

Second, two different nCubes can consist of exactly the same variables but not be the same. For example, both the census and the *Decennial Supplements* contain tables listing age by sex, but the census is tabulating the number of people alive on a given day while the *Supplements* are tabulating deaths over a period. The DDI standard records this via an nCube attribute called the *universe*, which is defined as a text string describing the sum of all values in the nCube: all people, all deaths, all persons in employment. Two other nCube attributes are *measurement units*, meaning what the numbers are counts of, such as 'persons' or 'acres', and *additivity*. Most basically, this last records whether the values within an nCube do add up to a meaningful total. For example, most historical British censuses published a single parish-level table, the only source of information on the most detailed administrative units. These tables are usually a mixture of information, the different columns not necessarily being logically connected. It would be possible to wholly define one of these tables as a single nCube, but clearly the result of adding together an acreage and a population total is meaningless. We generally avoided this, but it was useful to combine the current total population, the population 10 years previously and, occasionally, the population 20 years previously into a single non-additive nCube.

Thirdly, the detailed structure of an nCube is defined by its *Location Map*. The example below is the XML representation of a simple 2 × 2 nCube breaking down the total number of births by both gender and legitimacy. The nCube definition itself specifies the various attributes already mentioned as well as identifying the two variables used, defined elsewhere. The <text> attribute of the nCube definition could hold a much longer explanation of the data structure, and this is in fact where much of the text appearing on the *Vision of Britain* web site is held. The nCube definition is then followed by the Location Map, which records the actual location of the data. One key point is that in this terminology the number of legitimate male births is *not* a variable, it is a cell defined by the coming together of two different variables. This example was generated from the GB Historical GIS, and as explained below, the way that the physical locations of the data are specified is non-standard:

```
<nCube ID="N_BIRTH_LEG_SEX" dmnsQnty="2" cellQnty="4">
 <labl>Births by legitimacy and sex</labl>
 <universe>All Births</universe>
 <text>Births by legitimacy and sex.</text>
 <dmns rank="1" varRef="V_LEGIT"/>
 <dmns rank="2" varRef="V_SEX"/>
 <measure measUnit="Births" additivity="Y"/>
</nCube>

<locMap ID="LM">
 <dataItem ID="N_BIRTH_LEG_SEX_1" ncubeRef="N_BIRTH_LEG_SEX>
  <cubeCoord coordNo="1" coordVal="1"/>
  <cubeCoord coordNo="2" coordVal="1"/>
  <physLoc cellRef="L_M_Births"/>
 </dataItem>
 <dataItem ID="N_BIRTH_LEG_SEX_2" ncubeRef="N_BIRTH_LEG_SEX>
  <cubeCoord coordNo="1" coordVal="2"/>
  <cubeCoord coordNo="2" coordVal="1"/>
  <physLoc cellRef="I_M_Births"/>
 </dataItem>
 <dataItem ID="N_BIRTH_LEG_SEX_3" ncubeRef="N_BIRTH_LEG_SEX>
  <cubeCoord coordNo="1" coordVal="1"/>
  <cubeCoord coordNo="2" coordVal="2"/>
  <physLoc cellRef="L_F_Births"/>
 </dataItem>
 <dataItem ID="N_BIRTH_LEG_SEX_4" ncubeRef="N_BIRTH_LEG_SEX>
  <cubeCoord coordNo="1" coordVal="2"/>
  <cubeCoord coordNo="2" coordVal="2"/>
  <physLoc cellRef="I_F_Births"/>
 </dataItem>
</locMap>
```

Two concluding comments need to be made about the standard before discussing how we used it. Firstly, it is possible to include reporting geographies as DDI variables; for example, the set of counties within the USA could be defined as one member of a 'geography' variable group, and mapped to a simpler variable containing the 51 states via a set of category groups. Secondly, this is all a very abstract way of thinking about social statistics, and this in itself has inhibited adoption of the DDI standard by the data archiving community, and within the GB Historical GIS project.

13.3 Implementing the DDI within the GB Historical GIS

The original Great Britain Historical GIS was developed between 1994 and 2000 using a relatively conventional combination of ArcGIS and Oracle software: most of the intelligence lay in very complex AML code which assembled a large library of arcs (boundary segments), each with a beginning and end date, into boundary polygons for a particular date. Statistical reporting units – parishes, districts and counties – were defined via named label points, and there was no guarantee that the units named in the first column within each of the hundreds of statistical tables within the Oracle database matched the units defined in the label points. In many ways, this system was a triumph of making software do something it was not designed for, but it was almost impossible to ensure data integrity and it was clearly too slow to drive a high volume web site (Gregory and Southall, 1998).

New funding from the UK national lottery, discussed below, forced a major rethink leading to a completely new architecture in which almost everything was held in Oracle, ArcGIS playing no role. This would have been impossible when we started in 1994 because it requires a relational database with object extensions, enabling spatial data to be held alongside statistics and text. Figure 13.1 provides an overview, although it ignores how we hold text and historical maps.

The original GBH GIS had been able to create maps, using ArcMap, but it required users to select a particular statistical table from an ever-growing collection of many hundreds, and then to choose sensible combinations of columns from those tables to compute a rate. Our new approach began by saying that we were interested in the individual data values, and it was actively unhelpful to group them into 'tables' which generally came from census reports and each held data from a single year. Instead, we would hold *all* statistical data values in a single column of just one table, and give them meaning entirely through metadata.

Our main *data table* is almost as simple as Figure 13.1 suggests: one column holds the data values themselves and then other columns hold values recording location, meaning, date and source. The date column in fact holds an object, able to hold anything from a single calendar year, for a census, to a pair of calendar dates, e.g. for quarterly mortality data. The other columns hold identifiers whose meanings are defined in the three main metadata sub-systems. *Where* is recorded not by any kind of coordinate data but by a reference to a large *Administrative Gazetteer* of over 50 000 British local government units, modern and historical, each of which can have a set of associated variant names and, if available, one or more boundary polygons, using dates on the polygons to handle change over time. A central aim was to create a system which could hold information for units whose location was either unknown or had yet to be recorded through the expensive process of boundary digitization. Units are grouped into *Unit types* which have some of the characteristics of

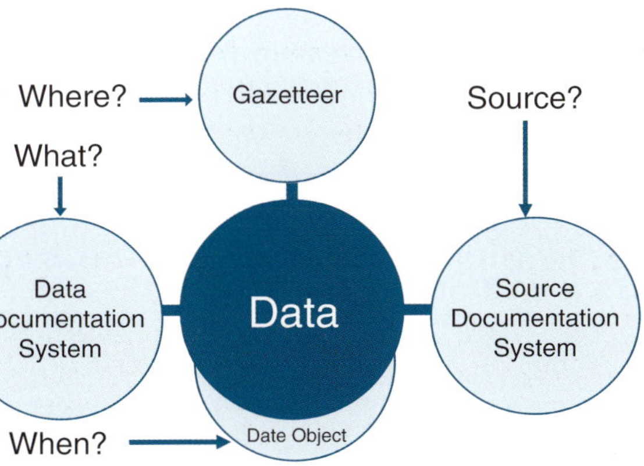

Figure 13.1 GB Historical GIS: overview

GIS coverages. The *Source Documentation System* (*SDS*) contains detailed information on the history of British censuses and their reports, and provides a unique identifier for each table within those reports. However, the current subject is our *Data Documentation System*, implementing the DDI.

13.3.1 DDI elements not implemented

Our implementation ignores large parts of the original Data Documentation Initiative standard, which identify *Collections* and the *Studies* of which they consist. At one level this is a quite obvious decision: the whole of the GB Historical GIS is a single data structure, and our data table is a single dataset or 'Study', so if need be we could provide this high-level documentation via a single block of static XML. From a data librarian's perspective, however, this approach is more questionable because the contents of our data table incorporate the results of a series of quite distinct projects, including much data computerized elsewhere and donated to us.

We are not data librarians and in the Great Britain Historical GIS two other metadata sub-systems record the provenance of the data: the SDS locates individual data values in particular rows and columns within named tables within the census reports, while a smaller 'Thanks' system allows us to associate each individual data value with a different set of 'credits' identifying the different roles of a whole series of contributors. Our aim with donated data is always to check back against the original statistical publication from which the data were transcribed, and if necessary modify the data so it matches the transcription we would have created, so distinguishing between donors and the original historical source is essential. However, none of this is the job of our DDS, which is about recording what the data *mean*, and giving them a new context supporting time series analysis.

We have also not directly implemented *Category Groups*, although they are implicit in the mappings we create between categories, or *Nested nCubes*, although some such mechanism is desirable in presenting very large nCubes such as the almost unclassified 1841 occupational data discussed earlier. We have also, obviously, not included either time or geography as DDI variables, but record those using the separate structures already described: British administrative geographies are too fluid to capture using one single hierarchy of units, and the US NHGIS does not treat geography as a simple variable (Southall, 2001).

13.3.2 Extensions to DDI: cellRefs and universes

The Location Map is one of the key innovations of the DDI Aggregate Data Extension, recording where data values occupying a particular cell within an nCube will actually be found. As originally implemented by the United States National Historical GIS, this meant identifying locations within conventional flat or hierarchical files held separately as ASCII. Our implementation is very different, holding everything in that single column; and we neither know nor care what order data values are within that column. We therefore defined a new concept, a 'cell reference' or *cellRef*, which is simply a unique identifier which appears in both the data table and the Location Map. Our cellRefs in fact have a readable structure: the first part, in upper case, usually identifies the nCube the cell is part of, and subsequent elements identify, in lower case, the categories within the different variables; these elements are linked by underscores. However, this is entirely to make them usable by the people building the system; the system itself would work just as well, and very slightly more efficiently, if these strings were replaced by arbitrary ID numbers.

Secondly, the DDI standard says that every nCube must have a *universe* attribute, specifying the population it covers, and a *measurement unit* attribute, specifying whether the numbers count people, or houses or hectares. One of our crucial changes was to re-define both universes and measurement units as entities. This was necessary as our rules for generating derived values do not look just for nCubes with related variables: two nCubes might both be based on two variables measuring sex and age, but it would be wrong to map from an nCube classifying total population to one classifying deaths. We therefore also check that the universes match, but this requires a fixed set of universe identifiers, not the simple uncontrolled string of text specified by DDI 2.0. DDI 3.0 has similarly introduced controlled vocabularies for universes.

13.3.3 Additional presentational elements

Our other extensions are designed to help organize a web site and assist in presentation. We replace *studies* and *collections*, as high-level organizing concepts, with *database* and *themes*. The *database* is the single entity in our particular system that everything else belongs to, directly or indirectly, and is the starting point for browsing the structure. *Themes* are used to divide up the statistical content and play a large role in the *Vision of Britain* web site, discussed below. We defined 10 themes, based on an exhaustive analysis of the contents pages of other people's social atlases: Population, Life and Death, Industry, Work and Poverty, Social Structure, Housing, Learning and Language, Roots and Religion, Agriculture and Land Use, and Political Life. All nCubes are assigned to a theme, and all variables are either directly assigned to a theme or belong to a *variable group* which is assigned to a theme.

Our last new entity is the *rate*, which currently defines what maps can be created. A major problem with earlier systems like the Great American History Machine was the ease with which naïve users could create totally illogical maps (Miller and Modell, 1998). To some extent, as discussed below, defining nCubes creates a set of ratios which are logically valid, but they are not necessarily interesting; and some important ratios are between numbers from very different structures, such as an infant mortality rate, which combines data on births and deaths. Our *rates* are defined in terms of a numerator, a denominator and a multiplier, the first two being cellRefs and the last a numeric constant. Even when the values come from a single nCube, rate definitions record key aspects of conventional practice. For example, we worry about the 'unemployment rate', defined as the number out of employment, divided by the number economically active, multiplied by 100; not an 'employment rate, based on the number in employment and multiplied by 1000'.

13.3.4 Implementing the DDS

Figure 13.2 is emphatically not a database structure diagram, as we further abstract the DDI and hold all entities – everything in Figure 13.2 except the location map and the

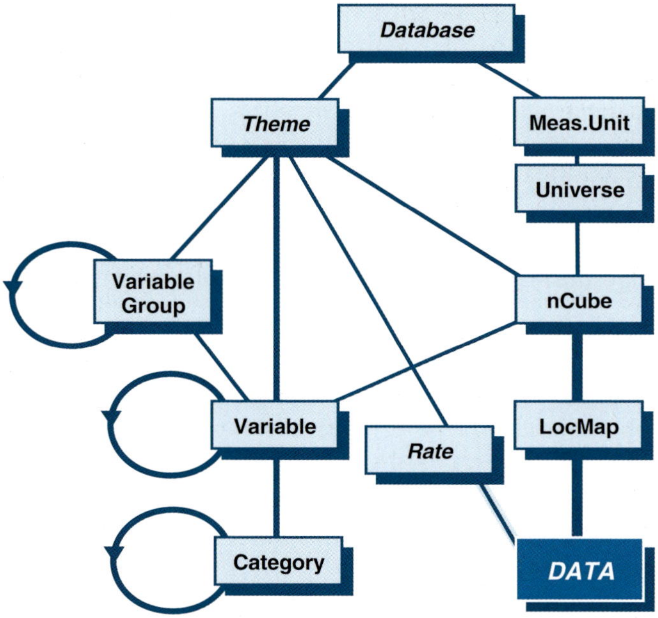

Figure 13.2 GBHGIS Data Documentation System: entities and relationships

data table itself – in a single table. Figure 13.3 shows the surprisingly simple database structure. *g_data_ent_type* defines the kind of entities which can exist in *g_data_ent*, while *g_data_legal_rel* defines what kind of relationships between them are allowed to exist in *g_data_rel*. Holding all entities in the one table again simplifies program code and will also assist in creating a planned search interface.

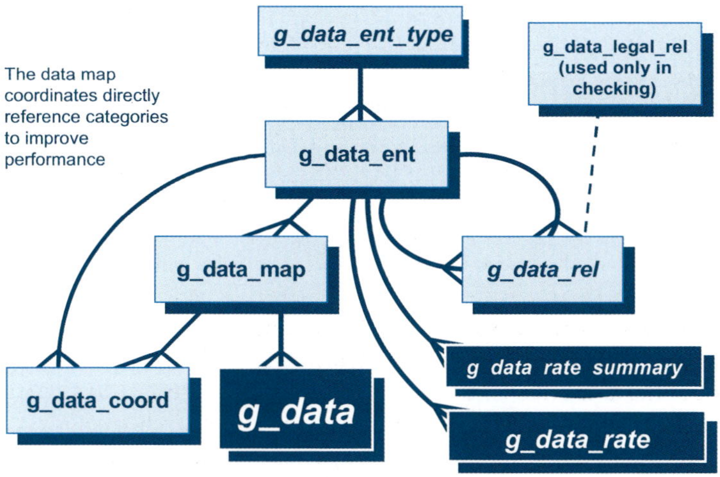

Figure 13.3 GBHGIS Data Documentation System: database implementation

The structure of the *location map* is fundamental to holding information about data structures with any number of dimensions in a fixed set of two-dimensional tables. The g_data_map table defines the cellRefs and associates each with the nCube to which it belongs. However, information held in *g_data_coord* locates it within the nCube: cells within a one-dimensional nCube, based on a single variable, have just one linked entry in *g_data_coord*, the value of coordNo always being '1' and the value of coordVal identifying the category; a cell in a three-dimensional nCube, on the other hand, has three linked entries in *g_data_coord*, with values of coordNo running from 1 to 3. The system would still work without also holding the ID of the relevant category in *g_data_coord*, but this substantially accelerates the system.

Another way we accelerate the system is by pre-computing all values of defined rates, and these are held in *g_data_rate*, with labelling information in *g_data_rate_summary*. There is one row in *g_data_rate_summary* for each combination of rate ID, unit type and date, so essentially each row describes a map. Information held includes values of sextiles, which define the keys for the default mapping. However, this is obviously to support a particular visualization interface, which we now discuss; it would be perfectly possible but slower to generate maps directly from the rate definitions and the raw counts in the main data table.

One obvious question is why this approach has not been taken before. The main answer is that it uses far more disk space than storing statistics in large numbers of separate tables: conventional methods of data archiving were originally developed in the 1970s, when disk space was very expensive. The world has changed, and the whole of the database behind *Vision of Britain*, including the scanned images of maps, which take up most of the space, could comfortably be held on an iPod.

13.4 Driving visualization in *Vision of Britain*

Thus far little has been said about visualization, other than as a justification for extensions to the DDI standard. The Great Britain Historical GIS is a data structure designed to support many activities, while visualization requires software and some kind of user interface. The particular interface to be discussed here, *A Vision of Britain through Time*, is just one of the many possible interfaces to the Great Britain Historical GIS, although it is the most complex that we have actually built. As an interface, it is very different from most of those discussed in this book, even those that are web-accessible (www.VisionOfBritain.org.uk). It was funded by the UK National Lottery, via the Big Lottery Fund, as part of their 'Digitization of Learning Materials' programme. A positive consequence was that we were, very unusually, able to obtain large-scale funding for a project centrally concerned with the visualization and mapping of social statistics, especially census data. However, the funding source imposed many constraints. One of them was simply that most of the budget had to go on 'digitization', i.e. the initial computerization of data, and programme rules tended to assume that the 'data' were scanned images which would be served to the public using an off-the-shelf content management system. New software development was something that had to be sneaked in, and one merit of the data architecture already described was that it transferred intelligence from software to the data structure: it enabled us to use a relatively small amount of code.

More specifically, all our content had to be accessible using a very basic web browser, perhaps on a mobile phone or gaming box rather than any kind of computer. We were not allowed to require any of the following: plug-ins like Flash or SVG viewers, Javascript,

ActiveX components or Java applets. Cascading style sheets were strongly promoted precisely because pages using them correctly should be usable even with browsers that ignore them. However, AJAX methodologies would clearly have been unacceptable had the concept been defined when the programme started. Strictly speaking, all of these technologies could be used to provide optional additional access methods, but our resources never permitted work on such extras. It rapidly became clear that all graphics generation would have to be done on the server, and this obviously limited interactivity. The screen layout had to work acceptably on an 800 by 600 display.

A further set of issues followed from the size and nature of the audience. It was extremely difficult to estimate likely numbers of users prior to launch, but the large and well-publicised problems encountered by the UK National Archives with their 1901 Census web site in the spring of 2002, coming as they did a few months into our own work, made the dangers clear and it was obvious we needed to plan for large numbers of simultaneous users. This proved justified. Funding issues meant we had to launch with a site that had not been fully stress tested, and strong publicity including a television appearance – for a statistical visualization project – led to the site crashing continuously for the fortnight following the launch. Figure 13.4 shows the results of that problematic launch, but also that we are now handling more users than, for example, the Office of National Statistics' *Neighbourhood Statistics* site. Serving a large general audience also meant that our repertoire of maps and graphs had to be immediately comprehensible, so nothing too avant garde.

13.4.1 Mapping rates

Our statistical maps are generated by a specially written servlet based around the Geotools Java class library (http://geotools.codehaus.org/) and also containing code from the

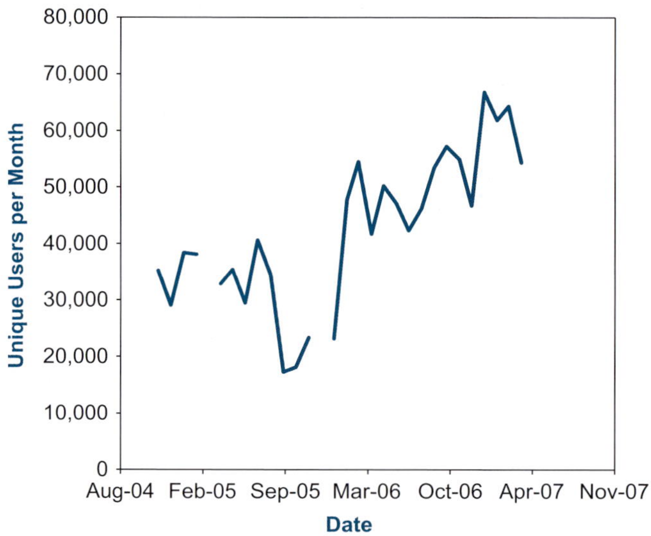

Figure 13.4 *A Vision of Britain through Time*: unique users 2004–2007

Geoserver project, enabling it to directly extract both polygons and statistical data from our Oracle data. So far, statistical mapping is limited to the *rates* defined in the DDS, and Figure 13.5 shows one of the resulting pages, slightly edited to save space. The map itself is an utterly unremarkable choropleth map of population growth rates, but the range of options the system is offering the user demonstrates its flexibility:

- The timeline bar above the map allows the user to request a map for a different date. Dates for which the same map cannot be drawn are greyed out, but this particular map requires only the most basic population count so it is available for every census year except the very first, 1801. The same timeline appears above time series graphs of the rate, and is used to request a map.
- Zoom in, zoom out and, hopefully, pan are self-explanatory even with a general audience. They are unlikely to have used even desktop GIS software, but Multimap, Streetmap and various on-line route planning sites should make these tools familiar. Choosing 'Select Unit' enables users to go to our 'home page' for the particular administrative unit they next click on.
- Once they start zooming in, few people can tell where they are looking at on a map consisting entirely of administrative boundaries. Another part of the project scanned three complete sets of historical one inch to one mile maps of Britain, plus less detailed maps from roughly the same dates. This material is held in a separate repository but is accessible using the Open Geospatial Consortium's Web Map Server standard. We present it independently in our 'historical mapping' section, but it is also available as an underlay for the statistical mapping, the statistical mapping servlet calling the historical WMS.
- The next set of options is the various rates which are available within the same theme for the same date and type of unit. An obvious limitation of this interface is that it does not show what additional rates might be available if the user changes the date or unit type.
- Finally, the user can request a map of the same rate for the same date but a different type of administrative unit. This particular map is an extreme case, because the 1911 census reported both on the nineteenth century system of Registration Counties and Districts as well as the then-new local government geography. The units actually being used in this map are modern district and unitary authorities, based on our redistricting work. The parish-level geography is greyed out but will become an option once the user zooms in sufficiently.

The greatest virtue of this system is how easily extensible it is: adding a whole new set of maps requires nothing more than a new rate definition within the DDS.

13.4.2 Graphing nCubes

The most easily accessible graphs in the system are also based on rates, presenting time series for the selected unit and an additional 'comparison unit'. By default, these comparisons are

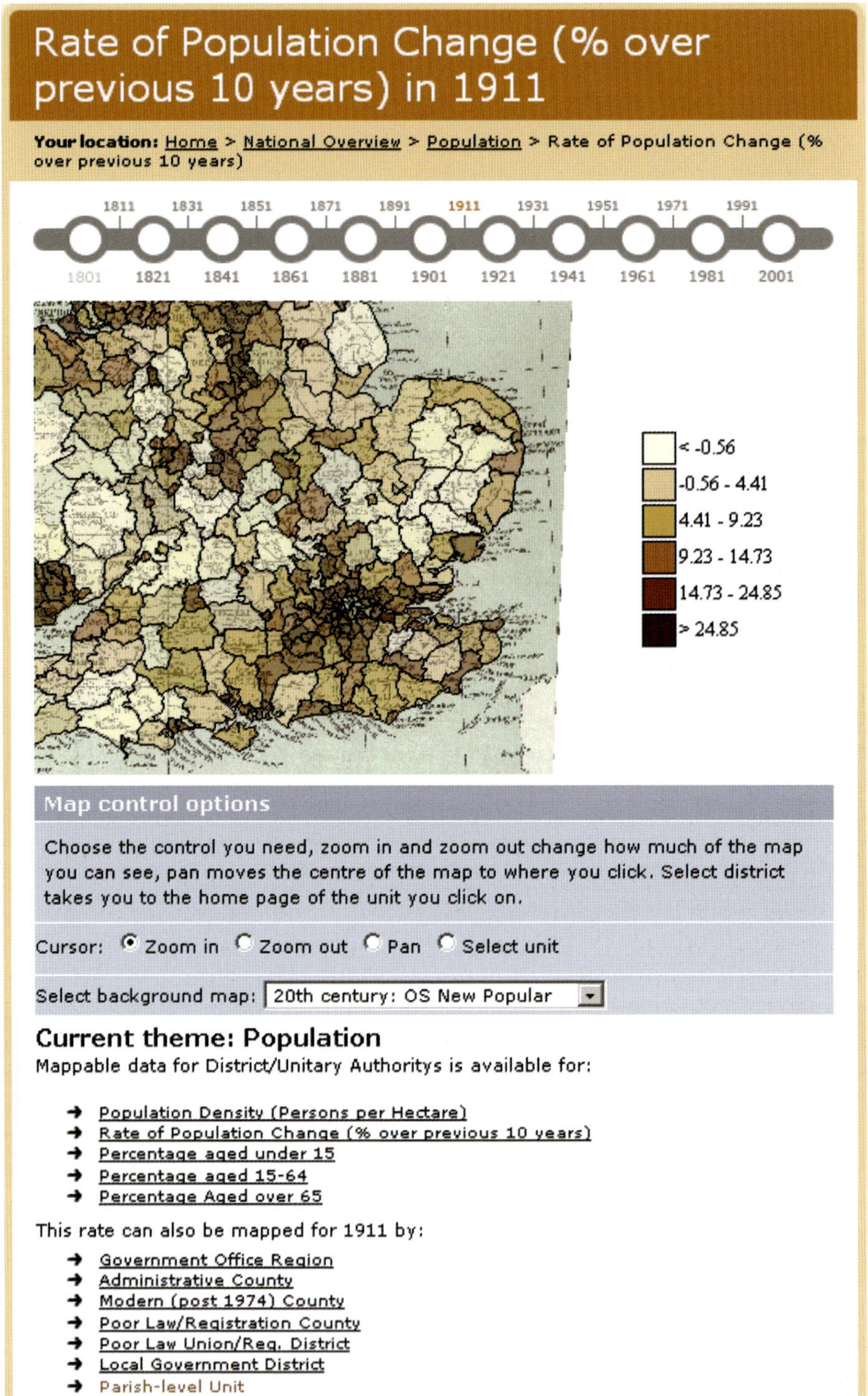

Figure 13.5 *A Vision of Britain through Time*: mapping a rate, with options

with national totals, and as noted above, clicking on points in these graphs takes users to a map of the rate at the relevant date, centred on the current unit. The graphs are either line or bar charts depending on the value of the 'continuous' attribute of the rate. Users are currently able to change the comparison unit only with the modern units covered by the 2001 census, as essentially the same data are available for all of these; so for example you can compare any pair of current local authorities.

However, most of the system's graphical repertoire is for presenting nCubes. Each administrative unit in the system, from the UK down to individual parishes, has its own 'home page', which lists the statistical themes for which data are available. The theme pages begin with links to the available rates, include some introductory text, and end with a list of available nCubes, including the period they cover and the variables they contain. Three alternative views are available for each nCube: a *Table View* lists the actual numbers; a *Source Info View* provides information on sources, and for census data may include links to reconstructions of the original table driven by the SDS; but the initial view of each nCube is a graph.

Figure 13.6 shows how the system decides what kind of graph to create based on what it knows about the relevant nCube. The very first case is slightly silly, as a one-dimensional nCube based on a single category variable holds just one number: we present it as a one-bar chart, always the same height but of course with a different scale. This graph, and all other graphs we generate, is created by an existing graphing servlet rather than one written especially for the project: we use Ce:wolf, based in turn on the Jfreechart class library. Unlike the statistical mapping servlet, this does not extract data from Oracle for itself but is passed an array of data by our software. Ce:wolf suffers from poor documentation but it is very fast.

This approach works well for a variety of different kinds of nCube based on single variables, whether for single dates or for time series, and also for two-dimensional nCubes where one of the variables has only two categories. Most often with social statistics that variable is sex, but *Vision of Britain* always generates 'population pyramids' for nCubes of that structure. With anything other than one-dimensional nCubes, time is always handled simply by creating a sequence of separate graphs, one for each date. Presentation of more complex two- and three-dimensional nCubes is less satisfactory, especially when data are available for a series of different dates, and the system currently makes no attempt to graph nCubes with four or more dimensions; if we had any, the current software would conceal their existence from users. However, this is a limitation not of the overall approach but of a system with limited interactivity where everything is done server-side.

13.4.3 Future potentials

The current graphical repertoire is severely restricted both by the rules imposed by the funding body and our own limited resources to develop software. New funding from the Joint Information Systems Committee is primarily aimed at extending our substantive coverage to include British parliamentary election results from 1832 onwards, a large part of the project being mapping the changing boundaries of constituencies. However, we will be able to make some additions to our graphing and mapping capabilities so it is useful to discuss the potentials.

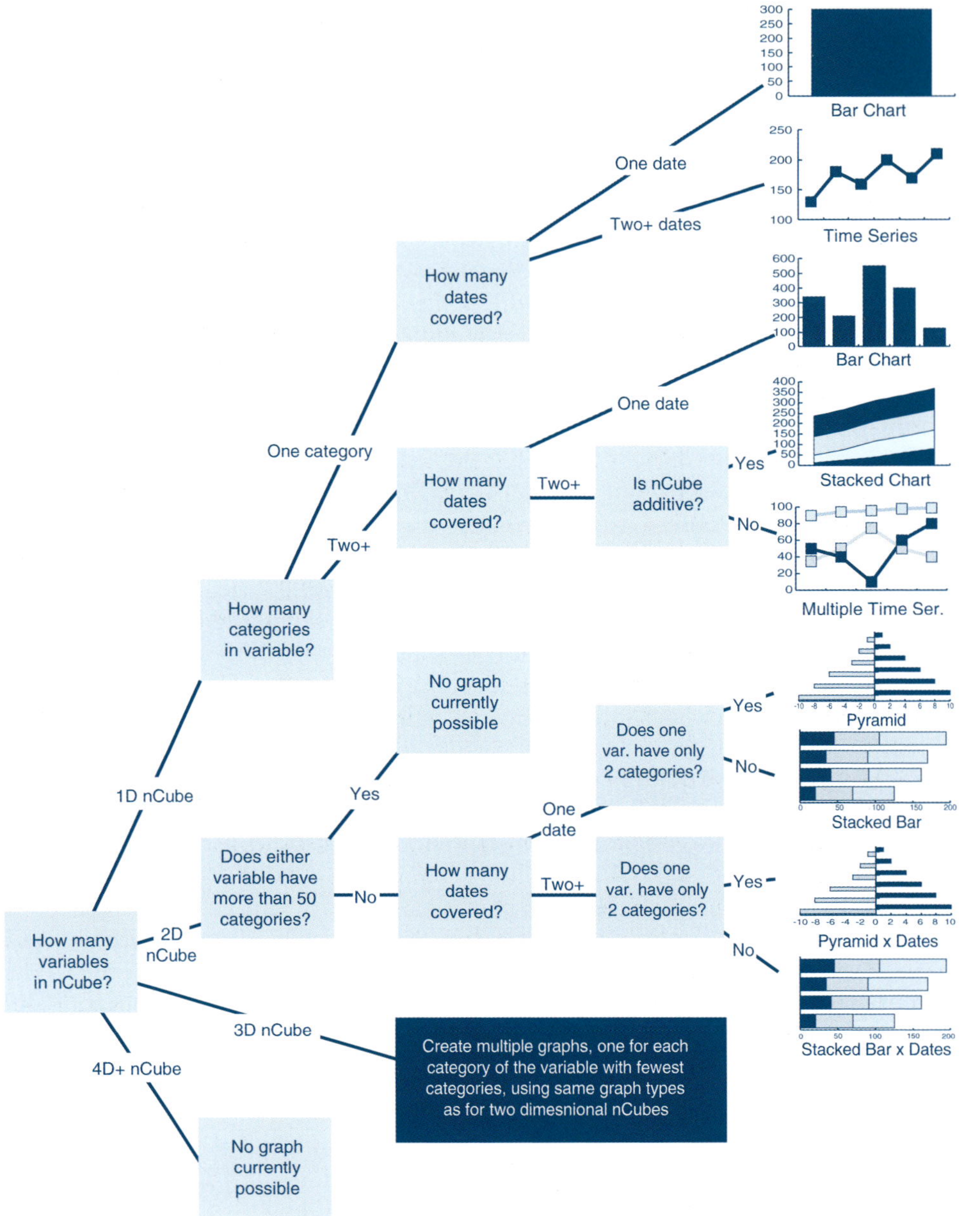

Figure 13.6 *A Vision of Britain through Time*: decision tree for graphing nCubes

Firstly, the current system maps only *rates*, but facilities to explore *nCubes* through maps are highly desirable. The problem here is partly one of providing a comprehensible user interface given the range of possibilities with higher dimension nCubes. Mapping one-dimensional nCubes is simple: a choropleth map showing any one of the categories within the single variable as a percentage of the universe, i.e. the overall total. With two-dimensional nCubes, there are five possible kinds of choropleth map: any cell as a percentage of the universe; any cell as a percentage of the row total for one dimension; any cell as a percentage of the row total for the other dimension; any row total on one dimension as a percentage of the universe; any row total on the other dimension as a percentage of the universe. With a three-dimensional nCube, there are 19 kinds of choropleth map, and we leave it as an exercise for the reader to work out the permutations. Providing a user interface to select between them is clearly non-trivial. A quite separate kind of option we hope to offer is cartograms as base maps instead of conventional boundary maps, and even animating them (Southall and White, 1997).

There are clearly large potentials for graphing higher-order nCubes, but these can only be realized within a software environment that offers three-dimensional graphics and animation. One aspect is that with one-dimensional nCubes there is usually one clear right way of presenting the data; with higher dimensions there are often a series of possibilities each of which will best reveal some facet of the data, and the user needs some quick mechanism for moving between them. Animation is arguably essential, as the only way of clearly presenting the time dimension, and as the web site's full name indicates, this resource is all about visualizing statistics *through Time*. The potential for presenting changing age and gender structure via a single animated population pyramid is very obvious. Some of the infrastructure for such developments is already in place, as until the impact of the lottery's technical rules became clear we planned to run GeoTools within users' browsers, and the software hooks for doing this remain in the system.

However, there are other potentials which could only be realized using much higher performance hardware. For example, the system holds geo-referenced historic maps for different dates, and could be easily extended to include digital elevation model data for the whole country. This would permit an interface in which users flew through a landscape in which the scanned maps were draped over a relief model. Animated three-dimensional graphics could then be located within this landscape to represent nCube structures associated with each geographical area. With the exception of the DEM data, all the information and metadata to support this vision is already within the system.

13.5 Conclusions

Academic visualization research is almost by definition about high-end visualization, because the low end is left mainly to the developers of Excel and various statistical packages. However, the effect of this is to prioritise helping people with PhDs to better understand their data over helping the general public, who perhaps need more help. Excel, in particular, knows far too little about the data it holds to ensure that users only create graphs that make some sense; the same is arguably true of the statistical maps that desktop GIS packages can create. The enormous strength of the DDI-based visualization outlined here is that the statistical metadata are not simply a source of labels for graphs, they capture the *logical* structure of the data, and constrain users' graphical choices to maps and graphs that make some sense.

Academic researchers looking at the current main outcome of this research, the *Vision of Britain through Time* web site, will be under-whelmed by the very unremarkable maps and graphs it creates, and it has already been noted that we were heavily constrained both by the very limited range of web technologies we were allowed to use and the need to use only types of maps and graphs which would be immediately understood by a very wide audience. However, some technically much more interesting possibilities were outlined above and we would be very interested in pursuing these collaboratively. Recent funding to extend the web site comes with fewer technical restrictions, and we could add more interactive vector-based visualizations using a combination of AJAX and Flash, or SVG.

The system also has some interesting analytic potential. Conventional statistical data libraries are very poorly suited to automated analysis because individual data values are spread across many different tables of every size and shape, and metadata are held separately and designed to be read only by humans. Conversely, our system holds both statistics and metadata in absolutely consistent locations, making it easily accessible to analytic software such as the Geographical Analysis Machine (Openshaw *et al.*, 1987). One interesting demonstration of how easily accessible the data structure is to automated exploration is its performance with Google, which is able to reach almost every page in the *Vision of Britain* web site: although we exclude them from our usage statistics, a third of all page views are for Googlebots, but in return the site features very strongly in search results.

One final caveat is necessary. The work reported here was very heavily influenced by data archivists, and especially the work of the DDI Alliance. However, making the data standard drive a visualization system required some extensions. Our ignoring the DDI concepts of *collection* and *study*, and introducing *database* and *theme*, are superficial. More importantly, our system works only because we have introduced a detailed controlled vocabulary for *universes*, and because almost all nCubes in the system are defined as *additive*; as already noted, the latest version of the DDI has adopted similar approaches. Both decisions complicate creating the metadata, but our generation of derived values depends on our software being able to identify which nCubes are built from related variables *and* share the same universe. Even more importantly, virtually any published statistical table can be defined as a non-additive nCube, but requiring additivity means that some individual census tables have to be defined as four or five separate nCubes; however, almost all our repertoire of visualizations are valid only for additive data.

The DDI Alliance is very much a creation of the data archiving community, but is increasingly seeking to develop a 'life-cycle approach', as illustrated in Figure 13.7 (Miller

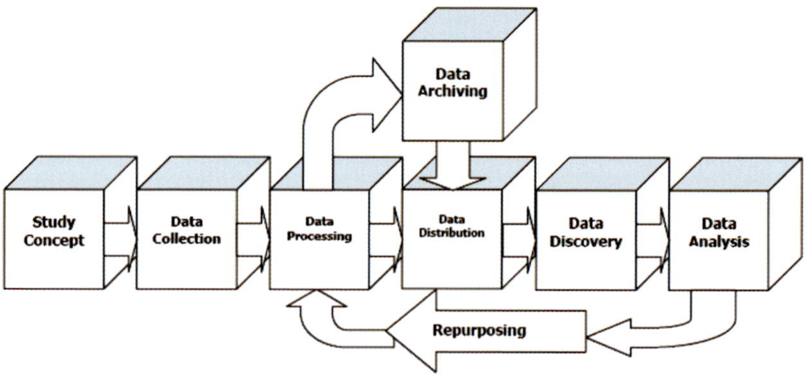

Figure 13.7 DDI: the life-cycle approach (© 2005, DDI Alliance)

and Vardigan, 2005). However, this approach can only succeed if the DDI framework is also adopted by those who originally create data, especially those who plan and carry out censuses and large-scale surveys. It will only *fully* succeed if it is also adopted by those who analyse and visualize social science data. So far, most geographers, even highly quantitative ones, know little of statistical metadata frameworks, but this chapter has argued we have much to gain from not just learning about this field but contributing to it.

Acknowledgements

Ian Turton and James Macgill, then of the Centre for Computational Geography at the University of Leeds, made a large contribution to the visualization facilities within *Vision of Britain*. Thanks to Wendy Thomas of the University of Minnesota and Mary Vardigan of the University of Michigan for comments on a draft version of this paper. Development of the *Vision of Britain* web site was supported by the Big Lottery Fund under award DIG/2000/147.

References

Blank, G. and Rasmussen, K. B. (2004) The Data Documentation Initiative: the value and significance of a worldwide standard. *Social Science Computer Review*, 22: 307–318.

Gregory, I. N. and Southall, H. R. (1998) Putting the past in its place: the Great Britain Historical GIS. In Carver, S. (ed.), *Innovations in GIS 5*. London, Taylor & Francis, pp. 210–221.

Miller, D. W. and Modell, J. (1998) Teaching United States history with the Great American History Machine. *Historical Methods*, 21(3): 121–134.

Miller, K. and Vardigan, M. (2005) How initiative benefits the research community – the Data Documentation Initiative, paper presented at the *First International Conference on e-Social Science*, Manchester, June 2005; www.ddialliance.org/DDI/papers/miller.pdf

Openshaw, S., Charlton, M., Wymer, C. and Craft, A. W. (1987) A mark I geographical analysis machine for the automated analysis of point data sets. *International Journal of GIS*, 1: 335–358.

Southall, H. R. (2001) Defining census geographies: international perspectives. *Of Significance... (Journal of the Association of Public Data Users)*, 3: 32–39.

Southall, H. R. and White, B. M. (1997) Visualising past geographies: the use of animated cartograms to represent long-run demographic change in Britain. In Mumford, A. (ed.), *Graphics, Visualization and the Social Sciences*. Loughborough, Advisory Group on Computer Graphics, Technical Report Series no. 33, pp. 95–102.

14

Making Uncertainty Usable: Approaches for Visualizing Uncertainty Information

Stephanie Deitrick and Robert Edsall

School of Geographical Sciences, Arizona State University

14.1 Introduction: the need for representations of uncertainty

Not long ago, we began the creation of an atlas of cultural geography of the Phoenix metro area. In one particular meeting with students for feedback on some of the draft maps, one student we noticed was concentrating very hard on a map of movie theatres in the region. We had told him the map was incomplete, but when we asked him about what he was seeing, he pointed to a spot on the map and somewhat reluctantly stated that he thought that there was a theatre at that location. Before we could remind him that these were purposely incomplete draft maps, he continued: he was *sure* he had been to a movie at that location, but he doubted himself because the theatre was not on the map. He must have been mistaken, he thought.

This all-too-common response to a map has haunted responsible cartographers – and delighted less scrupulous ones – for many years. We know that many incorrect conclusions and detrimental decisions are made from an unquestioning and uncritical use of maps, resulting in mistakes that range from the innocuous, such as missing a turn because the GPS map did not show the road, to the catastrophic, such as bombing a national embassy because the map used was not current (Reuters, 2006).

GIScience researchers, educators and students know that there is a multitude of reasons that our representations of geographic reality are incomplete and uncertain. Measurements cannot be taken at every location and can only obtain a certain level of accuracy. Observations

Geographic Visualization Edited by Martin Dodge, Mary McDerby and Martin Turner

of dynamic processes may exist only at specific temporal scales. The spherical surface of the Earth cannot be accurately represented on a flat piece of paper or on a computer screen. Other activities in the cartographic abstraction process lead to intentional and unintentional bias, and models in general, whether visual or computational, and no matter how sophisticated, can only approximate reality. Therefore, a fundamental and persistent discrepancy exists between geographic data and the 'reality' they are meant to represent.

In many applications and for many audiences, providing information enabling the understanding of this gap between reality and representation is critical. This chapter highlights research on communicating uncertainty and spatial data quality to GIS users, specifically focusing on uncertainty visualization. Providing information about uncertainty is considered by many users to be either irrelevant or detrimental for successful data communication and insight generation. Herein, we argue that uncertainty in geographic data should be made *usable* through innovative visualization techniques. In many cases, representing uncertainty has a positive effect in the visualization process. We provide a framework for increasing the usability of information about data uncertainty for decision-making.

14.2 The complexity of uncertainty

Uncertainty is, of course, a broad interdisciplinary subject, the infinite nuances of which cannot be explored in detail here (MacEachren *et al.*, 2005; Devillers and Jeansoulin, 2006; Goovaerts, 2006). Any discussion of uncertainty, however, should begin with a closer look at the many different terms in the literature that relate to, and are sometimes used interchangeably with, uncertainty.

Uncertainty broadly refers to incompleteness in knowledge, unknown or unknowable information about the discrepancy between an actual value and its representation in language, mathematics, databases or other forms of expression. Uncertainty can take many forms. The form of uncertainty that is relevant to a given problem or situation may change depending on the user of the data and the purpose of the data's use.

There are only a few forms of uncertainty that have been the focus of uncertainty visualization research. MacEachren, Brewer and Pickle (1998) represented the reliability (accuracy) of death rates for a map of mortality data. Cliburn *et al.* (2002) visualized the variability (error) in results of a water balance model. In their evaluation of fuzzy classifications, Blenkinsopp *et al.* (2000) were faced with uncertainty in the form of vagueness and ambiguity. Clearly by these examples, uncertainty is expressed in many different forms, many of which are summarized in Table 1.

The term *data quality* refers to the degree to which a dataset fulfils or conforms to specific requirements. It includes a range of attributes such as completeness, consistency and lineage (MacEachren, 1992; Evans, 1997; Veregin, 2005; Devillers and Jeansoulin, 2006). Data quality implies both an objective and subjective evaluation of the data. For example, for a particular study, a researcher might use a dataset that contains relevant information, but that was created for another study two years earlier. This dataset is comprehensive, with a large number of samples; the data source and producer are known, and all the data processing that was completed is well documented. In general, the dataset would be considered to have an objectively high level of quality. However, the time difference would render the same dataset as of subjectively low quality. Thus the criteria used to define quality may be very different for those producing and those using the data.

Table 14.1 Concepts of uncertainty (source: compiled from Duckham *et al.*, 2000; Goodchild, 2000; Zhang and Goodchild, 2002; Molden and Higgins, 2004; Fisher, 2005; MacEachren *et al.*, 2005; Veregin, 2005; Devillers and Jeansoulin, 2006)

Components	Definition
Data quality:	
Lineage	History of data, including sources, data processing and transformations
Completeness	Extent to which data is comprehensive
Logical consistency	Extent to which objects within the dataset agree, topology
Currency	Temporal gaps between occurrence, information collection and use
Credibility	Reliability of information source
Subjectivity	Amount of interpretation or judgment included
Accessibility	Format, availability, documentation of access or use restrictions
External quality	How well data meets the needs or specifications of the user
Internal quality	How closely data reflect or represent the actual phenomenon
Accuracy (positional, attribute and temporal)	Closeness of measured values, observations or estimates to the true value
Precision (statistical and numerical)	Number of significant digits of a measurement or observation (numerical precision)
	Conformity of repeated measurements to the reported value (statistical precision)
Fitness for use	Suitability of data for a particular use and user, subjective quality
Error	Difference between a measured or reported value and the true value, encompassing both precision and accuracy
Random error	Random derivation from the true value; 'noise': inconsistent effect across values (some results may be low, others may be high) and influences the *variance* of a measurement or sample
Systematic error	Systematic deviation from the true value; 'bias': consistent effect across values (errors occur in one direction, either low or high), influences the *average* of a measurement or sample
Vagueness	Poor definition of an object or class of objects
Ambiguity	Doubt in classification of an object; differing perceptions of the phenomenon

An important distinction should be made between the possibly diverging concerns of the producers of data and those of the users of the data, particularly with respect to uncertainty. Data producers are primarily concerned with *internal quality*, which corresponds to the similarity between the data and the actual phenomenon, which itself represents a 'perfect' dataset. Users, on the other hand, are typically more concerned with data's *external quality*, or its fitness for use in a given situation. For users, a data set's currency, scope and credibility are given high priority. Clearly, quality is a context-dependent concept, dependent on both the individual and the situation of the data's creation and use.

14.2.1 Uncertainty representation research in GIScience

The GIScience community has long identified uncertainty as an important research theme. Although uncertainty has been the focus of much research (Hunter and Goodchild, 1995;

Goodchild, 2000; Lucieer and Kraak, 2004; Deitrick and Edsall, 2006; Devillers and Jeansoulin, 2006; Goovaerts, 2006), current GIS technology and practice have done little to provide support for incorporating information about the quality of spatial data. Goodchild (2006) calls for the GIScience community to insist that GIS incorporate scientific standards and principles, including the reporting of results with a precision that does not exceed their accuracy or the precision of their source data. He notes a mismatch between responsible GIScience theory and practical GIS use concerning uncertainty. This mismatch, he writes, may stem from the comfort that GIScientists have with the necessary compromise between the conflicting objectives of representing the world with both accuracy *and* visual or computational simplicity, generalization and clarity. Users, on the other hand, are less likely to understand that GIS data and analyses are not necessarily accurate and reliable. In addition, many uncertainty models and concepts are based on difficult mathematics. The complexity of uncertainty added to the conflicting goals and differential training between GIScientists and the diverse community of users introduces the potential for misinterpretation of these concepts and the relationships they are meant to explain (Goodchild, 2006).

Geographic uncertainty in GIScience – and the importance of its communication – thus arises from a mismatch between the needs and goals of the user and those of the producer of geographic information. These needs include the match between the specifics of a user's request for data or information, the data format, the use of that data, and the phenomenon being investigated (Gottsegen, Montello and Goodchild, 1999).

Several recent efforts in cartography, visualization and GIScience research have sought to bridge this gap through the integration of uncertainty information into geographic representations. Researchers have sought the most appropriate and effective means of representing uncertainty to map readers, carrying out experiments comparing representational techniques. A common approach is to begin with the adaptation of Bertin's (1983) visual variables for the representation of uncertainty. In addition to these variables, additional graphic variables, such as transparency, saturation and clarity have also been proposed (MacEachren, 1992; Slocum *et al.*, 2004) specifically for uncertainty representation. Gershon (1998) proposed two general categories of representation strategy: intrinsic and extrinsic. *Intrinsic* techniques integrate uncertainty in the display by varying an existing object's appearance to show associated uncertainty, while *extrinsic* techniques rely on the addition of geometric objects to highlight uncertain information. Cliburn *et al.* (2002) utilized this distinction in an interactive environment which depicted the results of a water balance model along with the associated uncertainty of these results (Figure 14.1). Interactive computer environments have opened new possibilities for uncertainty representation, including the development of interfaces that allow users to manipulate the display of uncertainty by deciding how and when to display uncertainty information (Fisher, 1994; Davis and Keller, 1997; Ehlschlaeger, Shortridge and Goodchild, 1997; Cliburn et al., 2002; Aerts, Clarke and Keuper, 2003).

In most studies of uncertainty visualization in GIScience, when the importance of potential differences in users has been acknowledged, it is often included as an ancillary study, and not as the explicit and main focus of the study. The primary focus of most experiments has been on representational design. MacEachren, Brewer and Pickle (1998) developed and tested a pair of intrinsic methods for depicting 'reliability' of data on choropleth maps used in epidemiology. Newman and Lee (2004) evaluated both extrinsic and intrinsic techniques for the visualization of uncertainty in volumetric data comparing glyph-based techniques, such as cylinders and cones, with colour-mapping and transparency adjustments. Lucieer and Kraak (2004) developed an interactive environment for exploring the uncertainty of

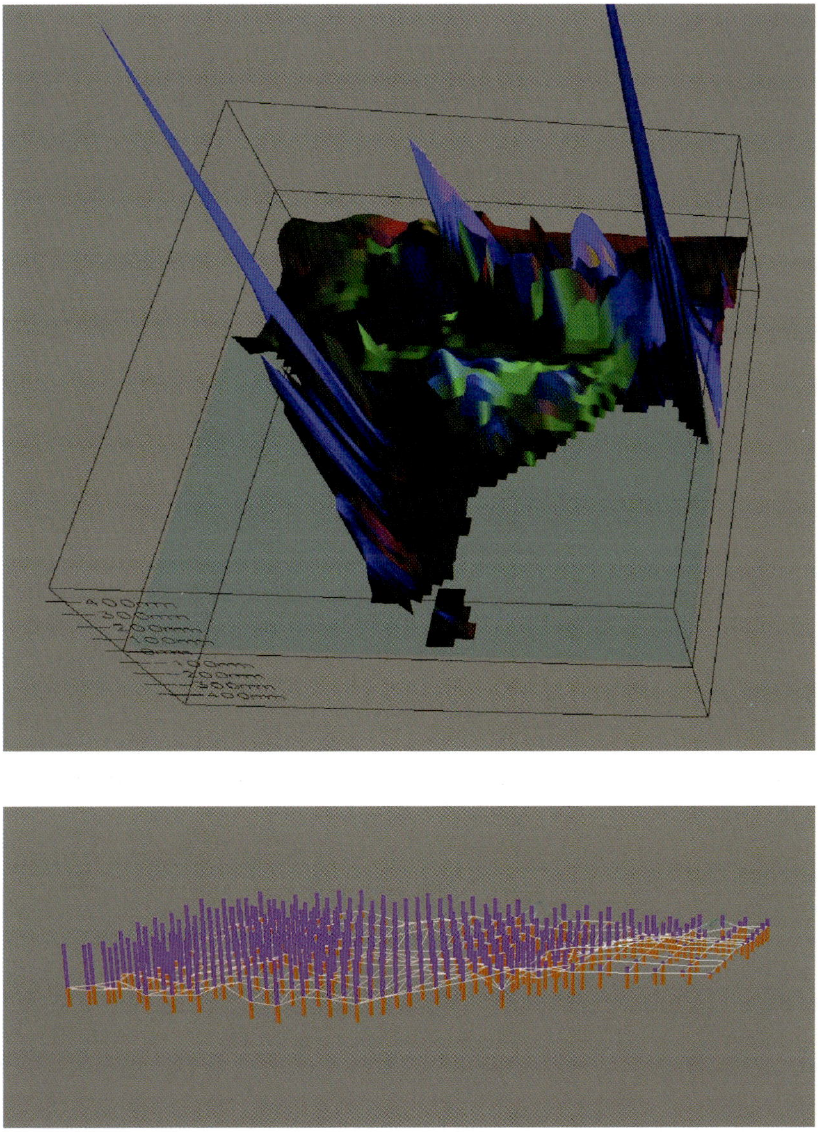

Figure 14.1 Example of intrinsic (upper) and extrinsic (lower) representations (*source:* reprinted from Cliburn *et al.*, 2002, *Computers and Graphics* **26**(6): 931–949, with permission from Elsevier)

classifications of remotely sensed images. In the case of Leitner and Buttenfield (2000), specific focus was on the alteration of the decision-making process by changing the representation, altering Bertin's visual variables systematically. In these cases, important insight for designing uncertainty representations was gained, but the influence of user diversity in the design of the representation was not a particular research focus.

However, research into uncertainty visualization in geography has also, on occasion, considered the influence of the user's experience and comfort with uncertainty as an independent variable. When differences in users are considered, the main focus has been on differences between novice and expert users, while other factors such as the users' comfort with uncertain information and their experience in making decisions are often downplayed. Blenkinsop *et al.* (2000) assessed user perception of a variety of uncertainty representations, including both interaction and animation, using results of a fuzzy classification of satellite imagery. The study examined the performance of two user groups, one expert and one novice, in determining classification uncertainty. The results were framed primarily in terms of effectiveness of the representation form (grey-scale images, histograms, animations, linked graphics and other methods) and not explicitly according to the differing profiles of the user groups, but the differences in users were discussed. Cliburn *et al.* (2002) developed a visualization environment to allow decision makers to visualize the results of a water-balance model. That study found that the complexity and density of some of the representation methods seemed to overwhelm novice users, while experts were able to use the detail more readily in decision making. For example, they suggest that intrinsic methods provide a more general representation of uncertainty data, which non-expert users may prefer over more-detailed extrinsic representations. Aerts, Clarke and Keuper (2003) evaluated two visualization techniques showing model results and the associated uncertainty, including a toggling animation and a side-by-side static comparison. Their focus was specifically on the end user, in this case urban planners, and on what uncertainty representations they found most useful.

A review of these studies and those from other disciplines suggests that the effectiveness of uncertainty communication may be as dependent on the user and context as it is on the representation method or the problem domain. Thus, we advocate the reorientation of research in uncertainty visualization to an approach that places the user at the centre of inquiry, and maintains a goal of making uncertainty information as usable as possible to the diverse community of users of geographic information.

14.2.2 The pervasiveness of uncertainty

The study of uncertainty is one of only a handful of themes that can be called truly interdisciplinary; almost all academic pursuits, from history to mathematics to philosophy to economics to linguistics, must consider uncertainty, whether it be variability, fuzzy categories, doubt, inaccuracy, vagueness or any of the other forms uncertainty takes on. Uncertainty is so much a fundamental piece of human experience that social anthropologists are discussing the possibility of an anthropology of uncertainty (Boholm, 2003; Anderson, 2006). Researchers in a broad array of disciplines – psychology, economics, political science, medicine, information science, as well as geography – have considered the processes by which decision making occurs under uncertain conditions (Molden and Higgins, 2004; Dequech, 2006). Although uncertainty takes on many different meanings throughout these disciplines, there are several concepts and theories that exist that may provide a foundation for uncertainty visualization and communication in GIScience. In order to inform a user-centred approach to the representation of uncertainty for decision making (a thankfully smaller but still daunting research theme than 'uncertainty' in general), we now turn to a review of uncertainty research in the psychology of decision making, information science (specifically information seeking) and risk communication.

Decisions: problems, frames and heuristics

Because one of the desired outcomes of an effort to make uncertainty more usable through visualization is the support of a more informed decision-making process, it would be beneficial to incorporate theories and research from decision science and information science into uncertainty visualization methods. Research in the psychology of decision making focuses on the means by which people merge beliefs (knowledge, expectations) and desires (personal values, goals) to make a decision (choose between alternatives or courses of action; Hastie, 2001). Decision making is often seen as the sequence of steps that people pass through in order to make a decision, including several components: a course of action, expectations about outcomes and processes, and beliefs (a decision maker's expectation) or utilities (what the decision maker wants) that define the consequences associated with the outcomes of each event or action (Hastie, 2001). Throughout this process, decision makers often attempt to identify or seek out information to support or identify a potential course of action. Information seeking refers to the variety of methods and actions that people employ to discover and gain access to information resources (Wilson, 1999). Not surprisingly, uncertainty is an essential and ubiquitous component of information seeking (Kuhlthau, 1991; Wilson, 1999; Wilson *et al.*, 2002).

Uncertainty is often considered to be negative and detrimental by information seekers, leading to a reduction in confidence (and an increase in anxiety). However, information science research has shown that the presentation of uncertainty can have a positive influence on problem-solving, encouraging individuals to take action and work through uncertainty to eventually generate knowledge (Anderson 2006). Information, in this case, has the potential to both reduce and produce uncertainty, depending on the context of the problem and the individual. The ultimate goal in problem solving is the reduction of uncertainty to acceptable levels, not necessarily the elimination of uncertainty. In the case of uncertainty visualization, consideration of the connotation of uncertainty (positive or negative) may play a role in the design of uncertainty representations. A simple semantic rewording of a legend, for example, from '50% uncertain' to '50% certain' may be an avenue by which some users would become more comfortable with uncertainty, allowing them to use it in productive and not detrimental ways.

The degree to which an information seeker can decide on what information to access (view, read or listen to), how long to access it, and in what order, is called *information control* (Ariely 2000). In website design, for example, information can be 'pushed' onto a user without the seeker's request, or it can be 'pulled' onto the display through a particular request for information. There are legitimate and productive reasons for both approaches, and it has been suggested that the degree to which an information seeker should have control of information is dependent on the experience, ability and/or knowledge of the user (Shapiro, Schwartz and Astin 1996; Wu and Lin 2006). This so-called *match theory* of information has provided a theoretical foundation to evaluate the influence of information control on the decision making of consumers.

For example, Wu and Lin (2006) found that experts should be given a high level of freedom to search for relevant information, while novices, who are less able to differentiate between relevant and irrelevant information, should have less information control. This suggests that there may be an optimal amount of information to present to users depending on their individual characteristics. Match theory should be adapted to visualization research, answering questions about the relationships between expertise, interactive affordances and information control. Instead of making uncertainty information available at detailed levels for all uses and users, it might be useful to evaluate whether experience should determine the extent and methods of uncertainty communication available for a given user, problem or use.

Any *decision problem* – the alternatives, consequences and probabilities involved with a particular decision, governed by the available data and the relative uncertainty of the data – is *framed* by the decision maker's concepts associated with a particular course of action (Tversky and Kahneman, 1981). Thus, the same decision problem can be framed in multiple ways – either by the same person who may have multiple or changing goals, or by many different decision makers, each of whom has a different perspective, expertise and conceptual model about the data and the phenomenon. When faced with decisions under uncertainty, individuals often revert to *heuristics*, or abstract mental rules, rather than statistics, to determine a course of action. In terms of time and information requirements, heuristics serve to efficiently generate satisfactory outcomes in situations that a decision maker frequently encounters. The individual learns to apply the heuristics that result in the most favourable outcomes; these repeatedly used rules reduce the complexity of assessing the alternatives and potential outcomes. Of course, there is no guarantee that, in any specific instance, heuristics will always generate the most favourable outcome, or that they are applicable for new situations or problems (Patt and Zeckhauser, 2000). Because they are used and reused in different decision problems, if they are incorrect they can result in systematic errors and bias in decision making (Tversky, 1974; Tversky and Kahneman, 1974).

Individual cognition, therefore, significantly informs the decision framework used to solve problems. Understanding the extent to which (and the problems for which) heuristics are used by decision makers when solving problems with uncertain information should be a consideration in the design of methods for communicating uncertainty. For example, individuals who, because of their particular decision frameworks, employ decision rules may be more likely to expect *discrete* alternatives, or 'scenarios'. Alternatively, others who prefer statistical explanations of uncertainty may be more likely to expect *continuous* alternatives, or 'ranges' of outcomes.

Risk management, uncertainty, and communication

Risk communication is defined as an interactive process of information and opinion exchange between individuals, groups, and institutions. These exchanges can include multiple 'risk messages', which are written, verbal or visual statements about a risk, and the expression of concerns, reactions and opinions about the risk (Patt and Dessai, 2005). The risk messages are specifically tailored to the parties involved; for example, a risk message does not necessarily involve statistical probabilities, which may be appropriate for a scientist, but not for less expert users of the information. Figure 14.2 presents several examples of potential risk messages.

Effective risk communication involves incorporating the potential cognition conflicts and knowledge of the users of the information into the risk message. Before scientists or organizations can effectively communicate risk, there needs to be an awareness of the decision frames and heuristics that individuals and groups use when evaluating alternatives and making decisions, as well as the potential actions that could be taken, and the consequences that may result (Grabill and Simmons, 1998). For example, Patt and Dessai (2005) evaluated the effectiveness of risk communication techniques for climate change data, based on a

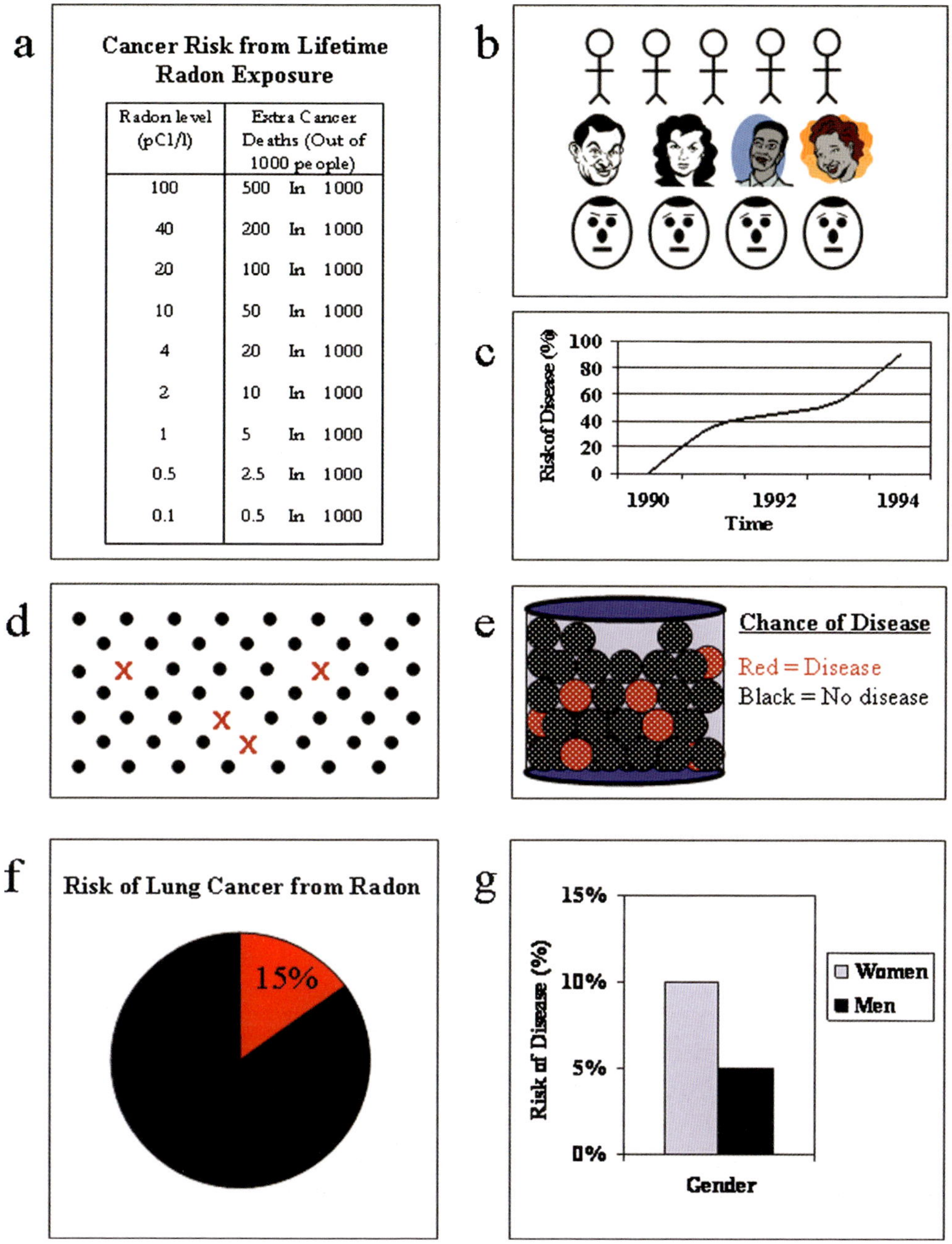

Figure 14.2 Examples of visual representations used to communicate risk. The representations include the following: (a) risk ladder; (b) Chernoff faces; (c) line graph; (d) graph with dots and Xs, with each X representing an area or item affected by a hazard; (e) marbles; (f) pie charts; and (g) histograms (source: reprinted from Lipkus and Hollands, 1999, *National Cancer Institute Monographs*, **25**:149–163 with permission from the author)

survey of both climate change experts and university students. The research identified the importance of using language that matches the cognitive framework of the decision maker in order to make the information easier to understand.

As with risk, a key barrier to the communication of uncertainty is how uncertainty is measured, described and ultimately perceived by individuals and groups. One method or one technique does not fit every user, situation or problem. In trying to support the integration of uncertain information into GIS, one approach apparent in much existing uncertainty visualization research is to develop a variety of tools and representation methods in order to provide users of GIS with an extensive selection of alternatives for incorporating uncertainty into their information display. An alternative (though not necessarily mutually exclusive) approach, more similar to that taken in risk communication, is to integrate the users of GIS early in the development process. This iterative approach requires identification of, and access to, a target user group from an early stage. Interaction with the user group would allow the identification of how decisions are made and what decision tasks would benefit from uncertainty visualization. GIScientists should be responsible for adapting uncertainty representation to the decision frames and heuristics identified during interactions with these target user groups. Based on this interaction, initial uncertainty representation methods would be developed and evaluated by the user group, allowing for feedback and potential refinement of the representations. Identifying the characteristics that make an uncertainty representation usable (and those aspects that do not) will aid in the development of usable uncertainty visualization tools.

14.3 Uncertainty visualization: a user-centred research agenda

Uncertainty is data- and context-dependent, but it is also very much user-dependent. Increasing the usability of uncertainty information in geographic visualization must be informed by user-centred approaches in other disciplines to handling uncertainty. Here, then, we propose establishing a research agenda in GIScience that places the needs, experiences and characteristics of the user at the centre. Theories associated with this approach utilize decision and information science research that seeks to understand the influence that context and experience have on the way in which users comprehend and incorporate uncertain information into their decisions.

In this section, we suggest several research themes that focus on making uncertainty in geographic visualization and communication usable, helpful and productive for users. A user-centred approach is rich with research opportunities that will serve to inform the design of uncertainty representations to maximize their effectiveness and utility based on the potential audience and the potential context of the representation and the analysis.

14.3.1 Different visualization designs for different levels, kinds of expertise

Domain expertise has been a factor that has been at least discussed in many studies of uncertainty communication. Expert-novice differences have been noted in visualization research in general (McGuiness, 1994) and it seems logical that one aspect of data visualization where expert-novice differences would be greatest is in the handling of uncertainty.

However, domain expertise is just one of several dimensions of expertise that could be considered in visualization design. For example, some individuals might have little expertise in the particular problem domain but might have a great deal of experience making decisions and solving problems under uncertain conditions. We might call these individuals decision experts – they might include policy makers, corporate executives and administrators who are frequently tasked with making difficult choices under less-than-ideal circumstances. While they are not necessarily domain experts, it is hard to call them 'novices' in the typical sense of the term. Another type of expertise that could play a role in determining appropriate design strategies is experience with complex visualization displays. Individuals who are not dazzled and overwhelmed by multiple detailed representations and sophisticated interaction will likely respond to a presentation of uncertainty information differently than those who do not encounter those types of displays regularly.

14.3.2 The influence of decision problem presentation and decision frame

The presentation of a decision problem can influence the way a decision maker approaches the problem and ultimately reaches a conclusion. For example, if model output is reported to an analyst using a specific level of 'certainty' rather than a specific (equivalent) level of 'uncertainty', different decision outcomes could result. More generally, the simple inclusion or omission of uncertainty information would also undoubtedly influence the decisions. Similarly, if a decision is presented as choosing between specific alternatives (choose A, B or C) compared with a ranking task (rank the alternatives from least to most favourable), different decisions could result (Deitrick, 2006). As discussed above, some individuals are able to consider statistical uncertainty more readily than others. Perhaps the design of uncertainty representations should be in part governed by these factors.

For example, treating uncertainty as a discrete variable (a piece of data is either 'certain' or 'not certain') is an approach that might be appropriate for one user group, while another group might be more likely to require a representation that treats uncertainty as continuous, with percentages and error ranges. Additionally, the representation of uncertainty to some may be more sensible if the possible outcomes are represented as discrete scenarios ('either this, this or this') rather than a statistical range of outcomes ('somewhere between this and this'). These differences can be addressed in terms of the decision framework within which an individual solves a problem. Interviews and cognitive walk-through methods with potential users could allow researchers to identify those components of the representation that are usable, the methods by which users reach decisions, and the most appropriate way to communicate uncertainty to that user group.

14.3.3 The influence of differing levels of uncertainty information control on decision making

The information-seeking literature suggests that individuals may vary (perhaps according to their relative expertise) in how much information they can use, and how much control they should be given in its access, in order to maximize the effectiveness of their search for knowledge, meaning or solutions. A relevant research question based on information seeking is whether increasing information control influences the search for knowledge among experts and novices. Do user characteristics have an influence on the amount and type of *interactivity* that should be given to the user? Given previous research we would expect that experts would be better able to 'request' information on uncertainty through interactive selections while novices would find 'pushed' information about uncertainty (i.e. as a default representation) more usable. We may also wish to investigate the extent to which novices (or other user groups) should be guided by uncertainty messages or other means of communicating the benefits and usefulness of uncertainty representations.

14.3.4 Encouraging action and discouraging anxiety due to uncertain information

In a user-centred approach to usable uncertainty visualization, visualization designers should study the *affective* nature of uncertainty. Interviews, focus group testing, and other usability testing could focus attention on both the reasons for negative connotations associated with uncertainty, and the strategies for creating uncertainty representations that promote creative and beneficial thinking. Affective computing research has led to user-friendly interfaces that facilitate understanding through, for example, the use of metaphor. As we saw in the reaction to the draft Phoenix atlas maps, uncertain information tends to evoke an emotional response, particularly in users who rarely need to handle uncertainty in geographic data. Understanding and being sensitive to that affective response to uncertainty should lead users to positive action rather than negative anxiety about the uncertain data.

14.4 Conclusion

With the globalization of users of geographic data, data has become available to a diverse group of users through numerous sources, including government data portals and internet mapping sites, and attention should be paid to different ways of communicating geographic information to this diverse user group. Many of these users have no experience with, or exposure to, the principles of GIScience, or to issues of uncertainty in geographic data. As access to geographic data and tools continues to grow, so does the urgency to communicate information about geographic uncertainty to a diverse community in a manner that is not only usable, but useful. Developing uncertainty communication methods in a manner that involves GIS users provides an opportunity for their specific experiences, knowledge and decision framing to be incorporated into the communication. If the goal is to support understanding and improved decision-making, user experience and expertise must specifically inform the development of uncertainty representation methods and tools.

References

Aerts, J. C. J. H., Clarke, K. C. and Keuper, A. D. (2003) Testing popular visualization techniques for representing model uncertainty. *Cartography and Geographic Information Science*, 30(3): 249–261.

Anderson, T. D. (2006) Uncertainty in action: observing information seeking within the creative process of scholarly research. *Information Research*, 12(1): 1368-1613; available at: http://informationr.net/ir/12-1/paper283.html (accessed 27 October 2006).

Ariely, D. (2000) Controlling the information flow: effects on consumers' decision making and preferences. *Journal of Consumer Research*, 27: 233–248.

Bertin, J. (1983) *Semiology of Graphics: Diagrams, Network, Maps*. Madison, WI, University of Wisconsin Press.

Blenkinsop, S., Fisher, P., Bastin, L. and Wood, J. (2000) Evaluating the perception of uncertainty in alternative visualization strategies. *Cartographica*, 37(1): 1–13.

Boholm, A. (2003) The cultural nature of risk: can there be an anthropology of uncertainty? *Ethnos*, 68(2): 159–178.

Cliburn, D. C., Feddema, J. J., Miller, J. R. and Slocum, T. A. (2002) Design and evaluation of a decision support system in a water balance application. *Computers & Graphics*, 26(6): 931–949.

Davis, T. and Keller, P. (1997) Modelling and visualizing multiple spatial uncertainties. *Computers & Geosciences*, 23(4): 397–408.

Deitrick, S. (2006) The influence of uncertainty visualization on decision making. Unpublished masters thesis, Geography, Arizona State University, Tempe, AZ.

Deitrick, S. and Edsall, R. (2006) The influence of uncertainty visualization on decision making: an empirical evaluation. In *Progress in Spatial Data Handling: 12th International Symposium on Spatial Data Handling*, Riedl, A., Kainz, W. and Elmes, G. (eds). New York, Springer, pp. 719–738.

Dequech, D. (2006) The new institutional economics and the theory of behaviour under uncertainty. *Journal of Economic Behavior and Organization*, 59: 109–131.

Devillers, R. and Jeansoulin, R. (eds) (2006) *Fundamentals of Spatial Data Quality*. Newport Beach: ISTE Ltd.

Duckham, M., Mason, K., Stell, K. and Worboys, M. (2000) A formal ontological approach to imperfection in geographic information. *GIS Research UK National Conference*, 5–7 April 2000, University of York.

Ehlschlaeger, C. R., Shortridge, A. M. and Goodchild, M. F. (1997) Visualizing spatial data uncertainty using animation. *Computers & Geosciences*, 23(4): 387–395.

Evans, B. J. (1997) Dynamic display of spatial data reliability: does it benefit the map user? *Computers & Geosciences*, 23(4): 409–422.

Fisher, P. (1994) Randomization and sound for the visualization of uncertain information. In *Visualization in Geographic Information Systems*, Unwin, D. and Hearnshaw, H. (eds). Chichester, Wiley, pp. 181–185.

Fisher, P. (2005) Models of uncertainty in spatial data. In *Geographic Information Systems: Principles, Techniques, Management and Applications*, Longley, P. A., Goodchild, M. F., Maguire, D. J. and Rhind, D. W. (eds). Hoboken, NJ, Wiley, pp. 69–83.

Gershon, N. (1998) Visualization of an imperfect world. *IEEE Computer Graphics and Applications*, 18(4): 43–45.

Goodchild, M. F. (2000) Introduction: special issue on 'certainty in geographic information systems', *Fuzzy Sets and Systems*, 113: 3–5.

Goodchild, M. F. (2006) Forward. In *Fundamentals of Spatial Data Quality*, Devillers, R. and Jeansoulin, R. (eds). Newport Beach: ISTE Ltd, pp. 13–16.

Goovaerts, P. (2006) Geostatistical analysis of disease data: visualization and propagation of spatial uncertainty in cancer mortality risk using Poisson kriging and *p*-field simulation. *International Journal of Health Geographics*, 5(7): 1–26.

Gottsegen, J., Montello, D. R. and Goodchild, M. F. (1999) A comprehensive model of uncertainty in spatial data. In *Spatial Accuracy and Assessment: Land Information Uncertainty in*

Natural Resources, Lowell, K. and Jaton, A. (eds). Chelsea, MI, Ann Arbor Press, pp. 175–182.

Grabill, J. T. and Simmons, W. M. (1998) Towards a critical rhetoric of risk communication: Producing citizens and the role of technical communicators. *Technical Communication Quarterly*, 7(4): 415–441.

Hastie, R. (2001) Problems for judgment and decision making. *Annual Review of Psychology*, 52: 653–683.

Hunter, G. J. and Goodchild, M. F. (1995) Dealing with error in spatial databases: a simple case study. *Photogrammetric Engineering & Remote Sensing*, 61(5): 529–537.

Kuhlthau, C. C. (1991) Inside the search process: Information seeking from the user's perspective. *Journal of the American Society for Information Science*, 42(5): 361–371.

Leitner, M. and Buttenfield, B. P. (2000) Guidelines for display of attribute certainty. *Cartography and Geographic Information Science*, 27(1): 3–14.

Lipkus, I. and Hollands, J. G. (1999) Visual communication of risk. *National Cancer Institute Monographs*, 25: 149–163.

Lucieer, A. and Kraak, M. J. (2004) Interactive and visual fuzzy classification of remotely sensed imagery for exploration of uncertainty. *International Journal of Geographic Information Science*, 18(5): 491–512.

MacEachren, A. M. (1992) Visualizing uncertain information. *Cartographic Perspectives*, 13: 10–19.

MacEachren, A. M., Brewer, C. A. and Pickle, L. W. (1998) Visualizing georeferenced data: representing reliability of health statistics. *Environment and Planning A*, 30: 1547–1561.

MacEachren, A. M., Robinson, A., Hopper, S., Gardner, S., Murray, R., Gahegan, M. and Hetzler, E. (2005) Visualizing geographic information uncertainty: what we know and what we need to know. *Cartography and Geographic Information Science*, 32(3): 139–160.

McGuiness, C. (1994) Expert/novice use of visualization tools. In MacEachren, A. M. and Taylor, D. R. F. (eds). *Visualization in Modern Cartography*. Oxford, Pergamon. pp. 185–199.

Molden, D. C. and Higgins, E. T. (2004) Categorization under uncertainty: resolving vagueness and ambiguity with eager versus vigilant strategies. *Social Cognition*, 22(2): 248–277.

Newman, T. and Lee, W. (2004) On visualizing uncertainty in volumetric data: techniques and their evaluation.*Journal of Visual Languages & Computing*, 15: 463–491.

Patt, A. and Dessai, W. (2005) Communicating uncertainty: lessons learned and suggestions for climate change assessment. *Comptes Rendus Geoscience*, 337(4): 425–441.

Patt, A. G. and Zeckhauser, R. (2000) Action bias and environmental decisions. *Journal of Risk and Uncertainty*, 21(1): 45–72.

Reuters. (2006) *Israel says UN deaths caused by map error* (cited 14 November 2006). Available at: www.boston.com/news/world/middleeast/articles/2006/09/14/ israel_says_un_deaths_caused_by_map_error/.

Shapiro, D. J., Schwartz, C. E. and Astin, J. A. (1996) Controlling ourselves, controlling our world: Psychology's role in understanding positive and negative consequences of seeking and gaining control. *American Psychologist*, 51(12): 1213–1230.

Slocum, T. A., McMaster, R. B., Kessler, F. C. and Howard, H. H. (2004) *Thematic Cartography and Geographic Visualization*, 2nd edn. Upper Saddle River, NJ, Prentice Hall.

Tversky, A. (1974) Assessing uncertainty. *Journal of the Royal Statistical Society. Series B (Methodological)*, 36(2): 148–159.

Tversky, A. and Kahneman, D. (1974) Judgment under uncertainty: heuristics and biases. *Science*, 185(4157): 1124–1131.

Tversky, A. and Kahneman, D. (1981) The framing of decisions and the psychology of choice. *Science*, 211(4481): 453–458.

Veregin, H. (2005) Data quality parameters. In*Geographic Information Systems: Principles, Techniques, Management and Applications*, Longley, P. A., Goodchild, M. F., Maguire, D. J. and Rhind, D. W. (eds). Hoboken, NJ: Wiley, pp. 177–189.

Wilson, T. D. (1999) Models of information behaviour research. *The Journal of Documentation*, 55(3): 249–270.

Wilson, T. D., Ford, N., Ellis, D., Foster, A. and Spink, A. (2002) Information seeking and mediated searching. Part 2. uncertainty and its correlates. *Journal of the American Society for Information Science*, 53(9): 704–715.

Wu, L.-L. and Lin, J.-Y. (2006) The quality of consumers' decision-making in the environment of E-commerce. *Psychology & Marketing*, 23(4): 297–311.

Zhang, J. and Goodchild, M. (2002) *Uncertainty in Geographical Information*. New York, Taylor & Francis.

15

Geovisualization and Time – New Opportunities for the Space–Time Cube

Menno-Jan Kraak

Department of Geo-Information Processing, International Institute for Geo-Information Science and Earth Observation

15.1 Introduction

Many of today's important challenges facing science and society not only have a fundamental geographic component but also involve changes over time as well, for example, understanding the impact of global environmental change, the effects of a potential outbreak and diffusion of avian flu, or preparing scenarios for disasters such as flooding. Maps offer interesting opportunities to visualize these dynamic phenomena. This is possible because the map is no longer the map as many of us know it. Traditionally it is known for it capacity to 'present' spatial patterns and relationships based on selection and abstraction of reality. Although this is still very much true, the map should also be seen as a flexible interface to geospatial data. Maps offer interaction with the data behind the visual representation and can be linked to other views that contain alternative graphic representations. Today's maps are instruments that encourage exploration and stimulate the user to think. The objective of this chapter is to see how the concept of the space–time cube can be extended beyond its original realm as defined by Hägerstrand in the late 1960s, and can support the challenges mention above (Hägerstrand, 1970).

Although time is a fundamental geographical notion, its definition is not straightforward (Vasiliev, 1997; Peuquet, 2002). Everyone seems to know what time is, but when one has to explain it to someone else difficulties start. Attempting to define time results in a variety

Geographic Visualization Edited by Martin Dodge, Mary McDerby and Martin Turner

of definitions (Peuquet, 2002). In that sense time is a phenomenon that is actually perceived by its consequences and can therefore be described through the changes that it induces. These changes can be broadly distinguished as occurring in the spatial domain and in the temporal domain (Blok, 2005). Changes in the spatial domain include: (a) appearance or disappearance, e.g. the emergence or vanishing of phenomena; (b) movement, e.g. change in position (location) or/and in geometry (shape) of the phenomenon; and (c) mutation (nominal: change in the character of a phenomenon; ordinal or interval/ratio: increase/decrease). Changes in the temporal domain refer to (a) moment in time, e.g. the instant (or interval) a change starts to occur; (b) pace, e.g. the rate of change over time; (c) sequence, e.g. the order of phases in a series of changes; (d) duration, e.g. the length of time during which a change takes place; and (e) frequency, e.g. the number of times a phase is repeated.

Some other aspects related to time should be also mentioned at this point, since they are closely related to spatio-temporal representations (such as the space–time cube) and to functionality and interaction with these representations. They are related to different descriptions or classifications of time (time granularities), interrelated varying resolutions and temporal uncertainties. Regarding the former, time can be described with varying different units, such as dates (absolute chronology) or historical/archaeological/geological periods (e.g. 'Roman', 'bronze age', 'Pleistocene', etc.). These different units for measuring (or describing) phenomena are known as 'granularities'. Examples of time granularities are calendar dates and hours, but also specialized units such as business days or academic years (Bettini *et al.*, 2000). The latter relates to space (spatial zoom) and time (temporal zoom), while it could also relate to thematic content ('semantic' zoom, as used in information visualization terminology; Ware, 2004). The third refers to inaccuracy in assigning a time-stamp in a phenomenon, event or record.

The changes in the temporal domain as described above can be represented graphically in many ways. These representations are based on alternative conceptions of time (Langran 1993); each one of them offers a different view of temporality and emphasizes the representation and effectiveness of different spatio-temporal aspects (Kousoulakou and Kraak, 1992). The main categories are (a) use of a single map (use of visual variables for understanding of changes); (b) use of small multiples ('spatial' deduction of changes, although in a discontinuous manner); and (c) animation (use of dynamic variables for the deduction of changes via memory). The space–time cube allows changes to be traced spatially, via the third dimension, a somewhat separate category combining elements from (b) and (c). Many other not strictly geographically oriented representations can be used in combination with map-based representations (Müller and Schumann, 2003).

In cartography the application of methods and techniques from other disciplines has led to geovisualization. It integrates approaches from scientific visualization, (exploratory) cartography, image analysis, information visualization, exploratory data analysis, visual analytics and GIScience to provide theory, methods and tools for the visual exploration, analysis, synthesis and presentation of geospatial data (Dykes *et al.* 2005). The visualizations should lead to insight that ultimately helps decision making. In this process maps and other graphics are used to stimulate (visual) thinking about geospatial patterns, relationships and trends, generate hypotheses, develop problem solutions and ultimately construct knowledge.

15.2 Hägerstrand's time geography and the space–time cube

At the end of the 1960s Hägerstrand introduced a space–time model to study individuals in their day-to-day activities. The model can represent people's physical mobility network, and is based on a space–time cube. From a visualization perspective the space–time cube is the most prominent element in Hägerstrand's approach. In its basic appearance these images consist of a cube with, on its base, a representation of geography (along the x- and y-axes), while the cube's height represents time (z-axis). It features elements such as a space–time path, and a space–time prism, respectively related to the movement and the potential movement of individuals. His model is often seen as the start of time-geography studies. Throughout the years his model has been applied and improved to understand our movements through space. Problems studied can be found in different fields of geography, and range from those on an individual movement to whole theories to optimize transportation.

During the 1970s and 1980s, Hägerstrand's time-geography was elaborated upon by his Lund group (Hägerstrand, 1970; Lenntorp, 1976). It has been commented and critiqued by for instance Pred (1977), who in his paper gives a good analysis of the theory, and summarizes it as 'the time geography framework is at one and the same time disarmingly simple in composition and ambitious in design'. As one of the great benefits of the approach, he notes that it takes away the (geographer's) over-emphasis on space, includes time and pays attention to people. Recently Miller (2002), in a homage to Hägerstrand, worded this as a shift of attention from a 'place-based perspective' (our traditional GIScience) to a more people-based perspective (time-geography). The simplicity is only partial though, because if one considers the space–time cube from a visual perspective it will be obvious that an interactive viewing environment is not easy to find. That is probably one of the reasons why the cube has been used only sporadically since its conceptualization. In addition, the approach has also been hampered by the difficulty in getting abundant data and methods and techniques to process data. This seems no longer to be the case, but it might have serious implications for privacy (Monmonier, 2002). Geographers see new opportunities to study human behaviour and this explains the revival in interest in Hägerstrand's time geography (Miller, 2002).

Although applications in most traditional human-geography domains have been described, only during the last decade have we witnessed an increased interest in the space–time cube. Miller (1999) applied its principles in trying to establish an accessibility measure in an urban environment. Kwan (2000) used it to study accessibility differences between the genders and among different ethnic groups. Forer developed an interesting data structure based on taxels ('time volumes') to be incorporated in the cube to represent the space–time prism (Forer, 1998; Forer and Huisman, 1998). Hedley *et al.* (1999) created an application in a GIScience environment for radiological hazard exposure. Improved data gathering techniques have given the interest in time-geography and the space–time cube a new impulse. Recent examples are described by Mountain and his colleagues (Mountain and Raper, 2001; Dykes and Mountain, 2003), who discuss the data collection techniques by mobile phone, GPS and location-based services and suggest a visual analytical method to deal with the data gathered. The application of sensor technology is also discussed by Laube *et al.* (2005). The space–time paths, or geospatial life lines as they are called by Hornsby and

Egenhofer (2002), are an object of study in the framework of moving objects. Analytical applications have been discussed by Andrienko *et al.* (2003) and the cartographics by Kraak and Koussoulakou (2004).

15.3 Basics of the space–time cube

Hägerstand's time geography sees both space and time as inseparable, and this becomes clear if one studies the graphic representation of his ideas, the space–time cube as displayed in Figure 15.1. Two of the cubes axes represent space, and the third axis represents time. This allows the display of trajectories, better known as space–time paths. These paths are influenced by constraints. One can distinguish between capability constraints (for instance mode of transport or the need for sleep), coupling constraints (for instance being at work or at the sports club), and authority constraints (for instance accessibility of buildings or parks in space and time). On an aggregate level time-geography can also deal with trends

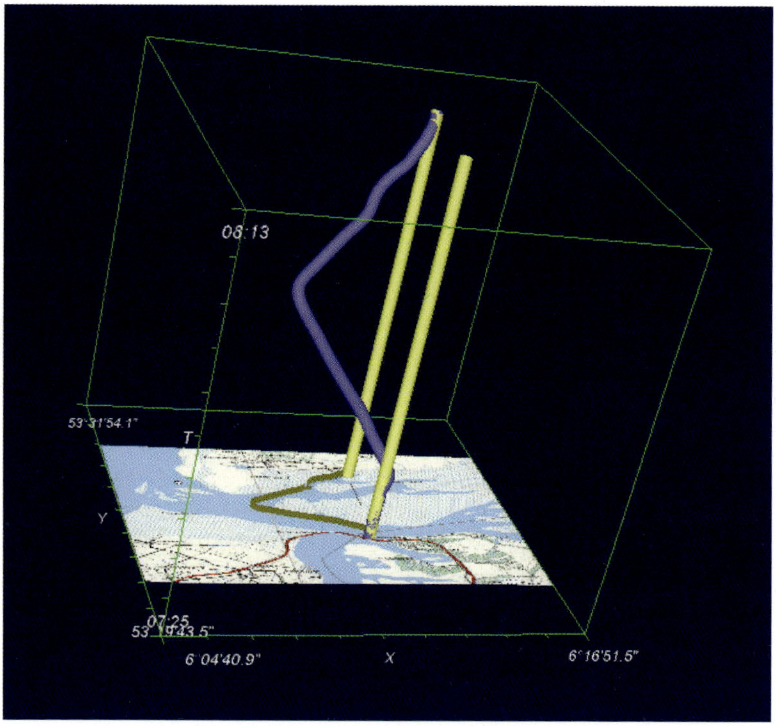

Figure 15.1 The space–time cube in its basic appearance with longitude and latitude along the x- and y-axes, and time along the z-axes. It shows a base map, a space–time path and two stations, as well as the footprint of the path. In this example a boat trip is shown. The trip is represented by the path starting at the bottom of the cube (Lauwersoog) and ending at the top (Schiermonnikoog). Both harbours appear as stations. The stations stand out as vertical lines from the cube's bottom to its top

in society. The example in Figure 15.1 displays a boat trip (the path) between two harbours (called stations). At this observation scale stations are equal to no-movement. However, if one zoomed in on those stations, it is very likely that one would observe movement, since people meet for activities such as sport or lectures. The time at which people meet at a station creates 'bundles'. The non-horizontal lines indicate movements. The space–time path can be projected onto the map, resulting in the path's footprint. Another important time-geography concept is the notion of the space–time prism. In the cube it occupies the volume in space and time in which a person can reach in a particular time-interval starting and returning to the same location (for instance: where can you get to during lunch time, and still be back on time). The widest extent is called the potential path space and its footprint is called the potential path area. It can be represented by a circle, assuming it is possible to reach every location at the edge of the circle. In reality the physical environment (being urban or rural) will not always allow this due to the nature of, for instance, the road pattern or traffic intensity; in the case of Figure 15.1's boat trip, the potential path space is limited to the size of the boat.

When the concept of the space–time cube was introduced, the options for creating the graphics were limited to manual methods and the user could only experience the single view created by the draftsperson. A different view of the cube would mean going through the laborious drawing exercise again. Today software exists that allows the automatic creation of the cube and its contents from a database. Developments in geovisualization allow one to link the (different) cube views to other alternative graphics. Based on the latest developments in geovisualization, this chapter presents an extended interactive and dynamic visualization environment, in which the user has full flexibility to view, manipulate and query the data in a space–time cube, while being linked to other views on the data. The aim is to have a graphic environment that allows creativity via an alternative perspective on the data to inspire the mind with new ideas and to solve particular geo-problems.

15.4 The space–time cube at work

In this section several traditional and non-traditional space–time cube applications will be discussed. For all cases it is assumed that a better (visual) exploration and understanding of temporal events is possible in the environment presented. Prominent keywords are interaction, dynamics and alternative views which each have their impact on the viewing environment proposed. Interaction is needed because the three-dimensional cube has to be manipulated in space to find the best possible view, and it should be possible to query the cube's content. Time, always present in the space–time cube, automatically introduces dynamics. The alternative graphics appear outside the cube and are dynamically linked. The combination should stimulate thinking and potentially lead to new insights and explanations. However, the question if this indeed works will be discussed in the last section of this chapter.

The available functionality can be distinguished in display and query functions and in viewing functions. The first include the options to switch content on or off, change content appearance and query display objects. The second include options to manipulate the cube in three-dimensional space (rotate and translate) and the possibility to zoom in and out. As a special function one can move the base map (by default at the bottom of the cube) along the time axis. This allows one to 'follow' the space–time path and keep track of its geographic position over time. The prototype has been developed in JAVA 3D and makes use of the Udig open source GIS environment (http://udig.refractions.net/confluence/display/UDIG/Home).

15.4.1 Sport (multiple linked views)

The first example is related to a sports event: running. Basic input is a GPS track which contains information on time and location. In a multiple linked view environment a map and a graph are linked to the cube. The first contains the track, and the second contains heart-rate data (the upper line) and speed data (the lower line) collected simultaneously with the GPS data. Figure 15.2 displays the views. The map in the cube has been moved along the time axis. This moment in time is also highlighted in both map and graph. Steep sections of the path indicate slower movement and correspond with drops in the graph. The viewing environment allows one to compare different runs, for instance by plotting

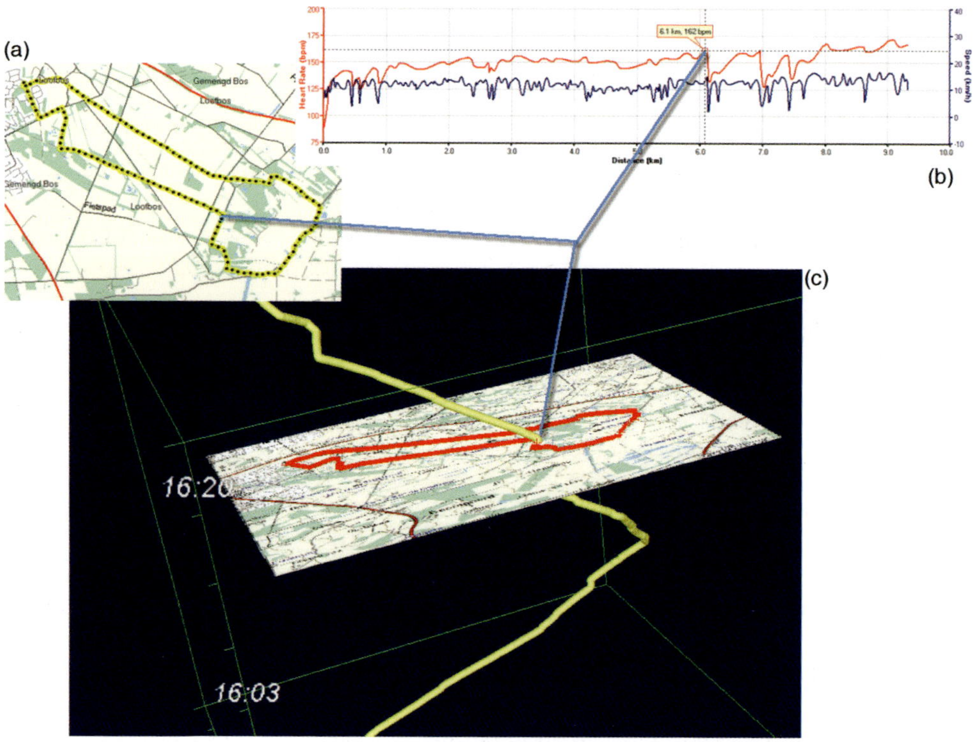

Figure 15.2 The space–time cube in a multiple linked view environment. A running event is displayed from different perspectives: (a) shows the track in the map; (b) shows a graph with the runner heart rate data; (c) displays the space–time cube with the track as a path and the base map with the path's footprint. The base map has been moved along the *T*-axis to a moment in time when the runner reached the 6.1 km point (see lines connecting the three views)

different runs in one graph and/or in the cube. With the cube, geography is included in the comparison, and one can judge when and where during the runs progress was slow or fast (steep or flat line segments).

Sport and time-geography have met before, for instance to analyse rugby matches (Moore *et al.*, 2003). The space–time cube is most suitable for the display and analysis of paths of (multiple) individuals, groups or other objects moving through space. However, other possibilities exist, and it could also be used for real-time monitoring. To illustrate this let us consider an orienteering event. The aim for participants in an orienteering run is to navigate (run) as quickly as possible via a set of check points from start to finish with the help of a map and compass. The participants all carry a special sensor that transmits every so many minutes their location. This approach allows the organization to monitor the runners' positions during the race. The check point can be expressed as stations, and the arrival at check points by the participants can be generated and displayed near real time. The base map displayed in the cube could be enhanced with a digital terrain model to be able to judge the influence of the terrain on the race. After the race the space–time cube viewing environment would allow the participants to analyse their run after the race and one could even use the cube as an animation environment to replay the orienteering event as a whole or for individual runners.

15.4.2 Sport in mountains (Google Earth)

The second example is also related to sport. The space–time cube in Figure 15.3 shows a running track. However, this time the environment of the run was the Alps. This meant that the terrain also had an influence on the run. How does one analyse this? The cube's base map has contours, but it is difficult to visualize the terrain from those lines. One option would be to introduce terrain in or below the cube. However, it was decided to export the space–time cube data to Google Earth. Here the space–time path can be seen in combination with the surrounding mountains. The path is not displayed as a single line but as a wall, otherwise the path would have disappeared into the landscape. The lowest point of this path/wall is the start of the run. An interesting question is how the viewer appreciates the combination of time and height along the same axis. Seeing the landscape will not be problematic but seeing the time path as a time path and not as a wall in the landscape is probably not as straightforward as it is in the space–time cube itself.

15.4.3 Napoleon's Russian campaign (spatial and temporal zoom)

In this example the options of spatial and temporal zoom are introduced, as well as adding attributes to the space–time path. Figure 15.4 illustrates Napoleon's Russian campaign of 1812. The data are based on Minard's famous map of this event (Kraak, 2003). The space–time cube in Figure 15.4(b) shows the path of the total campaign. The thickness of the path corresponds to the number of troops Napoleon had available. The path's footprint shows the geography of the campaign. Compared with Minard's map the space–time path clearly reveals that Napoleon and his troops stayed for a month in Moscow, something not so obvious from the original map.

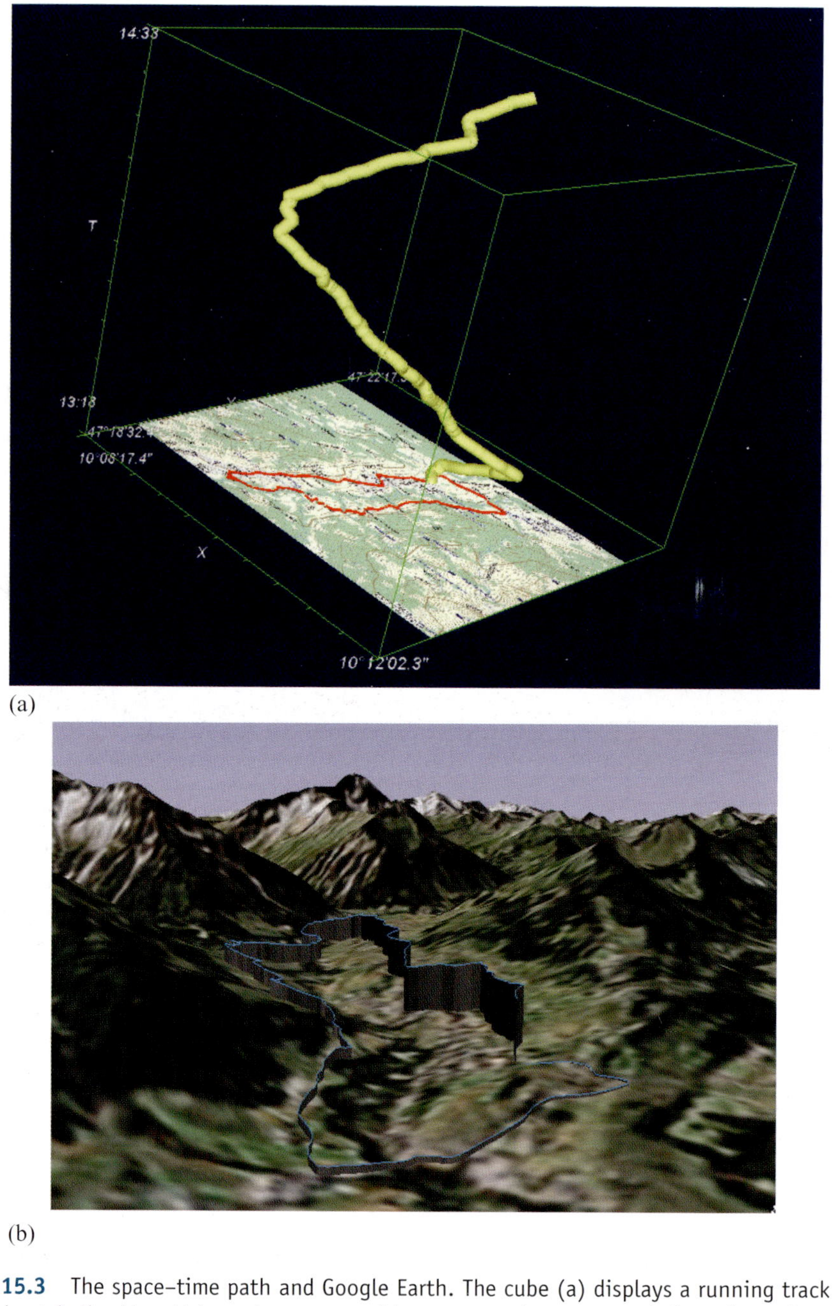

(a)

(b)

Figure 15.3 The space–time path and Google Earth. The cube (a) displays a running track in the Kleinwalsertal, the Alps. Using a base map, with contours, does not result in a three-dimensional impression. In (b) the path has been exported to the Google Earth environment. It is now display as a 'wall' among the mountains

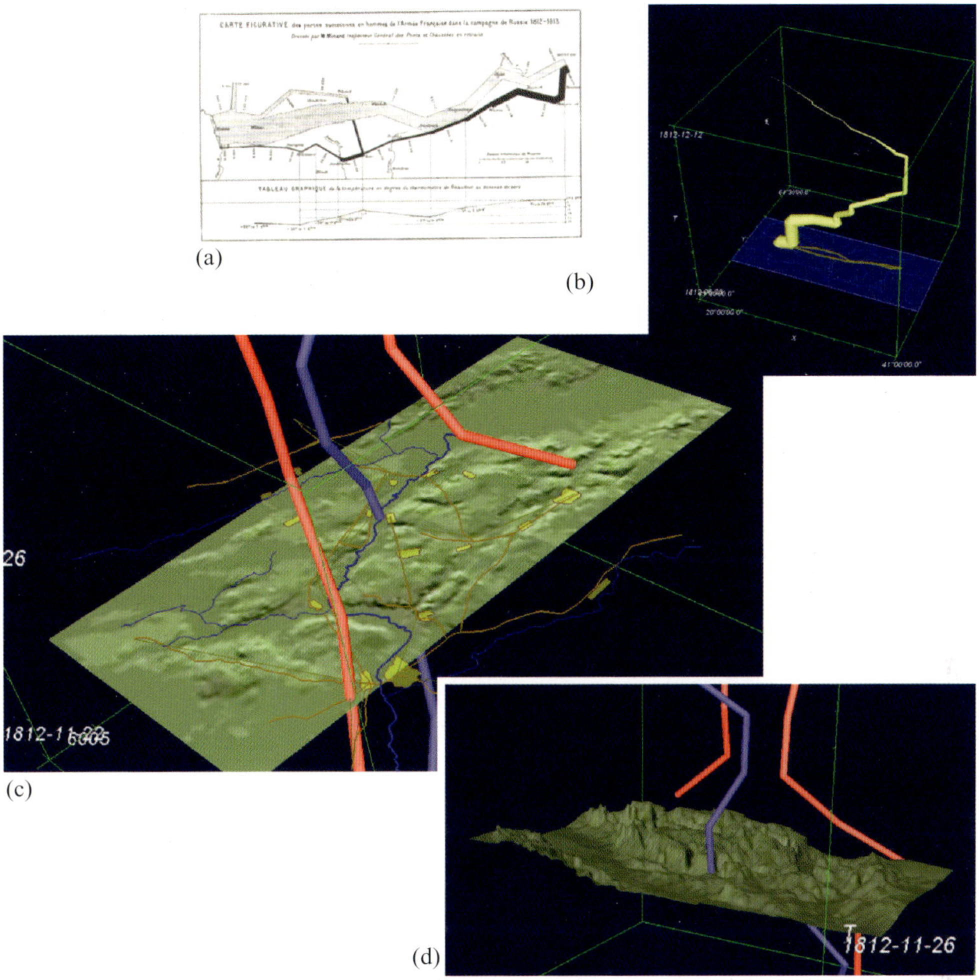

Figure 15.4 Napoleon's campaign to Moscow in a space–time cube. Napoleon's campaign as presented by Minard (a), has been converted into a space–time cube (b). The path's thickness corresponds to the number of troops. In (c) a detail of the campaign, the crossing of the Berezina River, is shown. The base map of the cube is moved to 26 November, and next to the path representing the French troops, two paths of the Russian troops are shown closing in on the French. In (d) the effect of the terrain is shown by replacing the base map with a terrain model

The cube detail shown in Figure 15.4(c) represents the situation on 26 November, the moment of the crossing of the Berezina River. The cube's base map, showing some topographic detail, has been moved along the time axis to this moment in time. Multiple paths are visible. The blue path, popping out of the river, represents the French troops. Both the red paths represent the Russian troops closing in on Napoleon. In the cube in Figure 15.4(d) the base map has been replaced with a terrain model to show the influence of the terrain on the movement of the troops. This model can be moved along the time axis just like the base map.

15.4.4 The Black Death display in the space–time cube (stations at work)

This example deviates from the basic time-geography approach. It is an experiment to see if the exploratory environment can also be used in epidemiological research. The Black Death is used as a case study. This has recently been discussed in quite some detail by (Christakos *et al.*, 2005). In this case study the objects of study are not moving humans but the Black Death epidemic that struck Europe in mediaeval times. Available data includes locations [see map in Figure 15.5(a)] for which there is a start and an end date of the epidemic and the number of deaths known or estimated. In the space–time cube terminology the locations are considered stations with attributes. The cube in Figure 15.5(c) displays them with their attribute values. This results in stations that vary in thickness depending on the number of deaths, and they appear only between the start and end dates of the epidemic at that particular location. The cloud of stations gives an impression of how the epidemic spread through Europe. Since the epidemic started in southern Europe and moved northwards, the locations in northern Europe are found in the upper half of the cube. If one looks at the cube from above, one can create a station-animation by moving the base map along the time axis.

The upper graph in Figure 15.5 contains all locations in alphabetical order with the starting date of the Black Death as an attribute. It is possible to sort the graph based on this starting date. The resulting sort order as seen in the lower graph can be used to create a space–time path. This would result in a path through the cube, giving an impression of the temporal spread of the epidemic [see Figure 15.5(f)]. Of course one has to realize that this is not necessarily the real spread of the disease across Europe. Similarly, one could swap the station's attributes in the graph to, for instance, the number of deaths, and generate a path starting with the lowest to the highest number of deaths. Such a path might not have real meaning and would result in a chaotic path going back and forth in time. However, let us not forget that new ideas come from unorthodox actions. Another such action would be to display the terrain model in the cube to see if the terrain had some influence on the spread of the disease. Back in that time the means of transport were much simpler and slower. A real interpretation of the view would be the task of an epidemiologist, but orthodox views on their data can also be of benefit.

15.4.5 Rotary Clubs in the Netherlands (splitting space–time paths)

Figure 15.6 shows the location and development of the Rotary Clubs in the Netherlands. The map shows their location (dots) and the spread (lines). In the cube the clubs are represented by stations. Stations appear when the clubs started. The horizontal lines in the cube link new clubs with their parents. The cube's base map can be moved along the time axis. It is possible to animate the movement along the time axis and see the clubs appear in both the cube and the map. The stations could be coloured according to specific club attributes.

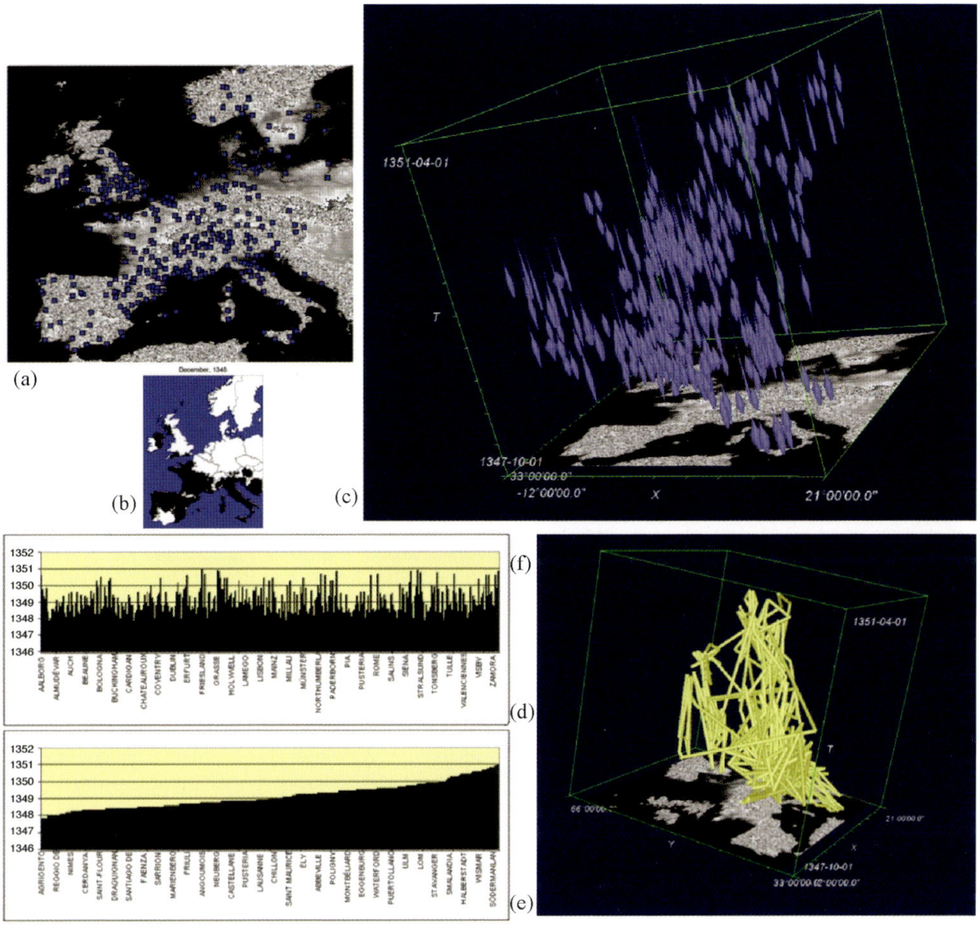

Figure 15.5 The Black Death in a space–time cube. The mediaeval Black Death epidemic is displayed in a map (a) and an animation frame (b). The map shows the location of data points and the animation frame the spread of the epidemic at the end of 1348 (data courtesy of Christakos *et al.*, 2005). The locations have been mapped as stations in the cube (c). Their size corresponds to the number of deaths over the years the plague was active at that location. The graphs in (d) and (e) show for each location the start date of the epidemic in alphabetical and chronological order, respectively. The chronological order has been used to create a path connecting all the station and is displayed in the cube in (f)

15.5 Discussion

From the above examples it seems that almost any kind of data can be displayed in the space–time cube, if not in one of the link views. However, there are some basic questions to be asked related to cognition and usability. Does the user understand what she sees? Are other mapping techniques more suitable? When does a certain space–time path make sense? Can we use time and the third spatial dimension along the same axis? These questions will

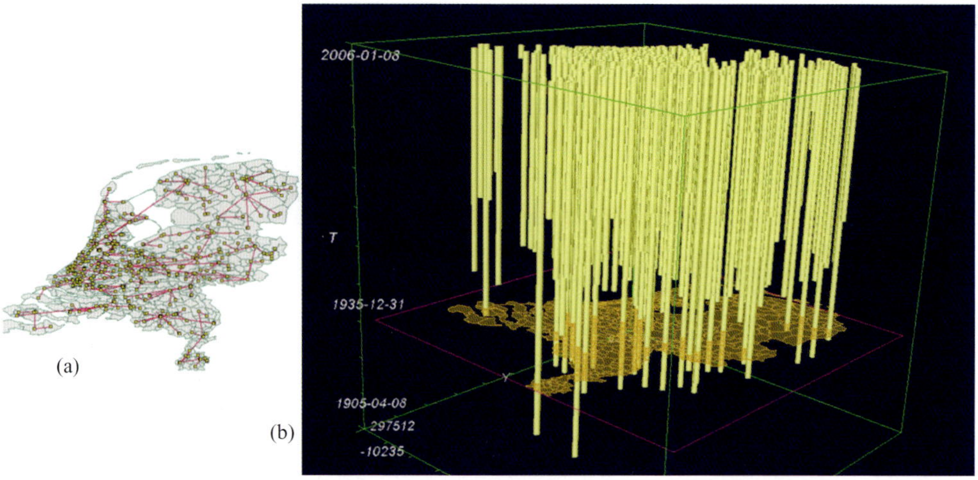

Figure 15.6 Rotary Clubs in the Netherlands. In the map (a) all Rotary Clubs in the Netherlands are displayed together with their history. In the cube (b) each club is represented by a station, which appears at the starting date of that particular club. The horizontal line in the cube links new clubs to their parent clubs. The base map has been moved to 1935

not have simple black-and-white answers. Although lots of testing has to be done, it is likely that the same graphic representation might work for some users with certain tasks in mind, while it might completely fail for others. For a start it would be interesting to see how a simple cube as shown in Figure 15.1 would compare with a map with the same data. The map could, for instance, have the start and arrival timed display on the harbour symbols. Probably these two come close, especially if both are interactive and easy accessible. An animation of the boat trip could be an alternative. The sport examples demonstrate that each view in a multiple view environment has its own strength. The path, map and graph each displaying and emphasizing different aspects of the data allow one to look at the data from different perspectives. This alone would, in an exploratory environment, be a good argument for having them all. For presentation purposes, depending on the message needing to be sent, one selects one of them. The data on Napoleon in Figure 15.4 clearly demonstrate the space–time cube's advantage in relation to the display of time. The stay in Moscow would be missed using the map alone. The crossing of the Berezina results is a more complex image. The cube gives the overview and a linked map could show the animated snapshot that corresponds with the cube's base map along the time axis. The relief model in the cube might look impressive; the combination of time and heights might be confusing. It could be better to have a separate view for the terrain model in which the paths are plotted on the terrain in real time. The Black Death example is the most complex application of the space–time cube. The station view with over 300 locations is reasonably clear. The path created from the start of the disease for each location seems complex and chaotic on its own. An animation through this cube with all other views linked and highlighting the current location and time might prove to be a possible solution. Looking at large amounts of data is not easy. A simplified mapping is always possible, but in an exploratory environment one would like to have a full overview, play with the data by selection and filtering and obtain details from the database on demand (Shneiderman, 1996).

References

Andrienko, N., Andrienko, G.L. *et al.* (2003) Visual data exploration using space–time cubes. In *21st International Cartographic Conference*, Durban.

Bettini, C., Jajodia, S. *et al.* (2000) *Time Granularities in Databases, Data Mining and Temporal Reasoning.* Berlin, Springer.

Blok, C. (2005) Dynamic visualization variables in animation to support monitoring. In *Proceedings 22nd International Cartographic Conference*, A Coruna.

Christakos, G., Olea, R. A. *et al.* (2005) *Interdisciplinary Public Health Reasoning and Epidemic Modelling: the Case of Black Death.* Berlin, Springer.

Dykes, J. A. and Mountain, D. M. (2003) Seeking structure in records of spatio-temporal behaviour: visualization issues, efforts and applications: computational statistics and data analysis (Data Viz II). *Computational Statistics and Data Analysis* 43(4): 581–603.

Dykes, J., MacEachren, A. M. *et al.* (eds) (2005) *Exploring Geovisualization.* Amsterdam, Elsevier.

Forer, P. (1998) Geometric approaches to the nexus of time, space, and microprocess: implementing a practical model for mundane socio-spatial systems. In *Spatial and Temporal Reasoning in Geographic Information Systems*, Egenhofer, M. J. and Gollege, R. G. (eds). Oxford, Oxford University Press.

Forer, P. and Huisman, O. (1998) Computational agents and urban life spaces: a preliminary realisation of the time-geography of student lifestyles. In *Third International Conference on GeoComputation*, Bristol.

Hägerstrand, T., (1970). What about people in regional science? *Papers in Regional Science Association* 24(1): 7–21.

Hedley, N. R., Drew, C. H. *et al.* (1999) Hagerstrand revisited: interactive space–time visualizations of complex spatial data. *Informatica: International Journal of Computing and Informatics* 23(2): 155–168.

Hornsby, K. and Egenhofer, M. J. (2002) Modeling moving objects over multiple granularities. *Annals of Mathematics and Artificial Intelligence* 36(1–2): 177–194.

Kousoulakou, A. and Kraak, M. J. (1992) The spatio-temporal map and cartographic communication. *Cartographic Journal* 29(2): 101–108.

Kraak, M. J. (2003) Geovisualization illustrated. *ISPRS Journal of Photogrammetry and Remote Sensing* 57(1): 1–10.

Kraak, M. J. and Kousoulakou, A. (2004) A visualization environment for the space–time cube. *Developments in Spatial Data Handling 11th International Symposium on Spatial Data Handling*, Fisher, P. F. (ed.). Berlin, Springer, pp. 189–200.

Kwan, M. P. (2000) Interactive geovisualization of activity travel patterns using three-dimensional geographical information systems: a methodological exploration with a large data set. *Transportation Research C* 8: 185–203.

Langran, G. (1993) *Time in Geographic Information Systems.* London, Taylor & Francis.

Laube, P., Imfeld, S. *et al.* (2005) Discovering relative motion patterns in groups of moving point objects. *International Journal of Geographical Information Science* 19: 639–668.

Lenntorp, B. (1976) *Paths in Space Time Environments: a Time Geographic Study of Movement Possibilities of Individuals.* Lund Studies in Geography B: Human Geography, Lund University.

Miller, H. J. (1999) Measuring space–time accessibility benefits within transportation networks: basic theory and computational procedures *Geographical Analysis* 31(2): 187–212.

Miller, H. J. (2002) What about people in geographic information science? *Re-Presenting Geographic Information Systems*, Unwin, D. (ed.). Chichester, Wiley.

Monmonier, M. (2002) *Spying with Maps.* Chicago, IL, University of Chicago Press.

Moore, A. B., Wigham, P. *et al.* (2003) A time geography approach to the visualisation of sport. *Geocomputation* 2003, Southampton (CD-ROM).

Mountain, D. M. and Raper, J. F. (2001) *Modelling Human Spatio-temporal Behaviour: a Challenge for Location-based Services.* Brisbane, GeoComputation.

Müller, W. and Schumann, H. (2003) Visualization methods for time-dependent data – an overview. *Proceedings of Winter Simulation 2003*, New Orleans, LA.

Peuquet, D. J. (2002) *Representations of Space and Time.* New York, The Guilford Press.

Pred, A. (1977) The choreography of existence: comments on Hagerstrand's time-geography and its usefulness. *Economic Geography* 53: 207–221.

Shneiderman, B. (1996) The eyes have it: a task by data type taxonomy for information visualization. In *Proceedings IEEE Symposium on Visual Languages*, Boulder, CO. New York, IEEE Computer Society, pp. 336–343.

Vasiliev, I. R. (1997) Mapping time. *Cartographica* 34(2): 1–51.

Ware, C. (2004) *Information Visualization: Perception for Design.* New York, Morgan Kaufmann.

16

Visualizing Data Gathered by Mobile Phones

Michael A. E. Wright, Leif Oppermann and Mauricio Capra

Mixed Reality Laboratory, University of Nottingham

16.1 Introduction

In an increasing number of research disciplines data gathering is an integral part of current research. Furthermore, data gathering is moving out of the traditional laboratory environment and into the 'wild'. For example, pollution monitoring as presented by Steed *et al.* (2003) and Rudman *et al.* (2005) uses mobile pollution monitoring equipment to gather data about pollution levels around a city. Other examples include Uncle Roy All Around You (Flintham *et al.*, 2003) where a range of data about movements and interactions between online and street players gave a rich data set from which player cooperation and emergent behaviour could be studied. This is further explored in mobile pervasive games such as Can You See Me Now (CYSMN; Anastasi *et al.*, 2002) and I Like Frank (Flintham *et al.*, 2001).

Visualization tools to graphically represent the data gathered by these systems are an important requirement. Ethnographers, for example, often collect a wide range of data such as text logs, audio and video. Tools such as Replayer (Tennent and Chalmers, 2005) allow the ethnographer to view this gathered data based on location tags attached to the data, which is often presented on a map. Similarly, the Digital Replay System (Crabtree *et al.*, 2006) presents spatial or map views of user movements through the physical environment, which they have combined with temporal views of the data such as, video, audio and text logs synchronized by time with the location data. Consequently, data presented in these visualizations is much more accessible to the ethnographer when compared with the raw record.

However, one possible limitation to these data gathering systems described above is the need to physically carry a number of different devices such as a PDA, GPS receiver and

Geographic Visualization Edited by Martin Dodge, Mary McDerby and Martin Turner

other sensors. These devices vary in weight and cost and so limit the number of possible participants that could potentially gather data. In contrast to this we propose that mobile phones, such as the Nokia Series 60 or Windows Enabled Mobile Phones, could be an excellent device by which multimedia data can be gathered by a large number of participants for a variety of research activities.

Indeed, mobile phones are the most ubiquitous piece of computing technology today. With over 54 million people owning and using one in the UK in 2004 (Ofcom, 2004) the shear numbers of mobile phones provides researchers with a readily available and relatively cheap piece of hardware to capture data (such as images, audio and text) which can be location-tagged. This location tag can either be spatial (i.e. GPS in newer mobile phones) or a non-spatial cell to which the participant's mobile phone is connected (i.e. the mobile phone mast they are receiving a signal from).

Capturing data in this way presents two main challenges: (1) the technical infrastructure to facilitate the user capturing and sending location tagged data; and (2) how to visualize this location-tagged data. Our focus in this chapter is on how to effectively visualize data captured by mobile phones; consequently, we present possible solutions to the second of these challenges.

16.2 What are we visualizing?

Mobile phones are a truly ubiquitous device. Modern mobile phones have the capability to capture multimedia data such as images, small videos, audio and textual data. The recent development of 2.5G and 3G mobile phone networks has increased data rates and brought about services that allow multimedia data to be more easily sent and received.

In a report exploring the state of the art in mobile devices, Strahan and Muldoon (2002) describe the then current direction of mobile phone services that are becoming the standard today. For example, GPRS (General Packet Radio Service) allows users to access resources and services from their mobile phone, including enabling internet access. MMS (Multimedia Messaging Service) is another feature of 3G networks where images (e.g. GIFs or JPEGs) and video (e.g. MPEG4s) can be sent and received. Additionally, video calling is yet another feature of the 3G networks as well as mobile TV.

From this we can see that current 3G mobile phone networks are built to enable support for multimedia data, and this feature is being taken up on by mobile phone manufacturers. For example, mid-range mobile phones such as the Nokia Series 60 (Nokia, 2007) enable mobile internet browsing as well as incorporating cameras to take photographs and small videos which can be uploaded to a PC or via GPRS to a server. Current high-end mobile phones such as the Nokia N93 (Nokia, 2007) have a built-in video camera and the O2 XDA (O2 XDA, 2007) offers PDA features on a mobile phone-size device.

Furthermore, mobile phone manufacturers are opening up the operating systems and development tools for their mobile phones. For example, the Symbian Operating System (2007) allows development of applications using Java and C++, and Windows Mobile Operating System (2007) allows development in C# and .NET.

All of these developments mean that, by either using existing software and services or developing new software, we are able to create new methods of data collection using mobile phones. The data that we can collect can use any of the integrated devices on the phone such as

- a camera to collect images and video;
- a microphone to capture audio data;
- a text editor or SMS to capture text data.

Furthermore, using Bluetooth we are able to connect additional sensors if needed. For example, we could connect pollution monitoring or heart rate sensors.

Meta data can also be captured about the participant and their location. This meta data can be captured using software such as Context Phone (Raento *et al.*, 2005). Context Phone is an open source software platform which uses four interconnected software modules to allow developers access to context information gathered using devices on and interactions with a mobile phone. For example, the sensor module captures information about location, user interaction, communication behaviour and physical environment.

One useful application of Context Phone is to capture meta data on the environment surrounding participants. To achieve this we can use the 'physical environment' sensor module and collect a log of Bluetooth connections surrounding the participant. Using this data we can give the analyst a sense of how busy or crowded the environment is.

All of these mobile phone features (integrated devices, software etc.) and network provider services allow the researcher to collect a wide range of data. Often we wish to visualize this data spatially. Therefore, location tagging data becomes an important requirement. In newer mobile phones, GPS is an integrated feature and exact coordinates of the phone can be captured and tracked. However, this is still not a standard feature of most mobile phones. A different approach is to use the mobile cell that the phone is connected to as a method for location tagging or getting the participant to self-report location information such as the street name, city or building where they are located.

16.3 How can we visualize this data?

Mobile phone data collection can enable researchers to collect various types of multimedia data. This data can be location-tagged as well as time-stamped and context information logged. All of this data can then be used to build up a rich picture of where and when the data was captured and the surrounding environment. Visualizing this data, therefore, becomes an important feature of data collection using mobile phones.

16.3.1 Visualizing location information

Visualization of location information is often map-based as it provides a natural and intuitive representation that can be quickly and easily interpreted. Map-based visualizations of environments are used in the pervasive games mentioned above (CYSMN, I Like Frank and Uncle Roy) as well as other pervasive games such as 'Ere be Dragons (Boyd Davis *et al.*, 2005).

In these pervasive games maps are used to present information to players, gam
and audience members and for later use in analysis. These map-based visualizati

movements to be viewed and the data collected at specific locations superimposed on this map. For example, in ethnographic studies of CYSMN, a map interface was used to select and view video, audio and text log data for a particular location.

Rudman *et al.* (2005) use maps to display to the participant the CO_2 levels measured as they walk around an urban environment. Sensors and a PDA are used to capture this CO_2 data, with a map on the PDA providing the user with a visual representation of where they have been and the CO_2 measured. In addition to this, photos could be taken at specific locations and attached on the map to document the levels of traffic at interesting locations. This information was then used to facilitate discussions and analysis about what the participants have encountered and the levels of CO_2 pollution in different locations.

These map-based visualizations allow the user to view data based on physical locations using accurate location positioning such as GPS. However, GPS is not always available if we use mobile phones for data collection; therefore, a more abstract representation of location information needs to be used. For example, Hitchers (Drozd *et al.*, 2006) uses mobile phone cell IDs as a positioning service and, as such, accurate positioning on a map is not always possible. Therefore, visualization of these cell IDs as nodes in a graph and the transitions between cells as edges provides a way of displaying a spatial representation of players' movements. This type of positioning is useful where GPS is not available and analysis of players' movements and/or context is needed.

The Hitchers visualization very much relies on the context information given by players about locations. These context labels allow the graph to become more meaningful as an often-travelled-through cell could be a popular café or a major road intersection on a player's route to work.

In the above examples we can see how location information could be visualized for the analyst. Despite the difference in the techniques for location-tagging data, the location information can be displayed to the analyst using the spatial nature of the data. In addition to this location information, context information can provide additional insight into the location data. Using Context Phone (Raento *et al.*, 2005), for example, as discussed in the previous section, we can capture data about a location, such as the number of Bluetooth connections, which could indicate how busy or densely populated the location is.

16.3.2 Visualizing context information

Context-aware systems are often associated with mobile ubiquitous computing as interactions between humans and computers in these environments can be made more meaningful by understanding the context of the participant.

Context in this case can be defined as where you are, who you are with and what resources are nearby (Schilit *et al.*, 1994). Additionally, context awareness can be defined as 'any information that can be used to characterise the situation of an entity where, an entity can be defined as a person, place, or object that is considered relevant to the interaction between a user and an application, including the user and applications themselves' (Dey and Abowd, 1999).

Both of these definitions are broad in their approach to what is context. Context can be gathered about vision, audio, motion and location, from bio sensors and other specialised sensors (Schmidt *et al.*, 1999) as well as many more. Visualization of these different combinations of variables can be a challenge and presenting multiple view visualizations of all these variables could lead to information overload for the analyst.

However, often we are only capturing a small sub set of these variables. For example, Cyberguide (Long *et al.*, 1996) only captures location and orientation context information. Even though this context information is, as the authors state, limited, it can still elicit meaningful information about the user's context that can be actively used.

Similarly, ActiveMap (McCarthy and Meidel, 1999) displays location information to support informal meetings between colleagues within a work environment. In this example the context information is not only the location but also the movement of different users, the 'freshness' of this information and groups of users at specific locations.

In the two examples above location is a major part of context awareness. However, location is not the only measurable context information we can gather, especially using mobile phones. As we have discussed, ContextPhone (Raento *et al.*, 2005) and other context-gathering software/hardware such as SenSay (Siewiorek *et al.*, 2003) allow a rich picture to be built up about a participant's movements, actions, environment and awareness. Another important aspect of context information is the notion of time and the history of a participant's context over time and how this changes.

16.3.3 Using time to provide additional insight

Time, as mentioned above, is an important part of visualizing the participants' context. Temporal visualizations allow the analyst to replay or view interesting events over time. Additionally, time visualizations can allow the analyst to view a snap shot or global picture of this information. For example, PeopleGarden (Xiong and Donath, 1999) visualizes user interaction on messages boards or chat rooms to allow new users to quickly gain an understanding of the other users' activities.

In the above sections we have explored the types of data we can capture using mobile phones as well as the types of visualizations that we could use to visualize this data. In the following section we shall look at two case studies where mobile phones have been used to capture data. These case studies focus on pervasive games from the Mixed Reality Laboratory (MRL) at the University of Nottingham, where text data from a mobile pervasive game was captured and location-tagged using cell ID data. This data was built up over the duration of the game and context information is gathered by inspecting the text gathered from the game play. Visualizations of this data were implemented to explore patterns of game play for future use in our exploration of pervasive games at the MRL.

16.4 Case studies

16.4.1 Hitchers

Hitchers (Drozd *et al.*, 2006) is a game for mobile phones that exploits cellular positioning to support location-based play. Players create digital hitch-hikers, and give them names, destinations and questions to ask other players which they drop into their current phone cell. By searching their current cell, players are able to view a list of hitchers currently in that cell and can pick them up, answer their questions, carry them to new locations and drop them again, providing location-labels as hints to where they can be found. In this way, hitchers pass from player to player, phone to phone and cell to cell, gathering information and encouraging players to label cells with meaningful place names.

During its runtime, the game builds up a collection of unique cell IDs and transitions between cell IDs. These unique cell IDs also have location labels given to them by one or more players. From this we are able to build up a connected graph whose nodes represent unique cell IDs and whose arcs represent direct transitions between pairs of cell IDs. This graph is visualized using a well-known force-directed placement algorithm (Fruchterman and Reingold, 1991) that is implemented in the JGraph library (Java Graph Library, 2007). Consequently, this is an abstract graph and the relative positions of the nodes bear no relation to geographical directions in the physical world.

As we have mentioned above, cell IDs can provide an alternative way of mapping participants' movements where more accurate location tracking techniques are not available. In this example, using this cell ID graph and location labels has enabled exploration of potential mappings between clusters and patterns of mapped cells and user-meaningful places and activities.

For example, Figure 16.1 shows how the data gathered from a Hitchers trial in Sweden is visualized in the Hitchers visualization tool. The figure reveals some distinctive clusters of

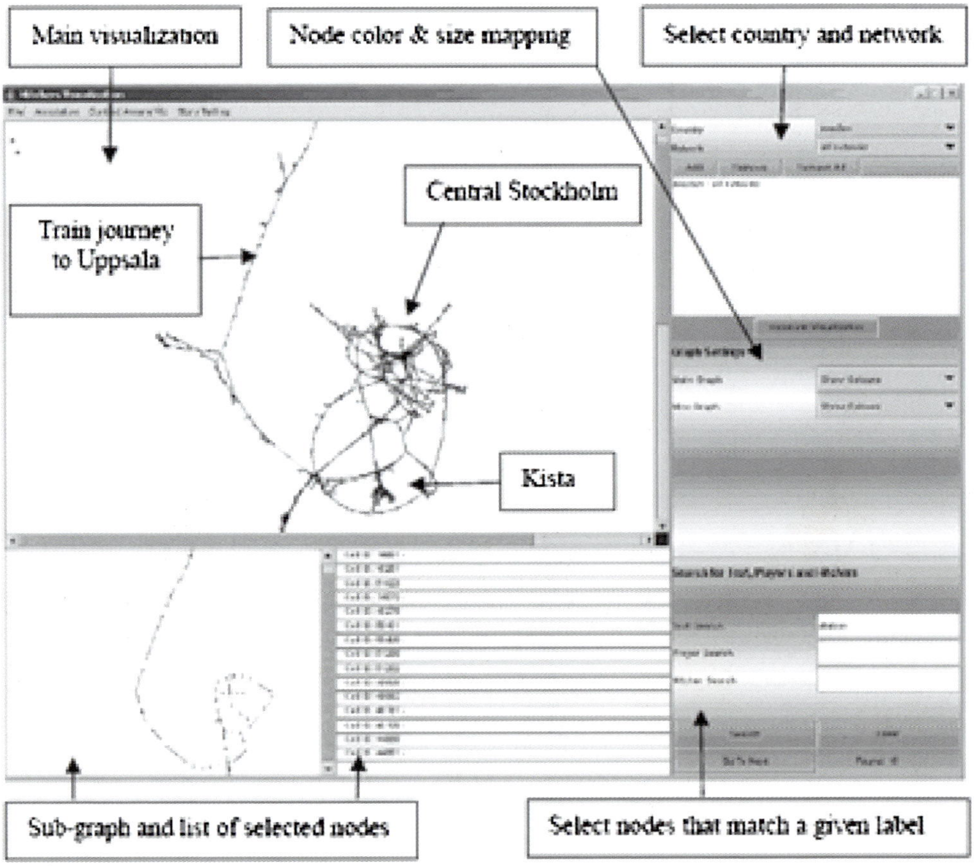

Figure 16.1 The Hitchers cell-data visualization tool

cells as well as some long thin trails of cells that connect them. Inspection of the player-given labels for these cells reveals likely physical locations to some of the key clusters (two of the larger clusters – the towns of Stockholm and Kista – are labelled in the figure). Furthermore, we have been able to determine that the long trails of cells correspond to train and car journeys made by players in the trial.

Filtering and zooming allows further exploration of this graph such that additional relationships and correlations can be identified and explored. For example, we are able to filter and display a sub-graph of all cells that have been labelled with the word 'station' along with intervening cells that connect these.

Another example of this filtering mechanism, shown in Figure 16.2, shows a series of views of a region of the graph that are highlighted according to three different location labels. The rightmost image highlights cells that have been labelled 'Kista', the middle image cells labelled 'SICS' (The Swedish Institute of Computer Science) and the left image cells labelled 'ICE' (Interactive Computing Environments Lab), suggesting that ICE is a place within SICS, which is a place within Kista.

16.4.2 Day of the Figurines

Day of the Figurines (Tandavanitj, 2005) is a slow pervasive game in the form of a massively multiplayer board-game that is played using mobile phones via the medium of text messaging. A first public test of Day of the Figurines took place in London in summer 2005 and involved 85 players playing over the course of a month. We engaged eight volunteers in a separate side-experiment to help us explore the potential of cell-positioning technology to support such a game.

As part of this trial we lent these eight players dedicated phones that could run local cell-logging software that recorded the sequences of cell IDs seen by the phone and uploaded

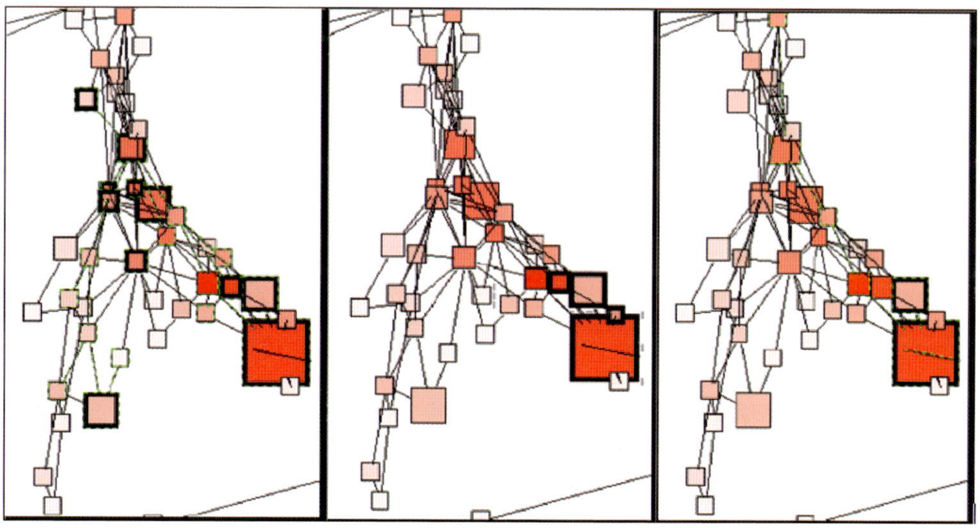

Figure 16.2 Highlighting cells that players have labelled 'Kista' (left), 'SICS' (middle) and 'ICE' (right) with a dark border. Is ICE a place within SICS which in turn, is within Kista?

this information back to a central server. Our aim was to explore whether such a mechanism might provide useful contextual information for managing a player's experience, e.g. automatically recognizing when and where they prefer to play and not to play and tailoring the delivery of text messages accordingly. The following statistical and chronological data-views helped in understanding the players' patterns of life and their times and places of interaction with the game.

Statistical visualization

A statistical diagram was created based on how long (in relative terms) each cell was visited throughout the game. This gives an idea of the player's most frequently visited places. The diagram also shows states of disconnection and complete misses, i.e. power off. An important feature in this statistical view is the option to apply colours to cell IDs.

Time line visualization

In order to represent the cell IDs visited by the phone along the days, we have created a time line visualization. In this one-dimensional graph we display the visited cell IDs. Through the use of colour it is easy to spot patterns of repetition or prolonged visits (see Figure 16.3). By using the same colours as in the statistical visualization, it is possible to make cross assumptions. For example, by identifying patterns for the most frequently visited cells, it is possible to guess work and home places for the typical nine to five worker.

The data displayed can be divided for example by days of the week, mapping the patterns of one person only, or compared with other players' data to show when they shared the same cell and at what time (see Figure 16.4).

16.5 Discussion

In this chapter we have explored the possibility of using mobile phones as data capture devices where data can be location tagged and visualized over time and supplemented with

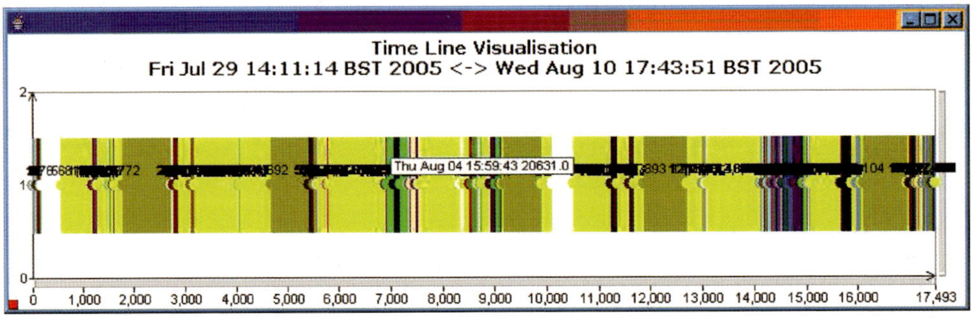

Figure 16.3 Time line visualization – single player's pattern

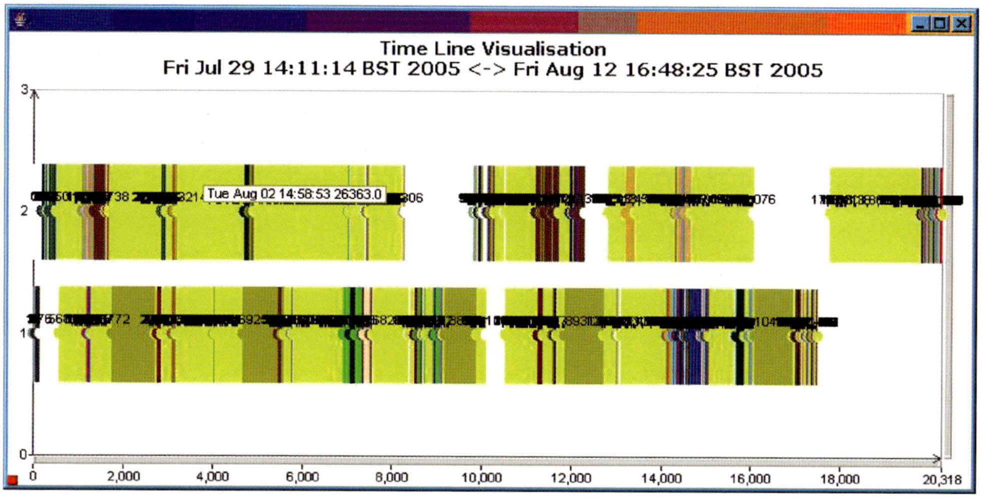

Figure 16.4 The pattern of two players

context information. Using mobile phones as a data collection device allows for a range of multimedia data to be collected. Collection of data using mobile phones gives the researcher a wide pool of potential participants for research activities such as exploration of social interaction, peoples' relationships with their environments and emergent behaviour in new types of pervasive gaming. The manner in which this data is collected is determined by the researcher. Simply asking the participant to collect data is one approach, or another could be to use games to prompt and direct participants to collect data.

Once we have collected this data we are then presented with the challenge of visualizing this data for analysis. Often we collect data which can be location tagged and therefore we are able to present this data spatially. In the hitchers case study we show how mobile phone cell IDs can be used to present a spatial representation of data collected during the course of the game. Although this spatial representation does not correspond to physical locations on a map, the analyst can view an abstract spatial representation of the data that can facilitate exploration and reveal clusters of interest.

Conversely, with newer mobile phones, GPS is becoming an integrated feature and therefore precise physical location information can be captured. This then allows the possibility of presenting data on maps through existing GIS software or other tools such as Google Earth (2007)

To supplement this location information it is often possible to collect context information about a location. This context information can include, for example, the number of Bluetooth connections, the number of recognized Bluetooth connections, WiFi access points, etc., which can be used to present information about the surrounding environment. For example, if an area has a large number of Bluetooth connections it could be assumed that the environment has a large number of people. If these Bluetooth connections are other mobile phones then this could be a crowded city centre; however, if they are Bluetooth connections from computers then it could be the participant's place of work.

Additional insight can be gained by inspecting this data over time. The advantage to allowing the analyst to view how the collected data was built up is that patterns in the data can be revealed and explored. For example, in the Day of the Figurines case study, location data is presented as a timeline displaying the cell IDs where a participant is located and how long they have spent in that cell. From this the analyst can view cells that a participant has most often visited as well as repetition of cell IDs at particular times of the day. Supplementing this timeline view with additional context data can allow a best guess of where that participant could be located in the physical world.

In these statistical visualizations we find another way to explore location-based data that focuses more on the activities of the participants. The advantage here is that often we wish to simply review the participants' activities over time. In this way the statistical visualizations used in the Day of the Figurines could, for example, provide a mechanism through which we could explore play patterns in mobile pervasive games.

16.6 Conclusion

Mobile phones are truly ubiquitous and pervasive in everyday life and can allow the researcher the potential for a large number of participants to take part in data gathering experiments. Coupled with effective visualizations of the data collected, we can allow researchers to collect more data and to effectively explore this data.

In this chapter we have presented two example case studies where we have collected data using mobile phones to specifically explore patterns of game play in mobile pervasive games. This data has been location-tagged and context information used to aid exploration. In both of the case studies we have used cell IDs as a way of location tagging data. However, as newer mobile phones become cheaper and more commonplace, researchers could have access to a pool of participants that have powerful data collection devices that can capture GPS location data as well as context information.

Furthermore, by utilizing the capabilities of different mobile phones, it could be possible to collect wider ranges of data as well as provide a way in which different location data could be merged. For example, by providing software that logs cell IDs and GPS coordinated to GPS-enabled mobile phone participants, the researcher can utilize this information to pin cell ID data collected by participants that do not have GPS-enabled mobile phones to certain areas on a map. Other examples could be to utilize the strengths of different mobile phones; for example, for PDA-style mobile phones the researcher could use questionnaire-style data collection, whereas mobile phones with video cameras could be used to video areas of interest.

References

Anastasi, R., Tandavanitj, N., Flintham, N., Crabtree, A., Adams, M., Row-Farr, J., Iddon, J., Benford, S., Hemmings, T., Izadi, S. and Taylor, I. (2002) Can you see me now? A citywide mixed-reality gaming experience. In *Ubicomp 2002*, Gothenburg, 2002.

Boyd Davis, S., Moar, M., Cox, J., Riddoch, C., Cooke, K., Jacobs, R., Watkins, M., Hull, R. and Melamed, T. (2005) 'Ere Be Dragons: an interactive artwork. In *MULTIMEDIA '05: Proceedings of the 13th Annual ACM International Conference on Multimedia*, Singapore. New York, ACM Press, pp. 1059–1060.

Crabtree, A., French, A., Greenhalgh, G., Benford, S., Heverst, K., Fitton, D., Rouncefield, M. and Graham, C. (2006) Developing digital records: early experiences of record and replay. *Journal of CSCW*, special issue on e-Research.

Dey, A. K. and Abowd, G. D. (1999) Towards a better understanding of context and context-awareness. GVU Technical Report GIT-GVU-99-22, College of Computing, Georgia Institute of Technology.

Drozd, A., Benford S., Tandavanitj, N., Wright, M. and Chamberlain, M. (2006) Hitchers: designing for cellular positioning. In *Ubicomp 2006*, Orange County, USA.

Flintham, M., Benford, S., Humble, J., Tandavanitj, N., Adams, M. and Row-Farr, J. U. (2001) I Like Frank: a mixed reality game for 3G phones; available at: www.amutualfriend.co.uk/html/papers.html

Flintham, M., Anastasi, R., Benford, S., Drozd, A., Mathrick, J., Rowland, D., Oldroyd, A., Sutton, J., Tandavanitj, N., Adams, M. and Row-Farr, J. U. (2003) Uncle Roy All Around You: mixing games and theatre on the city streets. In *Proceedings of Level Up Conference*, University of Utrecht, November.

Fruchterman, M. J. and Reingold, E. M. (1991) Graph drawing by force-directed placement. *Software – Practice and Experience* 21(11): 1129–1164.

Google Earth (2007) http://earth.google.com/ (accessed 14 August 2007).

Java Graph Library (2007) www.jgraph.com (accessed 14 August 2007).

Long, S., Kooper, R., Abowd, G. D. and Atkeson, C. G. (1996) Rapid prototyping of mobile context-aware applications: the cyberguide case study, *Proceedings of the 2nd Annual International Conference on Mobile Computing and Networking*. New York, ACM Press, pp. 97–107.

McCarthy, J. F. and Meidel, E. S. (1999) ActiveMap: a visualization tool for location awareness to support informal interactions, handheld and ubiquitous computing. In *Proceedings of the First International Symposium (HUC '99)*, Karlsruhe, September, pp. 158–170.

Nokia Mobile Phones (2007) www.nokia.co.uk/ (access date 14 August 2007).

O2 XDA (2007) www.my-xda.com/ (accessed 14 August 2007).

Ofcom (2004) www.ofcom.org.uk, Telecoms Research and Market Data.

Raento, M.., Oulasvirta, A., Petit, R. and Toivonen, H. (2005) ContextPhone: a prototyping platform for context-aware mobile applications. *Pervasive Computing*, 4(2): 51–59.

Rudman, P., North, S. and Chalmers, M. (2005) Mobile pollution mapping in the city. In *Proceedings of UK-UbiNet Workshop on e-Science and Ubicomp*, Edinburgh, May.

Schilit, B., Adams, N. and Want, T. R. (1994) Context-aware computing applications. In *First International Workshop on Mobile Computing Systems and Applications*, pp. 85–90.

Schmidt, A., Beigl, M. and Gellersen, H. W. (1999) There is more to context than location. *Computers & Graphics* 23(6): 893–901.

Siewiorek, D., Smailagic, A., Furukawa, J., Krause, A., Moraveji, N., Reiger, K. and Shaffer, J. (2003) SenSay: a context-aware mobile phone. In *Proceedings of the Seventh IEEE International Symposium on Wearable Computers*, pp. 248–249.

Steed, A., Spinello, S., Croxford, B. and Greenhalgh, C. (2003) e-Science in the streets: urban pollution monitoring. In *UK e-Science All Hands Meeting*.

Strahan, R. and Muldoon, C. (2002) State of the art mobile computing devices: report number $E = mC^2$. 1 December 2002, Department of Computer Science, University College Dublin; available at: http://emc2.ucd.ie/deliverables/e=mc2.1.1.2.2002.doc (accessed December 2006).

Symbian Operating System (2007) www.symbian.com/ (accessed 14 August 2007).

Tandavanitj, N. (2005) *IPERG Deliverable 12.3 First Phase Game Prototype for the first City as Theatre Public Performance*, Flintham, M. (ed.), August; available at: www.iperg.org (accessed December 2006).

Tennent, P. and Chalmers, M. (2005) Recording and understanding mobile people and mobile technology. In *Proceedings of the First International Conference on e-Social Science*, Manchester.

Windows Mobile Operating System (2007) www.microsoft.com/windowsmobile/ (accessed 14 August 2007).

Xiong, R. and Donath, J. (1999) PeopleGarden: creating data portraits for users. In *UIST '99: Proceedings of the 12th Annual ACM Symposium on User Interface Software and Technology*, Asheville, NC. New York, ACM Press, pp. 37–44.

Index

Geographic Visualization Edited by Martin Dodge, Mary McDerby and Martin Turner